ght A-7E Corsair from US Navy Attack Squadron VA-147 launches from the port bea
catapult of the USS *Constellation*. (*US Navy*)

UNITED STATES
NAVY AIRCRAFT
SINCE 1911

A Vo

UNITED STATES NAVY AIRCRAFT

SINCE 1911

SECOND EDITION

GORDON SWANBOROUGH

&

PETER M. BOWERS

NAVAL INSTITUTE PRESS

ANNAPOLIS, MARYLAND

BY THE SAME AUTHORS
United States Military Aircraft since 1908

By Peter M. Bowers
Boeing Aircraft since 1916

Library of Congress Catalog No. 76–12910
ISBN 0–87021–968–5

Published and distributed in the United States of America
by the Naval Institute Press, Annapolis, Maryland 21402
Printed in Great Britain
First published 1968
Second edition 1976
Second printing 1982

CONTENTS

PREFACE

From the time that the United States Navy took the first formal steps to create an air arm in 1911, to the present day, it has operated in parallel with, but separate from, the United States Air Force (and its predecessors). The special requirements of Naval aviation have called for the development and production of a wide variety of aeroplanes—and airships, which for many years figured in the Navy's operational forces.

This volume provides a chronicle of the aircraft which have served with the Navy, the Marines and the Coast Guard in 64 years—a period which opened with the Curtiss Pusher and comes up to date with the Grumman F-14A Tomcat variable-geometry fighter. Between these two extremes can be traced, through the contents of this book, the evolution of the carrier-based fighter, the torpedo-bomber, the anti-submarine patrol aircraft and other types with which the Navy and Marine Corps have fought in three wars, and, when the United States has not been actively engaged in hostilities, have helped to keep the peace in many parts of the world.

Following the pattern set by the companion volume *US Military Aircraft since 1908*, this volume includes primarily those types of aircraft which have reached operational service with the three subject Services. It excludes those prototypes, experimental types and research aircraft which have not served operationally. Between these two groups there is no clear division and of necessity some decisions as to which types to include or exclude have had to be made arbitrarily. As far as practicable, marginal aircraft types have been included, with the actual use to which the aircraft was put rating of greater importance than the quantity of a particular type built.

The main body of this volume is taken up with the descriptions of the most significant types of aircraft. Each of these is illustrated with one or more photographs and a three-view line drawing, often supplemented by additional drawings to show variants. The accompanying text sets out the development history of the aircraft type, refers to its operational service and makes specific reference to all known designated variants. Those types of aircraft which have been of relatively less importance are illustrated and described more briefly in the first appendix. Separate appendices cover, in a similar style, foreign aircraft types purchased for the use by the three Services, gliders and airships.

Maximum value can be derived from the information in this volume only if the reader has a clear understanding of the methods used by the Navy to designate its aircraft throughout the period under review. These methods are carefully described in a separate chapter, and a further appendix provides an index to every designated aeroplane, whether or not it is described in this book. Further interest in the nearly 500 photographs reproduced in this volume is created by the chapters on markings and colours of Naval aircraft.

Like its companion volume, this work is the result of a collaborative effort by the two authors and L. E. Bradford, who produced the line drawings. Its preparation has only been possible, however, by collaboration in a wider sense, involving many individuals over many years, and the authors acknowledge their special indebtedness to officials of the US Navy, the National Archives, the National Air Museum (Smithsonian Institution), USAF Photographic Library and the Wingfoot Lighter-than-Air Society; to the public relations staff of most of the US aircraft manufacturers; and to William T. Larkins, Lawrence S. Smalley, Harold Andrews and the late James C. Fahey. Other individuals who have contributed photographs are credited by name throughout this volume; their generosity in making these pictures available is gratefully acknowledged.

Hayes, Kent, England   GORDON SWANBOROUGH
Seattle, Washington, USA   PETER M. BOWERS
March, 1976

INTRODUCTORY NOTE

Best use of this book can be made if the reader clearly understands its scope and the arrangement of the material in it. Its scope is the aircraft—fixed and rotary wing, powered and unpowered, heavier- and lighter-than-air—which have served with the US Navy, the US Marine Corps or the US Coast Guard, or were being prepared for service when the book went to press. Aircraft of the US Army Signal Corps, the US Army Air Corps, the US Army Air Force and US Army Aviation are not included; nor are aircraft which have remained purely experimental throughout their life.

Arrangement of Material. Within the main section of the book and the three appendices covering heavier-than-air aircraft, the descriptions are arranged alphabetically by manufacturer, and chronologically within each manufacturer group. In those cases where a type was designed by one company but built by one or more others, it is located according to the name of the designing authority.

In the title of each aircraft description, the basic Navy designation and official popular name are included. In cases where a designation was changed during the production life of an aeroplane, the designation in use at the *end* of production has normally been given precedence in all references. This applies primarily to those types redesignated in 1962 when all Navy aeroplanes were brought into a unified tri-Service system. For simplicity and standardization, designations in the headings to aircraft descriptions exclude status prefixes and variant suffixes.

Technical Data. With some necessary exceptions, the data given throughout this volume are derived from official Navy performance charts. The exceptions are contemporary aircraft for which official figures are not available, or are given only in round terms.

Throughout the technical data, engines are referred to by their official designations. Popular names, such as Wasp and Cyclone, were applied by the engine manufacturers and a cross reference between these names and designations appears on page 527.

Serial Numbers. The serial numbers listed for each aircraft type are derived from unclassified US Navy documents giving information on contracts up to 1962. In some cases, subsequent changes to contracts result in inconsistencies between the serial numbers allocated and the numbers of aircraft known to have been built. Where such inconsistencies exist and cannot be resolved, the serial numbers are shown as 'allocations'.

Unit Information. Wherever possible, the number of the first squadron(s) to receive a particular aircraft type is recorded, but no attempt has been made to list all the user units of each type.

Photo Credits. Beneath each photograph appears an indication of its

ix

source, or the name of the private individual who took the photograph. The exchange of negatives between aircraft photographers, which is commonplace especially in the US, sometimes makes it difficult to identify the origin of a particular illustration as virtually identical negatives shot by one individual can be held by six or more collectors.

Every effort has nevertheless been made to give proper credit to the original photographer. Reference numbers are quoted only for photographs which can easily be obtained from official sources—particularly the National Archives, which now has possession of all US Navy photos prior to 1945 and has applied its own new reference numbers to these negatives.

US NAVAL AVIATION—
A BRIEF HISTORY

NAVAL interest in aviation as a potential aid to the fleet's fighting abilities can be dated back to 1898, the same year in which the US Army's interest was first aroused. Subject of this early interest was the work of Samuel P. Langley, whose model 'Aerodrome' flying machines were flying successfully in 1896. Under US Government contract in 1898, Langley undertook to build a full-size 'Aerodrome', and at the same time an inter-service board, including Naval officers, was appointed to investigate the possibilities for this aircraft. This was an inauspicious beginning, for Langley failed to demonstrate that the 'Aerodrome' could fly in two tests on the Potomac River at the end of 1903.

Five years elapsed after the 'Aerodrome' flopped miserably into the Potomac before the Navy again took official note of the aeroplane. Two officers were present as official observers in September 1908 when the Wright Model A was first demonstrated to the Army at Fort Myer, and, unofficially, others witnessed flying meetings in America and abroad. All reported enthusiastically on the possibilities of aircraft for naval purposes and although views within the Navy Department were diverse, the important step was taken on September 26, 1910, of designating an officer to whom all correspondence about aviation should be addressed. This officer was Capt W. I. Chambers, who served as officer in charge of aviation for the first three formative years.

In addition to setting up the necessary framework within which Naval aviation could evolve, Chambers selected Glenn Curtiss to build the first two Navy aeroplanes. The first of these, the A-1 Triad, was flown for the first time on July 1, 1911, by which time two Navy lieutenants (T. G. Ellyson and John Rodgers) had undergone pilot training. Also before the Navy accepted its first aeroplane, Chambers had arranged for Eugene Ely, a Curtiss demonstration pilot, to take-off from and land on special platforms aboard Navy ships—respectively on November 14, 1910, from the USS *Birmingham* in Hampton Roads and on January 18, 1911, aboard the USS *Pennsylvania* in San Francisco Bay.

Genuine ship-board operations were still a few years off, however, and the embryo Naval aviation unit was successively based at Annapolis; North Island, San Diego; Annapolis again and Pensacola, Fla, with an intermediate period at Guantanamo Bay, Cuba, for exercises with the fleet in January–February 1913. Naval aircraft were called upon a year later, during the Mexican crisis, to serve with the Atlantic fleet. In two detachments, aboard the USS *Birmingham* and *Mississippi*, seven pilots

1

and five aircraft participated; April 25, 1914, marked the first of a series of operational flights to observe Mexican positions.

Within the Navy Department, aviation became formally recognized on July 1, 1915, when an Office of Naval Aeronautics was set up. A Naval Flying Corps with a strength of 150 officers and 350 enlisted men was authorized in August 1916 and provision was also made for a Naval Reserve Flying Corps. Both the US Marines and the US Coast Guard had already become involved with aviation, with personnel from both services being trained as pilots under Navy auspices.

The 19 months during which America was an active combatant in World War I saw a dramatic expansion of the Naval Flying Corps. On April 6, 1917, the Corps had 54 aircraft, 48 qualified and student pilots and one base. By the time the war ended, the Navy had 12 air bases in the US and 27 in Europe; its flying-boats had attacked 25 enemy submarines and sunk or damaged at least half of them. The strength of Naval aviation in November 1918 totalled 6,716 officers and 30,693 enlisted men, plus 282 officers and 2,180 men in the Marine Corps; on hand were 2,107 aeroplanes and 15 dirigibles—a total which was not to be equalled in any single year until 1941.

While the size of the Naval Flying Corps inevitably diminished with the end of the war, the development of new applications of naval air power was pursued with great vigour in the 1920s. This period saw the evolution of the aircraft carrier as a major weapon and the acceptance of carrier-based fighter and observation squadrons as an integral part of the battle fleets. Less easily settled was the place to be occupied by Naval aviation in total defence strategy; this was the period of great inter-service rivalry between the Army and Navy, and within the Navy itself between proponents and critics of air power. The controversy was ended by the Pratt–Mac-Arthur agreement of January 9, 1931, which defined the naval air force as an element of the Fleet, to move with it and to help carry out its primary mission.

The Bureau of Aeronautics was created on August 10, 1921, to assume responsibility for all matters relating to naval aircraft, personnel and operation; BuAer thus became the aviation department in the office of the Secretary of the Navy and retained this role through World War II and on until the end of 1959. Marine Aviation remained under separate command, the responsibility of a Director of Aviation at Headquarters, Marine Corps. From its war-time peak, the strength of naval aviation fell to less than 1,000 aircraft in the 1920s but climbed slowly to reach 2,000 in 1938. That year, the Naval Expansion Act authorized the number of aircraft to be increased to at least 3,000. Two years later, in June 1940, the target was raised first to 4,500, then to 10,000 and a month later the ceiling was put at 15,000 'useful airplanes'.

By December 1941, when Japan attacked the United States, the air arms of the Navy and Marine Corps could muster 5,233 aeroplanes (including trainers and other non-combat types) and had 5,900 pilots and 21,678 enlisted men. Eight aircraft carriers were in commission, plus five land- or water-based patrol wings and two Marine aircraft wings. The attack on Pearl Harbor brought Naval aircraft into action immediately, when VS-6, flying from the USS *Enterprise*, somewhat fortuitously intercepted the attackers during the raid. Three days later, aircraft from the *Enterprise* sank a Japanese submarine—the first combat success of the war.

Fulfilling its primary mission of supporting the Fleet, naval aviation became heavily involved in the war against Japan in the Pacific area, and less so in the Atlantic battle with Germany and Italy. The record shows that Navy and Marine aircraft sank, in actions independent of ground or sea forces, 161 Japanese warships and 447 Japanese merchant ships, as well as 63 German submarines; and over 15,000 enemy aircraft were destroyed in the air or on the ground. The peak naval aircraft inventory was reached in 1945 with 40,912 aircraft on hand at July 1.

Operational forces were grouped in two major components, respectively in the Pacific and the Atlantic. In 1942–3 the Air Force, Pacific Fleet, and Air Force, Atlantic Fleet, were created as the administrative commands, these titles being given the additional prefix 'Naval' in 1957. In 1959 the Bureau of Aeronautics was merged with the Bureau of Ordnance to form the Bureau of Naval Weapons.

In the post-war rundown of forces, the strength of naval aviation was halved in 12 months, but the number of aircraft in the inventory remained well into five figures throughout the 1950s. Both Navy and Marine units were quickly in action in Korea, and in that three-year conflict these services actually flew 30 per cent of the total operational sorties mounted by all US forces.

Another major effort was made by Navy and Marine air forces in the war in Vietnam, where operations from task-force carriers steaming off the coast had become a daily routine by 1965. To meet the ever-increasing scale of operations in Vietnam, the Navy boosted its procurement of combat aircraft of all types and speeded up development schedules for new types to permit their rapid deployment in combat.

Procurement of aircraft by the US Marine Corps and their operations has been closely integrated with Naval aviation since 1912, as indicated. The US Coast Guard maintains close links with the Navy, particularly in respect of aircraft procurement, and almost without exception it acquires aircraft types already developed for or purchased by the Navy. During World War II the US Coast Guard was directly controlled by the Navy but both prior to that period, until 1941, and subsequently, from 1946, it has been administered by the Treasury Department.

3

US NAVAL AIRCRAFT DESIGNATIONS

EXCEPT for the short period from 1917 to 1922, the US Navy identified its aircraft up to 1962 by specific designation systems that conveyed a considerable amount of information concerning the origin and nature of the aircraft in simple and standardized form.

Since the aircraft described in this volume are presented by their official naval designations, the reader will benefit from an understanding of the various systems. These are detailed in this chapter. The extensive redesignation of existing naval aircraft in 1962 within the unified tri-service system then introduced presents a problem in the arrangement of the aircraft descriptions; the aircraft which were no longer in production are presented under their original designations with the new one following in parenthesis, whereas those produced after the change are given their new designation first with the old ones in parentheses. A complete cross-reference between the pre-1962 aircraft designations and those of the new system for every model in the US Navy inventory at the time is included in the Designation Index starting on page 527.

1911–1914 SYSTEM

Naval aircraft were originally identified by a letter to designate the manufacturer, followed by a number to show sequence of procurement, such as A-1 for the first Curtiss, A-2 for the second, etc. The second source, the Wright Brothers, was designated B.

A deficiency in this system became apparent as soon as different types of aeroplanes were obtained from the same manufacturer. Curtiss was then given another letter, C, to distinguish its flying-boats from its convertible landplanes–pontoon seaplanes.

This first system involved only five letters to identify manufacturers, as follows:

A—Curtiss (Land and Hydro)
B—Wright (Land and Hydro)
C—Curtiss (Flying Boat)
D—Burgess & Curtis (Hydroplane, no relation)
E—Curtiss (Amphibian)

1914–1916 SYSTEM

On March 27, 1914, a new system was adopted and all aircraft then on hand were redesignated. The aircraft were identified as to type and subtype and by sequence of procurement within each sub-type regardless of manufacturer in a manner similar to ship designations. The following designations were used:

AIRCRAFT TYPES OR CLASSES

A—Heavier than Air
B—Free Balloons
C—Dirigibles
D—Kite Balloons

AEROPLANE SUB-TYPES

AH—Hydro Aeroplane
AB—Flying Boat
AX—Amphibian

Published accounts of this system indicate that it applied to only the Navy's first 75 aeroplanes and was abandoned when a sequential serial number system, starting at 51, was adopted early in 1917. However, identification of later aircraft in official records indicates that it was carried on unofficially for a short while. In the case of AH-67 (properly A-67) the AH was used as a type designation and the 67 was the new serial number indicating the 67th naval aeroplane procured to that time rather than the 67th AH type.

1917–1922 SYSTEM

No standard system was used during these years until March 29, 1922. Aeroplanes were procured and operated under their manufacturer's names and model designations, as Curtiss N-9, Sopwith Baby seaplane and F-5L. The latter, an improved version of the Curtiss H-16 flying-boat developed in England as the F.5 and produced in the United States with the Liberty engine (hence the letter L), was produced by several manufacturers with no distinction made between them.

1922–1962 SYSTEM

On March 29, 1922, following the reorganization of naval aviation under the Bureau of Aeronautics, a systematic aircraft designation system was adopted. This recognized two classes of aircraft: aeroplanes, identified by the letter V, and airships, identified by the letter Z. These letters were used for administrative purposes and did not appear in the actual aeroplane designation although the Z was used in the rigid airship designations. The aeroplanes themselves were further identified as to manufacturer, type and model sequence by letters and numbers. (See page 503 for airship designations.)

5

This system was also used by the US Marines, which are part of the Navy, and was used by the US Coast Guard, starting in 1935. Up to 1935 the Coast Guard, which is under the jurisdiction of the Treasury Department in peacetime but which uses naval organization and equipment, had not used any specified system of designation. Individual aircraft were identified by serial number, sometimes preceded by initials identifying a specific design, as FLB-52 for Flying Life Boat. The 52 was the Coast Guard serial number.

The system was not applied to all aeroplanes operated by the Navy. Civilian aircraft taken into service during World War II, mostly transport types, were operated under their civil model numbers, such as B-314 for the Boeing Model 314. Captured enemy aircraft under test were flown under their original designations but with Navy serial numbers, as were a few isolated US Army models such as the Bell YP-59A. Army models adopted in numbers, even when transferred from the Army, were given new naval designations. Thus 48 B-17Gs transferred in 1945 became PB-1s; a single B-17F procured in 1943 operated under that designation until redesignated at the time the later models were acquired. In post-World War II years the standard system was by-passed in some cases where Air Force models were procured in quantity and operated under their original Air Force designations, such as T-28 and T-34. Minor changes made for naval use were reflected only in the series suffix letters, T-28B or C and T-34B.

There were very few cases of duplicate designations. Occasionally, strict adherence to the system was relaxed and duplication appeared. There was the Boeing PB-1 flying-boat of 1925, for example, duplicated by the Army Boeing B-17Gs transferred to the Navy in 1945 and

Type and model designations, adopted by the US Navy in 1922, did not appear on the aircraft until 1928. In 1925 use of the manufacturer's name was authorized and sometimes included the aircraft name bestowed by the manufacturer, as on the Vought O2U-1 Corsair (*left*). Both the manufacturer's name and Navy designation appeared together from 1928, as on the second XF8C-4 Helldiver (*right*) until the name was dropped in the early thirties. Note the use of black and white paint to contrast with the rudder stripes. (*Manufacturers' photos*)

6

redesignated PB-1, and the Vought UO-1 of 1923–8 duplicated by the Piper UO-1 of 1960.

In spite of being in use since 1922, the designation was not painted on naval aeroplanes until 1928, when it was applied in 3-in figures to the rudder. Early in 1925 the Navy had started the practice of painting the manufacturer's name across the top of the rudder. This short-lived practice overlapped the adoption of the letter-number system and for a short period the name and designation were both carried on some aeroplanes.

Seldom was a naval aeroplane redesignated. This was done when a designation symbol in use was discontinued, as T-for-Transport being changed to R, and when the Curtiss F11C-2 was redesignated BFC-2 upon adoption of the BF designation for Bomber-Fighter. In a few cases, the basic mission of the aeroplane was simply changed, as the Curtiss F8C-1 being reassigned to observation duties as OC-1. Oddly, an entirely new design, the Curtiss Helldiver dive-bomber, was given the designation XF8C-2. There was an F8C-4 production version, but the F8C-5 was more properly redesignated O2C-1 after delivery.

Aeroplanes were not redesignated when assigned new duties because of obsolescence. Observation types relegated to utility tasks (J) and training (N) retained their original O-designations.

At first, the identification emphasis was on the manufacturer, then the type of aircraft. As originally adopted in 1922, the letter identifying the manufacturer was placed first, then the aircraft type letter and finally the model number. The designation MO-1 identified the aeroplane as a Martin (M) and the first observation type (O) procured from Martin under the new system. The -1 indicated the initial configuration of that particular model. A recognizable change in the basic model would have become MO-2. A scout model procured from Martin was MS-1. An entirely new observation model from Martin, however, became M2O-1, the figure 2 identifying the second O-model from the same manufacturer.

This system grew as more detailed identification became desirable, and eventually contained six separate parts as detailed below, before being replaced by a common system for all US armed services aircraft in 1962.

A major change was made on March 10, 1923, when it became desirable to identify the aeroplane initially by type, so the type designation was placed first. Aircraft in service with the old arrangement retained it, however, as did aircraft of the same model procured under follow-up orders after the system was changed. A notable exception was the Martin SC-1, which was the Curtiss CS-1 built under licence by Martin but using the original Curtiss designation turned around to reflect the new practice.

While the letter sequence was reversed, the sequence of models was retained. The original Curtiss Navy racer design was the CR-1. Under the revised system, the second model was R2C-1. When a new Curtiss transport appeared in 1928, it was designated RC-1 with no conflict since the R-for-Racer had been dropped by that time.

7

The naval aeroplane designation system as used from March 10, 1923, to September 18, 1962, is discussed in detail below. The various portions are numbered in their sequence of adoption rather than by their sequence in the designation, thus:

$$
\begin{array}{cccccc}
\text{X} & \text{F} & 4 & \text{B}-1 & & \\
& \text{SB} & 2 & \text{C}-4 & \text{C} & \\
\text{H} & \text{O} & 2 & \text{S}-1 & & \\
(5) & (1) & (4) & (2)\ (3) & (6) &
\end{array}
$$

1	TYPE OR CLASS	4	MANUFACTURER TYPE SEQUENCE
2	MANUFACTURER	5	STATUS OR CLASS PREFIX
3	AIRCRAFT CONFIGURATION SEQUENCE	6	SPECIAL PURPOSE SUFFIX

TYPE OR CLASS DESIGNATION. Originally, only single letters were used to identify the aeroplane types but dual letters were adopted in March 1934 to identify specified dual-purpose types, as BF for Bomber-Fighter, PB for Patrol Bomber, etc. Some rather fine distinctions were made to emphasize the primary mission of the aeroplane, for example SO for Scout-Observation and OS for Observation-Scout. In one case only, three letters were used—for the experimental Hall Patrol Torpedo Bomber, XPTBH-2, of 1938.

In some cases, certain type letters were discontinued because of conflict with other types, as T-for-Torpedo and T-for-Transport, the latter being replaced by the letter R. When the torpedo designation was dropped after World War II, the letter T was readopted for new training-plane designs. Trainers then in service under the old N and SN designations retained them to the end of their service life or until the redesignation of 1962, when the surviving Beech SNBs became C-45s.

In other cases designation changes affected existing aircraft, the Consolidated PB4Y-2 Privateers still in service in 1948 becoming P4Y-2 when many dual designations were abandoned. This same type actually underwent a further change, to P-4, in 1962. There was no actual conflict be-between the P4Y-2 designation for the Privateer and the experimental XP4Y-1 flying-boat of 1942, which no longer existed. Using the PB4Y-2's original designation with only one letter dropped was a matter of convenience when strict adherence to the system would have dictated an entirely new designation, as P5Y-1.

There were also 'paper' type designations for political reasons. Some of the Navy's racers of the early 1920s, although flown with R-for-Racer designations, were camouflaged in the appropriations paperwork as fighters, so the CRs, R2Cs and R3Cs had equivalent fighter designations. Later racers flew with fighter designations. The first actual fighter that Curtiss built for the Navy under the new system became F4C-1.

Type designation letters assigned to all US Navy, Marine and Coast Guard aeroplanes from 1922 to 1962 are listed below, with dates from initial assignment to end of procurement of that type or its retirement from service. Assignment of two-letter designations was discontinued in 1946, but some aircraft carrying them retained them until the designation change of 1962. Letters in use to 1962 are indicated by an arrow in the right-hand 'year' column.

Letter	Class	Example	Year
A	Ambulance	AE-1	1943
A	Attack (formerly BT)	AD-1	1946 ⟶
B	Bomber	BT-1	1931–1943
BF	Bomber-fighter	BF2C-1	1934–1937
BT	Bomber-torpedo (to A)	BT2D-1	1942–1945
DS	Anti-submarine drone	DSN-1	1959 ⟶
F	Fighter	F6C-1	1922 ⟶
G	Transport, single-engine	GB-1	1939–1941
G	Flight refuelling tanker	GV-1	1958 ⟶
H	Helicopter (used in combination)	HO2S-1	1943 ⟶
H	Hospital (to A, 1943)	XHL-1	1929–1931, 1942
J	Transport (to R)	XJA-1	1928–1931
J	Utility	J2F-1	1931 ⟶
JR	Utility transport	JRF-5	1935 ⟶
L	Glider (used in combination)	LNS-1	1941–1945
M	Marine expeditionary	EM-2	1922–1923
N	Trainer	N2S-1	1922–1960
O	Observation	O2U-1	1922 ⟶
OS	Observation-scout	OS2U-1	1935–1945
P	Patrol	PB-1	1923 ⟶
P	Pursuit	WP-1	1923
PB	Patrol bomber	PBY-5	1935 ⟶
PTB	Patrol torpedo-bomber	XPTBH-2	1937
R	Racer	R3C-2	1922–1928
R	Transport	RR-5	1931 ⟶
RO	Rotorcycle	ROE-1	1954–1959
S	Anti-submarine	S2F-1	1951 ⟶
S	Scout	SU-2	1922–1946
SB	Scout-bomber	SBD-3	1934–1946
SN	Scout-trainer	SNJ-2	1939 ⟶
SO	Scout-observation	SOC-1	1934–1946
T	Torpedo	DT-2	1922–1935
T	Trainer	T2V-1	1948 ⟶
T	Transport	TA-1	1927–1930
TB	Torpedo-bomber	TBD-1	1935–1946
TD	Target drone	TD2C-1	1942–1946
TS	Torpedo-scout	XTSF-1	1943
U	Utility	UF-1	1955 ⟶
U	Unpiloted drone	UC-1K	1946–1955
W	Electronic search	W2F-1	1952 ⟶

MANUFACTURER'S DESIGNATION. A single letter was commonly used to identify the manufacturer. In the earliest applications this was logically the first letter of the manufacturer's name, as C-for-Curtiss. In only one

9

case was multiple lettering used to identify a US manufacturer, DW-for-Dayton-Wright, which built a scout version of the Douglas DT-2 under the designation of SDW-1. Dual identification letters were also used for some of the European designs procured for test in 1922–3.

Complications had arisen before adoption of the system due to the Navy's policy of having different manufacturers build the same design. The Navy's own plant, the Naval Aircraft Factory (NAF) in Philadelphia, built large numbers of outside designs under the original manufacturer's designations, for example Curtiss MF, Vought VE-7 and so on.

The Naval Aircraft Factory also built a number of experimental types designed by the Navy Bureau of Aeronautics in Washington, DC. When it proved desirable to put some of these Bureau designs into production, the work was farmed out to the established industry, which at first built them under the original arbitrary model designations selected by the Bureau. Thus the NAF built a few TS-1 carrier-based fighters, but the major production was by Curtiss, also as TS-1. There was no way of distinguishing between the two in service except by assigned serial numbers.

This problem was partially resolved early in 1923 when the Navy decided that the designation should reflect the actual manufacturer rather than the original designer. When a Navy-designed single-seat scout design was turned over to both Martin and Cox-Klemin for building, the Martin version became MS-1 and the Cox-Klemin became XS-1 (the letter C had already been assigned to Curtiss; also, Vought was assigned U because the logical V had already been selected to identify aeroplanes as a class). In World War II, most designs licensed to another manufacturer were given a new designation to reflect the actual builder, as the Vought F4U-1 built by Goodyear as FG-1 and by Brewster as F3A-1. A notable exception was the Grumman J2F built by Columbia under the original designation even though a new design by Columbia was identified by the letter L.

In some cases a letter used by one manufacturer was assigned to another after the first went out of business. In the heavy production period preceding and including World War II, several manufacturers used the same letter at the same time, e.g. Curtiss/Cessna, Boeing/Beech, Stearman/Sikorsky. In such cases the various models were not competitive. In one case where they were, Piper, identified by the letter E, started manufacturing gliders for the Navy after Pratt-Read, also identified by E, was already producing gliders of its own design. The Navy resolved this conflict by assigning Piper a second letter, P, to be used on its gliders while E was retained for Piper's aeroplanes. Piper was given still another letter, O, after being inactive as a supplier of naval aircraft for nearly 20 years.

The letters assigned to all suppliers of US Navy aircraft under the 1922–62 system are listed below with dates from initial assignment to the end of procurement from that manufacturer, retirement of his models from service, or change in the company. Not all of these were used on actual aircraft. Letters in use for one manufacturer from time of adoption to the change of 1962 are indicated by an arrow following the initial date.

10

MANUFACTURERS' IDENTIFICATION LETTERS

Letter	Manufacturer	Year
A	Aeromarine Plane and Motor Company	1922
A	Allied Aviation Corporation (glider only)	1941–1943
A	Atlantic Aircraft Corporation (American Fokker)	1927–1930
A	Brewster Aeronautical Corporation	1935–1943
A	General Aviation Corporation (ex-Atlantic)	1930–1932
A	Noorduyn Aviation, Ltd (Canada)	1946
B	Beech Aircraft Company	1937–1945
B	Boeing Aircraft Company	1923–1959
B	Edward G. Budd Manufacturing Company	1942–1944
B	Aerial Engineering Corporation (Booth, or Bee Line)	1922
BS	Blackburn Aeroplane & Motor Company (England)	1922
C	Cessna Aircraft Corporation (also E)	1943
C	Culver Aircraft Corporation	1943–1946
C	Curtiss Aeroplane & Motor Company	1922–1946
C	De Havilland Aircraft of Canada, Ltd	1955–1956
CH	Caspar-Werke G.m.b.H. (Germany)	1922
D	Douglas Aircraft Corporation	1922 ⟶
D	McDonnell Aircraft Corporation (to H)	1942–1946
D	Radioplane Corporation (drones)	1943 ⟶
D	Frankfort Sailplane Company (target drones)	1945–1946
DH	De Havilland Aircraft Co., Ltd (England)	1927–1931
DW	Dayton-Wright Airplane Company	1923
E	Bellanca Aircraft Corporation	1931–1937
E	Cessna Aircraft Corporation (formerly C)	1951 ⟶
E	Detroit Aircraft Corporation (to Great Lakes)	1928
E	Edo Aircraft Corporation	1943–1946
E	G. Elias & Brothers, Inc	1922–1924
E	Gould Aeronautical Corporation (gliders)	1942–1945
E	Hiller Aircraft Corporation (helicopters)	1948 ⟶
E	Piper Aircraft Corporation (aeroplanes) (also O, P)	1941–1945
E	Pratt-Read (gliders)	1942–1945
F	Fairchild Aircraft, Ltd (Canada)	1942–1945
F	Royal Dutch Aircraft Mfg. Works (Fokker)	1922
F	Grumman Aircraft Engineering Corporation	1931 ⟶
G	Gallaudet Aircraft Corporation	1922
G	Bell Aircraft Corporation (for Bell-built Great Lakes)	1935–1936
G	Eberhart Aeroplane & Motor Company	1927–1928
G	Globe Aircraft Corporation (target drones)	1946–1959
G	Goodyear Aircraft Corporation	1942 ⟶
G	Great Lakes Aircraft Corporation (ex-Detroit)	1929–1935
G	A.G.A. Aviation Corporation (gliders)	1942
H	Hall-Aluminum Aircraft Corporation (into Consolidated, 1940)	1928–1940
H	Howard Aircraft Company	1941–1944
H	Huff, Daland & Company (became Keystone, K, 1927)	1922–1927
H	Stearman-Hammond Aircraft Corporation	1937–1938
H	Snead and Company (gliders)	1942
HP	Handley Page, Ltd (England)	1922
J	Berliner-Joyce Aircraft Corporation	1929–1935

11

Letter	Manufacturer	Year
J	General Aviation Corporation (ex-Atlantic)	1935
J	North American Aviation Corporation (successor to General)	1937 —→
JL	Junkers-Larson Aircraft Corporation	1922
K	Fairchild Aircraft Corporation (Kreider-Reisner)	1937–1942
K	Kaiser Cargo, Inc (Fleetwings Div.)	1943–1945
K	Kaman Aircraft Corporation (helicopters)	1950 —→
K	Keystone Aircraft Corporation (ex-Huff, Daland)	1927–1930
K	Kinner Airplane and Motor Corporation	1935–1936
K	Kreider-Reisner Aircraft Company, Inc (Fairchild)	1935
K	Martin, J. V.	1922–1924
K	Nash-Kelvinator Corporation	1942
L	Bell Aircraft Corporation (helicopters since 1946)	1939 —→
L	Columbia Aircraft Corporation	1945
L	Grover Loening, Inc	1932–1933
L	Langley Aviation Corporation	1942–1943
L	Loening Aeronautical Engineering Corporation (into Keystone)	1922–1932
L	L. W. F. Engineering Corporation	1922
M	Glenn L. Martin Company	1922 —→
M	General Motors Corporation (Eastern Aircraft Division)	1942–1945
M	McCulloch Motors Corporation	1953–1954
N	Gyrodyne Co of America Inc (drones, rotocycles)	1960
N	Naval Aircraft Factory	1922–1945
O	Lockheed Aircraft Corporation (Plant B)	1931–1950
O	Piper Aircraft Corporation (also E, P)	1960
O	Viking Flying Boat Corporation	1929–1936
P	Piasecki Helicopter Corporation (ex- P-V Engineering)	1946–1960
P	Piper Aircraft Corporation (gliders)	1942–1943
P	Pitcairn Autogyro Company	1931–1932
P	P-V Engineering Forum (became Piasecki, Vertol)	1944 —→
P	Vertol Aircraft Corporation (ex-Piasecki, P-V; to Boeing 1960)	1956 —→
P	Spartan Aircraft Company	1940–1941
PL	George Parnall & Company (England)	1922
Q	Bristol Aeronautical Corporation (glider)	1941–1943
Q	Fairchild Engine and Airplane Corporation	1928 —→
Q	Stinson Aircraft Corporation (into Convair, 1942)	1934–1936
Q	Chas. Ward Hall, Inc	1926
R	Aeronca Aircraft Corporation (Army TG-5 glider)	1942
R	American Aviation Corporation (glider)	1942
R	Brunswick-Balke-Collender Corporation	1942–1943
R	Ford Motor Company	1927–1932
R	Interstate Aircraft & Engineering Corporation (drones)	1942 —→
R	Maxson-Brewster Corporation	1939–1940
R	Radioplane Div, Northrop Corporation (target drones)	1943 —→
R	Ryan Aeronautical Company	1941–1946
RO	Officine Ferroviarie Meridionali Romeo (Italian-built Fokker C-V)	1933
S	Schweizer Aircraft Corporation (glider)	1941
S	Sikorsky Aviation Corporation (helicopters since 1943)	1928 —→
S	Stearman Aircraft Company (became Boeing-Wichita, 1939)	1934–1945
S	Stout Engineering Laboratories	1922

Letter	Manufacturer	Year
S	Sperry Gyroscope Company (target drones)	1950
S	Supermarine (Spitfire as FS-1)	1943
T	New Standard Aircraft Corporation	1930–1934
T	The Northrop Corporation (became Douglas-El Segundo)	1933–1937
T	Taylorcraft Aviation Corporation (Army TG-6 glider)	1942
T	Northrop Aircraft, Inc (Army P-61 as F2T-1)	1944
T	Temco Aircraft Corporation	1956
T	Timm Aircraft Corporation	1941–1943
T	Thomas-Morse Aircraft Corporation	1922
U	Chance Vought Corporation (now Ling-Temco-Vought)	1922 —→
V	Lockheed Aircraft Corporation (Vega, Plant A)	1942 —→
V	Canadian Vickers, Ltd	1942–1945
V	Vultee Aircraft, Inc (into Consolidated as Convair, Y, 1942)	1941
VK	Vickers, Ltd (England)	1922
W	Waco Aircraft Corporation (gliders since 1942)	1934–1945
W	Wright Aeronautical Corporation	1922–1926
W	Canadian Car & Foundry Co, Ltd	1942–1945
X	Cox-Klemin Aircraft Corporation	1922–1924
Y	Consolidated Aircraft Corporation (to Convair, 1942)	1926 —→
Z	Pennsylvania Aircraft Syndicate	1933–1934

AIRCRAFT CONFIGURATION SEQUENCE. The initial configuration of a naval aeroplane was identified by the figure 1 separated from the basic designation by a dash. Minor changes were identified by sequential numbers: -2, -3, etc. In some cases there were much greater changes between an experimental prototype, as XSB2C-1, and its first production version, the SB2C-1, than there was between the SB2C-1 and the improved SB2C-3 (the XSB2C-2 was a seaplane version). However, it was generally recognized that there were significant differences between prototype and initial production models.

TYPE SEQUENCE NUMBER. This reflects the procurement sequence of different models of a similar type procured from the same manufacturer, as FB-1 for Boeing's first fighter (no number used in this case), F2B-1 for the second Boeing fighter model, F3B-1 for the third, etc.

Of necessity, there was a little skipping around in the early years of the system because of conflict with existing designations. For instance, there was no F5C designation assigned because of the large numbers of wartime F-5L flying-boats, simply called F-5, still in service.

The Navy took care to avoid duplication of designations when different manufacturers using the same letter produced the same types of aircraft years apart. The XFA-1 was an experimental fighter produced by General Aviation in 1930. This was a successor firm to Atlantic, the American Fokker organization, and retained the original A-for-Atlantic letter. In 1935, after General became North American and adopted the letter J, the letter A was assigned to Brewster. Brewster's first design for the Navy was a scout bomber, the XSBA-1. Since XFA-1 had already been used, Brewster's first experimental Navy fighter design became the XF2A-1.

13

Prefix letters to indicate the special status of an individual aeroplane were not added to the basic designation until 1927. At that time the letter X, as already used by the Army Air Corps, was adopted to identify experimental prototypes or, in some cases, standard models withdrawn from normal service and modified for experimental purposes. In some cases prototype aircraft that had not originally carried the X prefix when first flown were given it after it was adopted, as in the case of the Boeing FB-5 prototype of 1926 becoming XFB-5 in 1927.

For a short period following World War II the letter N as a prefix indicated a first-line combat aircraft assigned to a reserve training squadron. At the time of the Korean War the letter Y was added to indicate a prototype aircraft in other than straight experimental status, roughly equivalent to the Army/Air Force letter Y indicating service test status. Extensive use of prefixes was made after adoption of the unified 1962 system (see page 16).

During World War II, gliders and helicopters were identified as to class by prefix letters in conjunction with otherwise standard designations, e.g. HO2S-1 for Sikorsky's second observation helicopter (H) and LNS-1 for Schweizer's first training glider (L). Although these glider and helicopter letters, as well as others used to identify target drones, were really prefixes to the designation, they were widely regarded as part of a two-letter type designation like SB or OS.

SPECIAL PURPOSE SUFFIX. Suffix letters were used sparingly in the between-war years. During World War I and shortly thereafter they were used as part of the manufacturer's designation to indicate various series of a basic model, special configuration, etc. The H in N-9H and JN-4H did not indicate sequential development of those models but variants fitted with the American-built Hispano-Suiza, or Hisso, engine. The letter L in connection with the HS-1L flying-boat similarly identified the Liberty engine version. The Vought VE-7 was normally unarmed; the VE-7G was fitted with guns and the VE-7SF was converted to a single-seater with flotation gear.

The first suffix used under the new (1922) system was the letter C to indicate a modification of a standard aeroplane to withstand catapulting when such work was necessary, such as UO-1C for those seaplane versions based on battleships and launched by catapult (surprisingly, there was no distinction made between land- and seaplane versions of the same aeroplane).

Later, limited use was made of suffix letters A and B following the complete designation to indicate a minor alteration not considered sufficient to change the dash number of the aeroplane, e.g. F4B-1A for a standard F4B-1 converted to a two-seater for the Assistant Secretary of the Navy and TBD-1A for a standard TBD-1 tested as a seaplane. The earlier NY-1B had been an NY-1 trainer fitted with longer NY-2 wings.

Suffix letters with specific meanings came into wide use during and after

World War II to identify aeroplanes used for other than their designated mission, use of special equipment, etc. Occasional multiple use of the same letter was somewhat confusing. For example, by World War II the letter A identified a normally unarmed aeroplane fitted with armament (J2F-2A); a machine not normally based on aircraft carriers fitted with arrester gear (SOC-3A); former Army aircraft, basically similar to standard Navy aircraft, transferred to the Navy (F4B-4A, formerly Army P-12); and an amphibious version of a flying-boat (PBY-5A). Other examples of duplicate use will be found in the following list.

During World War II separate use was made of some suffix letters to indicate similarity of Navy-designated aeroplanes to their original Army versions, as PBJ-1J for the Navy equivalent of the North American B-25J, used in numbers by the Navy. This use was rare, however.

The nomenclature applying to some of the following suffixes changed from year to year while the meaning remained essentially the same, such as 'Special Search' and then 'Special Radar' for a TBM-3D equipped with search radar. In other cases, changes or improvements in equipment produced a second sub-variant of the same type, identified by a number 2 following the suffix, as in TBM-3S2 and F3D-2T2.

NAVY SUFFIX LETTERS

Suffix	Meaning	Example
A	Miscellaneous modification	OL-9A
A	Armament on normally unarmed aeroplane	J2F-2A
A	Arrester gear on normally non-carrier aircraft	SOC-3A
A	Built for Army or obtained from Army	SBD-3A to A-24 F4B-4A from P 12
A	Amphibious version	PBY-5A
A	Land-based version of carrier aircraft	F4F-3A
B	Miscellaneous modification	NY-1B
B	Special armament	F8F-1B
B	British version (lend–lease)	JRF-6B
C	Arrester gear added	SNJ-5C
C	Reinforced for catapulting	UO-1C
C	Cannon armament	SB2C-1C
C	Navy equivalent of Army C-series aircraft	SBN-2C = AT-7C
CP	Trimetrogen camera	TBM-1CP
D	Drop tanks	F4U-1D
D	Drone control	JD-1D
D	Navy equivalent of Army D-series aircraft	PBJ-1D = B-25D
D	Special search	TBM-3D
D	Special radar	TBM-3D
E	Electronic equipment	SB2C-4E
F	Flagship conversion	PB2Y-5F
F	Special power plant	P2V-5F
G	Search and rescue (later Coast Guard)	PBM-5G
G	Coast Guard version	PB-1G
G	Gun on normally unarmed aircraft	UO-1G
G	Navy equivalent of Army G-series aircraft	PBJ-1G = B-25G

15

Suffix	Meaning	Example
H	Hospital conversion	PB2Y-3H
H	Navy equivalent of Army H-series aircraft	PBJ-1H = B-25H
J	Special weather equipment	TBF-1J
J	Navy equivalent of Army J-series aircraft	PBJ-1J = B-25J
K	Drone conversion	F6F-5K
L	Winterized	TBF-1L
L	Searchlight carrier	P2V-5L
M	Missile launcher	F7U-3M
N	Night fighter	F6F-5N
N	All-weather operation	AD-4N
NA	Night version stripped for day attack	AD-4NA
NL	Night, winterized	AD-4NL
P	Photographic	F7F-3P
Q	Electronic countermeasures	AD-3Q
R	Support transport	R4D-6R
R	Transport conversion	PBM-3R
S	Anti-submarine	P5M-2S
T	Two-seat trainer version	F9F-8T
U	Utility	TBM-3U
W	Special search	AF-2W
W	Early warning	AD-5W
Z	Administrative version	R4D-5Z

1962 SYSTEM

The Navy mission–manufacturer–number system survived until September 18, 1962, when the US Department of Defense combined all Air Force and Navy aircraft designations into the existing Air Force system. When aeroplanes of directly comparable make and model were involved, the Navy models were given established Air Force designations with the next available series letter, as the Douglas R4D-8 becoming C-117D. Where there were no equivalent Air Force types, new series were started under the old Navy type letters, as the Douglas AD-6 becoming the A-1E and the A3D-1 becoming the A-3. Electronically equipped aeroplanes like the Grumman W2F-1 became the E-2 in a new E-for-Electronic Search class. Some duplication of older Army Air Corps designations resulted because some high-numbered series were started from 1 again. The USAF C-for-Cargo numbers had reached C-142, so a new series was started almost 40 years after the original. The same thing was done to fighters after F-111. This resulted in some duplication of the old (and discontinued) Army F-for-Foto designations, but did not duplicate earlier Navy F-for-Fighter designations because of detail differences in the system. The McDonnell F4H-1 became F-4B in its Navy version and F-4C for the Air Force without quite duplicating the old F4B-1 and F4C-1 designations.

The transition to the new system was eased for naval personnel and others well versed in the old system by using dual designations on the aeroplanes for a year or so. When the new designation was first applied,

16

For a year after the unified designation system was introduced in 1962, old and new designations were both carried, normally on the rear fuselage above the serial number. (*Peter M. Bowers*)

the old one was reapplied directly below or alongside in parenthesis. A cross-reference between all old and new designations for aircraft affected by the 1962 change is provided in the Designation Index starting on page 527.

BASIC MISSION AND TYPE SYMBOLS. The basis of the tri-service designation system used since 1962 is the use of a Basic Mission Symbol and a design number, as F-14 for the 14th design in the Fighter series, or S-3, the third anti-submarine aircraft design. Twelve basic mission symbols are in use, together with two type symbols—H for helicopter and V for aircraft having VTOL or STOL capability. The H and V symbols are used immediately preceding the series number and are therefore sometimes confused with basic mission symbols, but in most cases they are used in conjunction with the latter, as in AV-8, a VTOL attack aircraft, or SH-3, an anti-submarine helicopter. The basic mission and type symbols are as follows:

Symbol	Mission	Remarks
A	Attack	Joint use by Navy and USAF.
B	Bomber	Primarily USAF. One Navy model, JD-1, reverted to original USAF B-26 designation.
C	Cargo/Transport	Joint use; replaces old Navy R, SN and some W types.
E	Special electronics	Joint use; former search types WF and W2F.
F	Fighter	Joint use.
H	Helicopter	Joint use by Navy, USAF, US Army (Type symbol).
K	Tanker	Joint use; aircraft designed for refuelling of others.
O	Observation	Joint use with US Army; replaces USAF L, continues Navy O.
P	Patrol	Navy only; continues P series.
S	Anti-submarine	Navy only; continues S series.
T	Trainer	Joint use.
U	Utility	Joint use by Navy, USAF, US Army.
V	VTOL and STOL	Joint use by Navy, USAF, US Army (Type symbol).
X	Research	Joint use.

MODIFIED MISSION SYMBOLS (PREFIX LETTERS). The use of prefix letters to indicate modification of an aircraft for use in other than its basic mission is adapted from USAF procedures and is equivalent to the old Navy use of suffix letters for the same purpose. In some cases, two prefix

17

letters are used, one being a status prefix (see below) or in designations including a type symbol, as in TAV-8A or YOV-1D. The 16 letters used for modified mission symbols in the 1962 system are as follows:

A	Attack	O	Observation
C	Cargo/transport	Q	Drone
D	Drone director	R	Reconnaissance
E	Special electronic installation	S	Anti-submarine
H	Search/Rescue	T	Trainer
K	Tanker	U	Utility
L	Cold weather operations	V	Staff transport
M	Missile carrier	W	Weather reconnaissance

STATUS PREFIX SYMBOLS. This continues the established practice in use by both the Navy and USAF of indicating the special status of prototype, service test and standard aircraft withdrawn from regular service for experimental purposes. Additional letters were picked up from USAF practice. The use of special test prefixes eliminated the practice, sometimes necessary, of assigning X-for-Experimental to a standard model to remove it from standard category in order to protect a test programme from delays incurred by mandatory compliance with Technical Orders or other directives pertaining to in-service modification of the standard aircraft or improvement of its combat effectiveness; items entirely inapplicable to the operation of the particular aircraft used for test work. The six status prefix letters in use are as follows:

Letter	Status	Remarks
G	Permanently grounded	Old USAF 'Class 26'; obsolete or damaged aircraft used for ground instruction and training. Navy previously had no designation for aircraft so used.
J	Special test, temporary	Aircraft modified for test work that can easily be reconverted to their original operational status.
N	Special test, permanent	Major alterations to test aircraft that preclude their ever being returned to original configuration.
X	Experimental	Usually first prototypes or other models not yet adopted for standard service.
Y	Prototype	Model procured in limited quantity to develop the potentialities of the design; old Air Corps/USAF 'Service Test'.
Z	Planning	Used for identification during the planning or pre-development stage; formerly used by Air Corps/USAF to identify obsolete types still being flown.

SERIES LETTER. This too has been adapted from Air Force practice to indicate sequential improvement or alteration of the basic model. There is no direct correlation of new suffix letters with older Navy dash numbers under the new system since direct substitution of letters for numbers is impossible due to the Navy practice of adding special purpose suffixes to the dash number under its old system.

US NAVAL AIRCRAFT COLOURING

THE standard colouring of US Naval aircraft has changed considerably over the years, so much that it is often possible to date a photograph fairly accurately from a study of the paint on the aircraft involved.

Other than providing a protective finish, overall aircraft colouring serves two primary purposes—as camouflage in areas apt to be under enemy surveillance and to identify aircraft used for special purposes. In each case, single colours or a combination of colours have been used. The most commonly used colours are described in the following paragraphs, which have been grouped under three different headings: STANDARD COLOURING, OVERALL SPECIAL COLOURING and PARTIAL SPECIAL COLOURING.

US Navy combat and support aircraft have appeared in camouflage colours during two periods—late 1917 to 1920 and continuously from February 1941 to the present. In some cases the camouflage is applied with a non-specular, dull non-reflecting finish; at other times the paint is glossy.

Non-standard colour schemes have existed in great profusion and detailed descriptions are not justified in this book. Many have been the result of special tests or assignments or the transfer of aircraft to the Navy from other services. In World War I and again in World War II the Navy operated significant numbers of aircraft obtained from the US Army and the Allied powers in their original colours.

STANDARD COLOURING

Colour was not a consideration for US Naval aircraft until America's entry into World War I in 1917. Up to that time the cloth-covered areas of the aircraft had been clear-doped, which imparted an all-white or light tan shade according to the nature of the material. If varnish was applied over the dope, a burnt-umber shade was imparted. Wood areas were usually clear-varnished, while sheet metal areas were left natural or were given a variety of non-standard coatings. Overall painted colouring was introduced late in 1917 as the impending operations in the war zone dictated the use of protective colouration. In every case where a change in basic colouring was decreed, there was a period of overlap which sometimes lasted several years. The basic schemes are noted below in chronological order of use, followed by notes on some special colour schemes adopted for various purposes.

Grey. A shade named in the specifications as *stone grey* was used on all surfaces—fabric, wood and metal alike—from late 1917 into 1920. After

19

that time the basic colouring changed to silver and the grey was retained for the wooden hulls of flying-boats until 1928.

This shade of grey, which photographed fairly dark, was also specified for metal parts throughout the 1920s, but was often replaced by silver. In 1930 a much lighter shade of grey, practically an off-white, was specified for metal parts and appeared on the entire fuselage of the new smooth-skin semi-monocoque designs introduced at that time. Grey for such purposes was replaced by silver in 1936.

Overall light grey was readopted as camouflage for naval aircraft in February 1941, and remained in use as a solid colour on some types into early 1942. From the summer of 1941 the grey was used as an underside colour when grey-green top and side colouring began to appear. The grey was replaced by white in a new camouflage scheme adopted early in 1943.

Light grey again appeared in 1955, this time on top surface and side camouflage in connection with white undersides. This was still the standard colouring for Navy combat aircraft in 1974. Over the same period considerably darker shades of grey have been used for patrol aircraft and some scouting and patrol helicopters and flying-boats.

Silver. Overall silver became the standard colouring for all US Navy aeroplanes, combat and non-combat types alike, from 1920 until the adoption of warpaint in 1941. It was retained after that time for most uncamouflaged or specially coloured types and remains in use to the present. From the early 1920s until 1942 the silver was supplemented by chrome-yellow on the entire upper wing surface (top wing only on biplanes) and the top of the horizontal tail surfaces. Only a few transport types of the 1930s flew without the yellow during this period. All-metal uncamouflaged trainers were in natural finish throughout World War II to 1955, and most transports remain unpainted except for white cabin tops today.

Sea Green and Grey. A non-specular grey-green shade called *sea green* was adopted as side and top surface camouflage in mid-1941 for aircraft operating primarily over water areas. Undersurfaces were light grey. By early 1942 it had become standard for all camouflaged aircraft and remained so until the end of 1942. On carrier-based types with upward-

Douglas TDB-1s of VT-7, showing the red-and-white tail stripes used from January to May 1942. The green finish for upper surfaces is carried on the underside of the folding wing panels; the remainder of the under surfaces is grey. (*IWM photo A9377*)

folding wings, the undersurfaces of the folding portions were painted with the topside grey-green to maintain the camouflage when the machine was viewed from above.

Sea Blue and White. In 1943 the green camouflage was replaced by a blue scheme graduated from a dark non-specular blue on top surfaces to a light grey-blue at lower sides where white undersurface colouring began. This combination remained standard into 1946.

Midnight Blue. Early in 1944 overall midnight blue was adopted as the standard colouring for all carrier-based aircraft. It was adopted for other types after the war and remained the standard camouflage until 1955. Lettering and markings were applied in white except for reserve unit markings which were in yellow for a short period after the war.

Brown and Green. A three-colour, brown and two-tone green, camouflage pattern, somewhat similar to the famous 'Sand and Spinach' used by the Royal Air Force early in World War II, was adopted for top and side surfaces of some naval aircraft operating in Vietnam in 1965. This followed the similar scheme adopted by the US Air Force, and involved either white or light grey undersurfaces.

OVERALL SPECIAL COLOURING

Yellow. The most widely used special colouring, starting in the 1920s, was overall chrome yellow. The purpose was to make the aircraft highly visible to others in the vicinity, and the colour remained in use until 1956. Its major use was on primary trainers, but was also applied to some advanced and instrument trainers during World War II. Yellow was also used for some utility and target-towing types starting in 1938, on some ambulance types during World War II, and on some radio-controlled test aircraft at the end of the war.

Black. Some types, especially flying-boats, were painted a dull black all over for night operations during World War II. Some night fighters obtained from the Army were glossy black.

Red. Overall red was used during World War II and post-war years for radio-controlled aircraft used as targets. More recently, solid red or red-orange 'Day Glo' colouring has been used for special test and some rescue aircraft.

White. Limited use of all-white colouring was made as Arctic camouflage and for some patrol flying-boats during World War II.

White and Orange. From 1956 a dual colour scheme of white and orange has been used for naval training aircraft, including obsolescent combat types used for training. It also appears on some utility and light transport types. White, with varying areas of orange 'Day Glo' according to aircraft configuration, has been standard for Coast Guard amphibians since 1960.

A red-orange band round the fuselage of this Chance Vought NFG-1D in 1949 indicates its use by a Navy Reserve unit. (*Peter M. Bowers*)

Blue, Yellow and Red. In 1946 the basic dark blue standard camouflage was modified for target tugs. The fuselage remained blue, but the wings and fixed tail surfaces were painted chrome yellow while movable surfaces and engine cowlings were painted red and chordwise red bands 3 ft wide were painted on each wing one-third the distance from fuselage to wingtip.

PARTIAL SPECIAL COLOURING

From the late 1920s, additional colours have been applied to localized areas of naval aircraft already carrying standard colour schemes to identify them for special purposes. These are not to be confused with the various stripes, chevrons and area colours used to identify individual aircraft as to organization (see page 32).

Command Blue. From the early 1930s until World War II, aircraft assigned to flag officers have had their fuselages, and sometimes engine nacelles and vertical tails, painted a solid blue. Aircraft assigned to commanders or to those officers with the rank of Captain had only the area of the fuselage from the nose to a point aft of the cockpit or cabin painted blue. When the flag officer is actually aboard the aircraft, a placard representing his 'flag' and carrying the appropriate number of stars is fitted in a special rack provided for the purpose on each side of the aircraft.

White Cabin Tops. Since World War II the armed forces have picked up a trick from the airlines and painted the fuselage tops of most personnel-carrying aircraft white. The purpose of this is to reflect sunlight and reduce cabin temperatures. This is applied without regard to existing colour schemes, and is even seen on some blue or dark grey patrol types.

Instrument Trainer Stripes. In the 1930s the Navy applied red bands 3 ft wide around the rear fuselage and chordwise on the upper and lower wing surfaces of aircraft used as instrument trainers. This was a warning to other pilots that the pilot was 'under the hood' and could not be expected to see other aircraft. In 1943 the colour of the striping was changed to light green and was still in use in 1968.

22

Target-Towing Stripes. In addition to the special colouring applied to the aircraft used to tow sleeve targets for gunnery practice since the end of World War II (see above), 3-ft-wide red bands were applied chordwise to the upper and lower wing surfaces midway between the wingtips and the fuselage.

Yellow Rescue Markings. From the end of World War II until 1960 Navy and Coast Guard aircraft used in air–sea rescue work have carried a standardized marking shared with the Air Force. This consisted of a 3-ft yellow band around the rear of the fuselage or hull just ahead of the tail and outlined with 6-in black stripes, and yellow wingtips set off with 6-in black stripes. In addition, some flying-boats and landplane patrol or transport types carried a large yellow rectangle on the nose with the word RESCUE painted on it in black.

'Barber Pole' Recall Stripes. A unique practice was followed at some of the large training centres along the Gulf Coast during World War II. The primary trainers operating from these bases did not carry radio and there was no way to recall them in an emergency, such as to avoid an imminent storm. An aeroplane at the station would be painted all over with equally spaced paralleled stripes contrasting with the basic colour. This was the 'Recall' machine, and its appearance in the area where students were flying meant that they should return to their base at once.

Reserve Unit Stripes. After World War II, aircraft assigned to squadrons manned by reserve pilots were distinguished from those flown by the regular Navy by means of a 2- to 3-ft yellow band painted around the aft fuselage ahead of the tail. Some of these aircraft operated without national marking on the fuselage. Aircraft identification numbers were also yellow.

This scheme was soon replaced by standard white markings, the national insignia was reapplied to the fuselage, and a 3-ft-wide red-orange band was painted around the fuselage as a background to the insignia. In the 1950s, the fuselage insignia was moved forward but the band remained approximately 3 ft forward of the horizontal tail. The band was then frequently broken to allow the service name and the squadron number or station name to be painted against the basic fuselage colour.

High gloss white and red-orange finish indicates the training role of this Grumman TF-9J of VT-23 at NAAS Kingsville, Texas, in 1967. (*Harry B. Adams*)

MARKINGS OF US NAVAL AIRCRAFT

NATIONAL MARKINGS

Prior to its entry into World War I the United States did not have an internationally recognized military aircraft marking. However, from late 1916 the Navy had used a blue anchor—a marking universally associated with nautical operations—painted on the rudder.

On May 19, 1917, an official marking was adopted for both Navy and Army aircraft. Based on the markings in use by other Allied powers, this consisted of a white five-pointed star backed by a blue disc. A smaller red disc was centred on the star. Initial application of this marking was full-chord on the wing, inboard of the ailerons. There was great variation in size and location into the early 1920s, when the size of the wing marking was reduced to a maximum of 60 in, centred between the aileron and the leading edge, and usually located one diameter inboard of the wingtip. Tail marking was three equal-width vertical stripes, red, white and blue with the red at the trailing edge.

In January 1918 the wing marking was changed at the request of the Allies to three concentric circles, red outside, then blue and a white centre. At the same time, the order of tail stripes was reversed. This arrangement remained official until August 19, 1919, but many 1917 aircraft operating

Photographed in February 1917, this Sturtevant AH-24 (later A76) is unpainted and bears the blue anchor insignia on rudder and under wing. (*US Navy*)

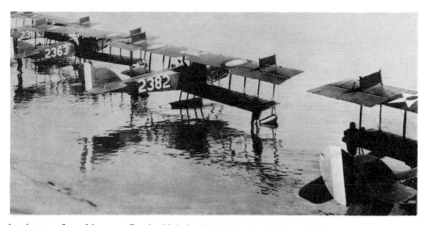

A mixture of markings on Curtiss N-9s in 1918. Number 2363 at left has correct markings; 2382 has the order of tail stripe colours reversed, as in 1917, and the aircraft extreme right has the 1917 star insignia on the wings but 1918 tail stripes. (*US Navy*)

in the United States carried the old markings throughout the war or had interesting mixtures of old and new. Although the 1917 star and order of tail stripes were readopted in August 1919, replacement on existing aircraft was slow, some carrying the circles into the early 1920s.

The tail stripe began to disappear from naval aircraft in 1926 when solid tail colours were adopted for unit recognition. By 1940 only a few utility and reserve aircraft used them. Even the majority of the aircraft without colour coding on the tail, such as trainers and transports, had their stripes deleted, beginning in the early 1930s. The Marine Corps, however, retained tail stripes right up to the adoption of camouflage in 1941.

The Coast Guard developed distinctive tail striping of its own. Starting in 1931 it expanded the vertical striping from the rudder only to cover the entire vertical tail surface. In 1934 this was changed to a new design, based on the Coast Guard flag, that was applied to the rudder only. The top quarter of the rudder was painted blue, with five alternating red and white vertical stripes beneath it. This was retained until the Coast Guard was taken over by the Navy upon US entry into World War II. The Coast Guard did not use the star marking on its aircraft until 1942. It used standard Navy applications during the war and has retained the star since regaining its status as a separate service after the war.

The adoption of camouflage in February 1941 brought changes in the application of the national marking. The star was added to each side of the fuselage and one was deleted from each wing, leaving one on the upper port surface and one on the lower starboard. Uncamouflaged types were not affected until 1943, when the new fuselage and wing locations were standardized. A few trainers operated to the end of the war with the old arrangement.

25

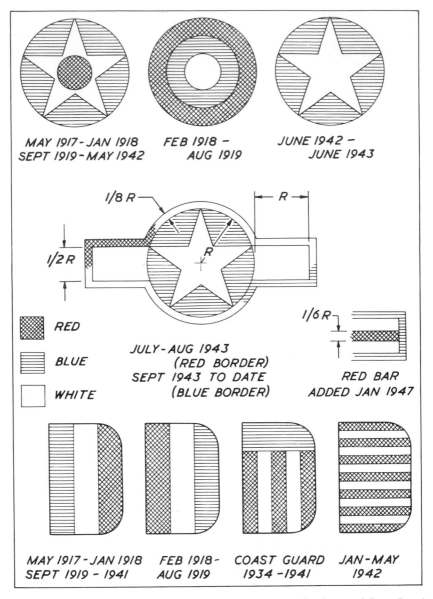

MAY 1917 - JAN 1918
SEPT 1919 - MAY 1942

FEB 1918 -
AUG 1919

JUNE 1942 -
JUNE 1943

1/8 R

R

1/2 R

R

RED

BLUE

WHITE

JULY - AUG 1943
(RED BORDER)
SEPT 1943 TO DATE
(BLUE BORDER)

1/6 R

RED BAR
ADDED JAN 1947

MAY 1917 - JAN 1918
SEPT 1919 - 1941

FEB 1918 -
AUG 1919

COAST GUARD
1934 - 1941

JAN - MAY
1942

The national insignia used on aircraft flown by the US Navy, Marine Corps and Coast Guard has undergone a number of changes in the period since 1917. This chart shows the colours and proportions of the various insignia used on wings and rudders from then until this volume went to press.

Grumman JF-2 in US Coast Guard markings. In October 1936 the serial number changed to V144. (*Gordon S. Williams*)

Earlier, the star had been applied to the fuselages of uncamouflaged aircraft engaged in the Neutrality Patrol, which had been established in the Atlantic Ocean by the Navy in 1940. This was not an additional national marking as such, but served to identify the aircraft engaged in this patrol. It was applied to the nose, not to the aft portion of the fuselage.

Further changes that proved to be temporary were made on January 5, 1942. Thirteen horizontal red and white rudder stripes similar to the pre-war Army markings but without the vertical blue stripe, were added to all camouflaged Navy, Marine and Coast Guard aircraft, and remained to the end of May 1942. Stars went back on both wingtips, where they remained into 1943. In some early 1942 Pacific combat areas, the stars were painted as large as possible on both wings and fuselage for positive identification.

Starting on June 1, 1942, the national marking underwent several changes in form (see page 26). The first was deletion of the red centre because of

Neutrality patrol star on the forward fuselage of a Grumman F3F-1 in 1940. (*US Navy*)

similarity to the Japanese marking. The next was addition of white rectangles and a red border in July 1943. This border was soon changed to blue because a flash of red could be mistaken for Japanese colouring in the heat of combat. Red was reintroduced in the form of a red bar through the white rectangle in January 1947. The markings of 1947 were still in effect in 1974.

SERIAL NUMBERS

From the beginning, individual US Naval aircraft have been identified by serial numbers. For the first 75 aircraft procured, various designations and procurement sequence numbers were combined according to the systems described on pages 4–5. Although assigned numbers confirm 75 aircraft procured or redesignated as a result of rebuilding, it is certain that fewer than 50 existed in the Navy inventory by 1917.

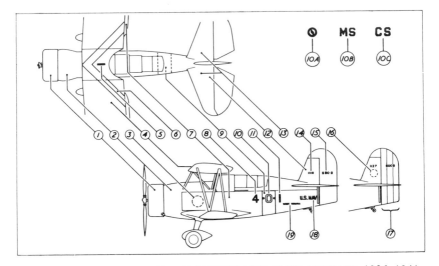

REPRESENTATIVE MARKINGS—UNCAMOUFLAGED FLEET AIRCRAFT, 1926–1941
1. Cowling in section colour and fractions (see page 34). 2. Metal parts light grey to 1935. 3. Navy Squadron Insignia or Marine Corps Emblem (location variable). 4. Upper surface of top wing chrome yellow. 5. Number of aeroplane in squadron (black). 6. Upper wing chevron in section colour (see page 34). 7. Fabric areas silver. 8. Squadron number (this and all other figures black on silver background). 9. Section leader stripe in section colour (on USMC since 1934). Usually carried completely around fuselage. 10. Squadron type letter (black, except white against red, blue, or black). 10A. Slash identified Navy Observation types into early 1930s; USMC Observation through 1936. Circle around Squadron type symbol identified USMC to July 1937. 10B. Prefix M added to Squadron type letter for USMC, July 1937. 10C. Prefix C added to cruiser-based scouts, July 1937. 11. Number of aeroplane in squadron. 12. Solid or partial-colour empennage for identification of squadron, ship or station, (seldom used by USMC). 13. Upper surface of horizontal tail chrome yellow on uncoloured tails; decreasing use in late 1930s. 14. Navy Serial number (also called Bureau number or BuNo.) Colour as Item 10. 15. Type–Model–Manufacturer designation. Colour as Item 10. 16. USMC Squadron Insignia. 17. Vertical tail stripes in decreasing use by Navy from 1926 through 1940; used on most USMC types into 1941. 18. Service name, U.S. NAVY or U.S. MARINES. 19. Ship name for battleship and cruiser-based aircraft (location variable).

28

This system of numbers was abandoned and serials were thereafter assigned in strict sequence of procurement, taking number 51 as a convenient starting-point. The old prefix letters were retained for a short time as a general type designation but no longer had relationship to the serial numbers. On May 19, 1917, these were officially abandoned and replaced by the letter A (for aeroplane). This new arrangement was applied to each side of the fuselage or hull in large block figures, a dash frequently separating the letter from the numbers (this dash has been omitted from serials quoted throughout this volume for the sake of a standardized presentation). The serial number remained in this location until 1924, when it was applied in 3-inch figures to the vertical fin to make room for unit markings on the fuselage. From 1918 to 1924 the serial was also painted in black at the top of the white rudder stripe in 3-inch figures. The letter A was abandoned at the end of June 1930. The serial numbers related to the aeroplanes described in this book include the letter where applicable.

Serials were assigned by the purchasing agency (Bureau of Aeronautics since 1921) at the time an unbuilt aeroplane, balloon, non-rigid airship or glider was contracted for or upon acquisition of an existing aircraft by purchase, capture or transfer to the Navy from the Army or Air Force. When assigned serials were unused, as in the case of a contract cancellation, they were seldom reassigned.

Procurement of naval aircraft was on a relatively small scale compared to the Army's purchases, and did not reach 9,999 until 1934. At that time, rather than expand the serial number to five digits, the Navy started a new series at 0001. The rapid increase in procurement immediately preceding World War II soon produced a situation where numbers in the second serial series threatened to duplicate late numbers in the first series that were still in use. To avoid this, the second series was discontinued at 7,304 in 1940, and a third series of five digits was started at 00001. This was allowed to continue into six digits during World War II, and by late 1974 the series was approaching 160,000. This, of course, does not reflect the actual number of aeroplanes delivered, because of large-scale cancellations of contracts at the end of World War II and other programme changes in later years.

Marine Corps aeroplanes were procured by the Navy and fitted into the Navy serial number system, but the Coast Guard used three different systems of its own from 1927. The first was a simple numerical system that reached 55 but covered only 16 aeroplanes because of grouping different models in numerical series, with a 20-series for amphibians and a 50-series for flying-boats, for example.

In 1934 the Coast Guard adopted a new three-digit grouping series: 100s for amphibians, 200s for flying-boats, 300s for landplanes and 400s for convertible designs that could operate on wheels or floats. A third system was adopted in October 1936 when every aeroplane ever assigned to the Coast Guard was given a three-digit serial number prefixed by the letter V.

During World War II the Coast Guard was taken over by the Navy and operated a mixture of its own pre-war V-serialled aircraft and later models procured with Navy serial numbers. The V was dropped from those aircraft still carrying them at the end of 1945. These were all three-digit numbers, and the single digit 1 was added as a prefix in January 1951 to make a four-digit number for use as a radio call-sign; at the same time five-digit Navy serials on Coast Guard planes were cut down to their last four digits. (The Coast Guard was returned to Treasury Department jurisdiction in June 1946.)

The standard location for the serial number remained near the top of the vertical fin until 1947, when it, along with the model and service designations, was moved to a point on the fuselage below the leading edge of the horizontal tail. The serial number was usually in figures a little larger than those used for the model designation and name of the service.

During and after World War II, aircraft not identified by a unit number frequently carried the serial number, or at least the last three digits of it, on the nose. In 1957 the full serial number, or the last four digits, began to reappear in figures 12 or more inches high on the vertical tail surfaces.

SERVICE DESIGNATORS

1924–1941. Starting in 1924, when the large serial numbers were deleted from the fuselage, the words U.S. NAVY in block letters were applied to each side of the fuselage or hull to identify naval aircraft. To make room for organizational markings, the location became standardized towards the rear of the fuselage, sometimes almost entirely under the horizontal tail. Up to World War II the preferred height was 8 in.

The Marine Corps followed the Navy marking system and carried the lettering U.S. MARINES in the same locations used by Navy planes. The Coast Guard, not using Navy/Marine organizational markings, was able to apply the lettering U.S. COAST GUARD along the fuselage centreline midway between the wings and tail.

From 1929, the letters U.S. NAVY were added to the undersides of the wing (lower wing only on biplanes). The Marine Corps followed suit, but also applied its lettering across the top wing surface at times. In the case of sweptwing aircraft such as the Curtiss F8C Helldiver, the top wing lettering sometimes followed the sweep and was sometimes applied straight across. The Coast Guard used only the initials USCG under the port and starboard wings. The Navy and Marines deleted the wing lettering on July 8, 1931. The Marines retained it on the top wing in some special applications for several years. The Coast Guard retained wing lettering until the adoption of camouflage.

World War II. Upon the adoption of camouflage in 1941, the service designations of Navy, Marine and Coast Guard aircraft were removed from the fuselages and hulls (and the lower wings of Coast Guard aircraft) and relocated in 1-in letters above the serial number on the vertical fin (up to 4-in on large bombers and flying-boats). The old sizes and locations were retained for uncamouflaged aircraft. On camouflaged and un-camouflaged types alike the prefix letters US were deleted.

Post-war Years. After the war, the model designations and serial numbers of camouflaged types were moved to a rear-fuselage position beneath the leading edge of the horizontal tail, and the service designation, in the same or slightly larger-sized lettering, moved with it. This soon became the standard location for non-camouflaged types as well. It soon became desirable to emphasize the name of the service and the words NAVY or MARINES were applied farther forward on the fuselage in 12-in letters, starting in February 1950. In some cases, the words appeared on the vertical fin of delta wing or tailless types and on some transport types the full words UNITED STATES NAVY were spelled out above the line of cabin windows. The NAVY or MARINES designation was authorized for use on the

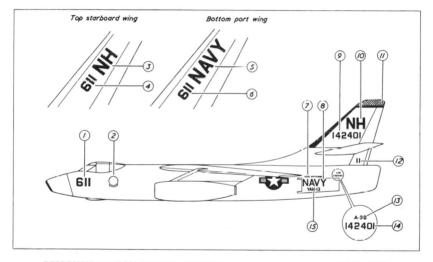

REPRESENTATIVE MARKINGS—CAMOUFLAGED FLEET AIRCRAFT, 1947–1968

1. Unit aircraft numeral: up to three digits. Sometimes last three digits of serial number. Minimum height: 16 in. 2. Squadron insignia: location variable. 3. Unit identifying letter(s) from January 1947: 36-in on upper starboard surface near tip, 16-in on lower starboard between national insignia and fuselage. Sometimes on lower port wing to 1950. 4. Unit aircraft numeral: 24-in on upper starboard wing, 18-in inboard of (3). 5. Branch of service: lower port wing from 1950 on all but white-painted carrier-based or patrol types. Minimum: 24-in for NAVY, 12-in MARINES. 6. Unit aircraft numeral: 16-inch on lower surface of port wing inboard of (5). Not on white-painted carrier or patrol aircraft wings since 1960. 7. Name of aircraft carrier: location variable. 8. Branch of service: largest practicable of 12, 16, 20, 24, 30, or 36-in since February, 1950. Two-inch figures above serial number, 1947–1950. 9. Serial (BuNo) number used in whole or last four digits as radio call number in addition to standard application at (14). Minimum height: 12 in. 10. Unit identifying letter(s): 36-in high single, 30-in double. 11. Squadron identification colour at top 7-in of vertical tail. Other unit identification variable in design and location. 12. Last two digits of (1). Location on vertical tail variable. 13. Model designation: 2-in letters. 14. Serial Number (BuNo): 4-in figures. 15. Squadron designation added 1950: half size of (8); location variable.

31

undersides of port wings, but the practice was not uniform. Some reserve aircraft shared by Navy and Marine squadrons carried both designations on the side of the same aeroplane.

SQUADRON IDENTIFICATION SYSTEM

From 1925 until early in 1942, all US Naval aeroplanes assigned to regular squadrons were painted with a standardized number–letter–number system that identified each individual machine as to number and class of squadron and position of the machine in the squadron. For example, the marking 2-F-5 on the side of the fuselage (nose of flying-boats and patrol landplanes), in block figures, identified the aircraft as belonging to Squadron VF-2 and showed that it occupied position number 5 within the squadron. For this purpose the number of the squadron appeared before the letter to avoid duplicating aeroplane designations, as F2B.

The letters painted on the aircraft correspond to the primary aircraft type designations listed below. Dual designations of aircraft, as SB for Scout-Bomber, were not reflected in the squadron class letter. For administrative purposes, all squadron designations were preceded by the letter 'V' to identify heavier-than-air units.

VB   Bombing
VF   Fighting
VJ   Utility
VM   Miscellaneous
VN   Training
VO   Observation
VP   Patrol
VS   Scouting
VT   Torpedo
VX   Experimental (not to be confused with the X-for-Experimental used in designations of prototype aircraft).

Additional letters, not applied to the aircraft, were used to identify a secondary mission or particular status of the squadron. VF-3B would indicate that Fighting Squadron 3 also had a bombing role while VS-2R identified the second scouting squadron in the reserves which was not to be confused with Scouting Squadron 2 in the fleet. Scouting squadrons attached to battle cruisers had their designations followed by the letter 'S', as VS-5S, while observation types flying from battleships were further identified by the letter 'B', as VO-3B. The letter 'M' identified aircraft of the Marine Corps, as VF-9M.

On July 1, 1937, all Navy and Marine squadrons were redesignated. The distinguishing letters were retained but the order was changed, e.g. VF-9M becoming VMF-1 with the 'M-for-Marines' preceding the squadron class letter. This changeover brought changes in the painting of some aircraft. Since 1926, the Marines had distinguished their fleet aircraft from those of the Navy by enclosing the squadron class letter within a circle, without the separating dashes used by the Navy. In the 1937 change, the circle was eliminated and the letter 'M' was applied to the aircraft as a prefix, VF-4M becoming VMF-2 and the fifth aeroplane in this unit being painted 2-MF-5. Similarly, the cruiser-based scouts adopted the letter 'C-for-Cruiser' to distinguish them from scouts based on aircraft carriers. Such distinction was not necessary for the Observation squadrons since they operated only from battleships (the plain O-for-Observation designation had been dropped for new procurement in 1934 and had been replaced by SO for Scout-Observation and later OS for Observation-Scout, but the observation squadrons were still identified by 'O' alone).

After squadron designations were reapplied to Navy and Marine aircraft in 1950, the full administrative designation was painted on, such as VAH-13 for Heavy Attack Squadron 13.

The standard naval squadron was made up of 18 aircraft of the same make and model, and was divided into six sections of three aeroplanes each. The number of each individual aeroplane tied it to a fixed position in the rigid squadron organization, which was further indicated by special markings and colours.

The leader of each section was identified by a 20-in band in the section colours painted around the fuselage and centered on the squadron designation letter. This colour was repeated on the ring cowling of radial engines and was painted completely around the ring on the forward portion of deep-chord cowlings on leader aircraft only. The second and third machines in each section did not carry the fuselage stripe and had only the top or bottom of the cowling painted, as shown in the chart on page 34.

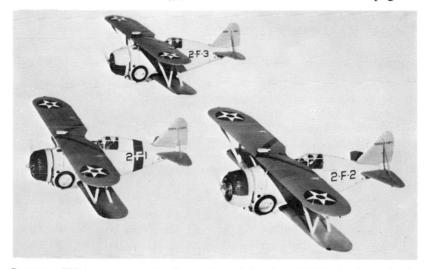

Grumman F2F-1s comprising the first section of Squadron VF-2B. Section leader has fuselage stripe and full-colour cowling; aircraft 2 and 3 have partial colour cowlings. Aircraft 4 in this squadron would be leader of second section, with cowling pattern repeated in different colours (see page 34).

33

Section Number	Section Colour	Aircraft Number	Leader Stripe	Cowling Painted
1	Insignia red	1	Yes	Full
		2	No	Top half
		3	No	Bottom half
2	White	4	Yes	Full
		5	No	Top half
		6	No	Bottom half
3	True blue	7	Yes	Full
		8	No	Top half
		9	No	Bottom half
4	Black	10	Yes	Full
		11	No	Top half
		12	No	Bottom half
5	Willow green	13	Yes	Full
		14	No	Top half
		15	No	Bottom half
6	Lemon yellow	16	Yes	Full
		17	No	Top half
		18	No	Bottom half

The section colour was also used in a chevron painted full-chord across the topmost wing surface and the individual aeroplane number was painted in black within the chevron, which was usually pointed forward but was sometimes pointed aft.

There were exceptions to the 18-plane squadron organization. Sometimes squadrons had additional utility aircraft assigned, and these were given continuing numbers past 18 although they did not carry section colours. Scout aeroplanes attached to cruisers in the fleet were organized in four-plane sections, with all four assigned to one cruiser. This changed the numbers of the section leaders, number 5 being the leader of the second section, for example, and introduced an additional cowling colour pattern—a longitudinal stripe one-sixth the diameter of the cowl painted along the top and bottom. Some cruiser sections used only two aircraft, so number 3 was the leader of the second section. Marine Squadron VF-9M operated in the 1934–7 period as a 24-plane squadron with only individual aeroplane numbers for identification and solid-colour cowlings, tails and wheel discs, for each machine in an eight-plane section.

The number and letters were normally in black. When the command stripe was red, blue or black, the squadron letter was white. When grey warpaint was adopted in 1941, all lettering became white. When grey-green was adopted, the lettering was sometimes white and sometimes black.

The designating system remained in use during the war years, but for security reasons the full designation was seldom painted on the aeroplane —only the individual number. The section colours were deleted with the adoption of camouflage.

From late 1946 squadrons were identified as to group or carrier by a two-letter code appearing on the vertical tail, the upper starboard wing, and sometimes the lower port wing. Reserve units had a single number to identify the section of the country and a letter to identify the individual organization or base. Individual carrier-based aircraft have been identified by numbers within blocks of one hundred assigned to the several squadrons within a carrier group. This appeared on the nose or near the cockpit. Other aircraft use sequential numbers within their units or the last two or three digits of the aircraft serial number, usually on the nose. This number was repeated just ahead of the squadron letters on the wing.

In February 1950 squadron numbers and letters reappeared on the fuselage in figures at least 12-in high. These were still in use, along with the unit code letters, in 1974.

UNIT IDENTIFICATION

From 1926 to 1941 coloured tail assemblies were the principal means of identifying aircraft at a distance as to unit. Different squadrons using the same type of aircraft had their entire horizontal and vertical tail surfaces painted in one of six solid colours—insignia red, white, true blue, black, willow green and lemon yellow. This tail colouring did not immediately replace the standard vertical rudder stripe national marking. The stripe gradually disappeared, however, until only a few carrier utility planes and land-based reserve units used them by 1940. The Marines retained the stripes on most aircraft until 1941 and applied colour only to the vertical fins of fighter types as a squadron identification in the late 1920s (a notable exception was VF-9M in the 1934–7 period, which used solid colours on the entire tail for different sections within the squadron).

Originally, the solid colour tails distinguished squadrons using similar aircraft. From March 15, 1937, the colour identified all the aircraft of a particular aircraft carrier, as follows:

LEMON YELLOW—LEXINGTON (CV-2)
WHITE—SARATOGA (CV-3)
WILLOW GREEN—RANGER (CV-4)
INSIGNIA RED—YORKTOWN (CV-5)
TRUE BLUE—ENTERPRISE (CV-6)
BLACK—WASP (CV-7)

35

Solid colouring also identified some land-based aircraft as to home station. When the number of squadrons of a given aircraft type began to exceed the six basic colours, it became necessary to alter the solid-colour scheme by using patterns, such as horizontal stripes or diamonds. To distinguish the catapult seaplanes used by cruiser scouting squadrons from similar models used by battleship observation squadrons, the scouts used coloured bars on their tails while the observation types used solid colours. This colouring system was abandoned with the adoption of camouflage in 1941.

In 1944, when overall midnight blue became the standard colour for carrier-based aircraft, various white patterns were added to the wingtips and vertical tails to identify the aircraft of a particular carrier.

This system was abandoned at the war's end and the aircraft were identified as to unit (wing or group, and not carrier) by one or two large letters on the vertical tail. This system was still in use in 1974, but has been supplemented by a great variety of tail colour patterns or fuselage decoration. In addition to the orange stripe around the rear fuselage, aircraft of the reserves were identified as to regional unit by a two-digit number–letter combination on the vertical tail and the upper starboard wing.

Aircraft assigned to ships or naval air stations as utility or special-purpose types were usually identified as to unit by simple lettering and a number, such as U.S.S. SARATOGA–5 or N.R.A.B. OAKLAND–10 (Naval Reserve Air Base). Aircraft of the commanders of large units into early World War II carried the name of the unit on the fuselage, as COMMANDER, RANGER AIR GROUP, or COMMANDER, CARRIER DIVISION ONE.

A well-known marking in the 1930s was the single word ANACOSTIA painted on the fuselage of experimental and utility aircraft assigned to NAS Anacostia, Maryland. After World War II, test aircraft carried the letters NATC, meaning Naval Air Test Center. Also in post-World War II years, the station names of reserve units frequently appeared against the reserve stripe, with or without the prefix letters NAS.

In the late 1920s and early 1930s, some stations and organizations used systems similar to operational squadron markings, as 2-HR-7 for the second of several units at Hampton Roads, Virginia, or 2-Xdi-3 for the Second Experimental District.

SQUADRON INSIGNIA

The use of individual squadron insignia became prevalent in the mid-1920s. Standard location for naval aircraft was close to the cockpit on conventional landplanes and on the nose of flying-boats. The Marine Corps used the Marine emblem in this position and carried the squadron insignia on the vertical fin. The practice was abandoned during World War II, but squadron insignia have reappeared in limited applications since the readoption of light grey camouflage.

36

Aeromarine 39-B making the first landing on USS *Langley* on October 26, 1922. (*US Navy*)

Aeromarine 39-A, 39-B

In 1917 the Navy placed with the Aeromarine Plane and Motor Company of Keyport, NJ, what was at that time the largest single order for Navy aircraft—50 of the Model 39-A and 150 39-B trainers. These were conventional two-bay wood and fabric biplanes and could be fitted with wheels or floats. The 39-As used the four-cylinder Hall-Scott A-7A engine of 100 hp and the seaplane versions had twin wooden floats. The 39-B was powered by the 100 hp Curtiss OXX-6 engine, the seaplanes having the single main pontoon with small wingtip floats for stability which the Navy preferred for its training and service seaplanes and was to retain until seaplane trainers were dropped from the inventory in 1960.

A number of the 39-Bs survived World War I, and two were used for the Navy's early experiments in deck landing. Various types of arrester gear were tried on a dummy carrier deck at Langley Field, Virginia, in 1921. The aeroplane was fitted with the forerunner of the modern hook that engaged the cross-deck arrester cables, while alignment hooks were fitted to the undercarriage to engage longitudinal wires on the deck to keep the machine running straight. In anticipation of forced landings at sea in the course of later operations from shipboard, a hydrovane was fitted ahead of the wheels to prevent nosing over on alighting.

On October 26, 1922, a 39-B piloted by Lt Cdr Geoffrey DeChevalier, Naval Aviator No. 7, made the first landing on the deck of the Navy's first aircraft carrier, the USS *Langley*, a converted collier. This was nine days after the first take-off had been made in a Vought VE-7.

TECHNICAL DATA (Aeromarine 39-A)

Manufacturer: Aeromarine Plane and Motor Company, Keyport, NJ
Type: Training biplane.
Accommodation: Two pilots.

37

Power plant: One 100 hp Hall-Scott A-7A.

Dimensions: Span (upper), 47 ft, (lower) 36 ft.; length, 26 ft 3 in; height, 12 ft 8¾ in; wing area, 494 sq ft.

Weights: Empty, 1,650 lb; gross, 2,220 lb.

Performance: Max speed, 73 mph at sea level; climb, 27 min to 5,000 ft; service ceiling, 7,500 ft; endurance, 4·6 hr.

Serial numbers:
 A450–A499.

TECHNICAL DATA (Aeromarine 39-B)

As Aeromarine 39-A except:

Power plant: One 100 hp Curtiss OXX-6.

Dimensions: Length, 30 ft 4¼ in; height, 13 ft 2 in.

Weights: Empty, 1,939 lb; gross, 2,505 lb.

Performance: Max speed, over 73 mph at sea level; climb, 30·6 min to 5,000 ft; service ceiling, over 5,000 ft; range, 273 st miles.

Serial numbers:
 A500–A649.

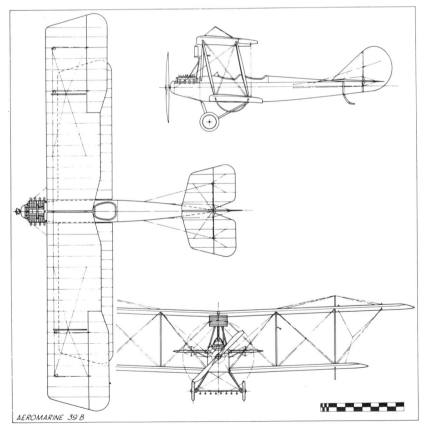

AEROMARINE 39 B

The Atlantic RA-2 serial A8018 with a spare OC-1 wing under the fuselage, in Nicaragua, 1929. (*US Navy*)

Atlantic TA, RA Series

In 1926 the Navy followed the lead of the Army Air Corps, which had evaluated a new commercial airliner, the Fokker F.VII trimotor, following its sensational participation in the 1925 Ford Reliability Tour. After testing the new machine, which had been built in Holland, the Army ordered three of a modified version designated the C-2. The Navy followed suit and ordered an additional three under the designation of TA-1 for Transport, Atlantic. The Navy also had some direct involvement with the prototype, as Edsel Ford, son of the motor magnate, bought it for use by Navy Commander Richard E. Byrd's 1926 Arctic expedition. On May 9 Byrd and Navy pilot Floyd Bennett made a successful round trip flight over the North Pole from their base in Spitzbergen.

Atlantic Aircraft Corporation was formed late in 1923 to manufacture Fokker aircraft in the United States at a time when Anthony Fokker's name was clouded by his war-time activities on behalf of Germany. The TA-1s were virtually identical with the F.VIIA trimotor except for a wider fuselage better suited to carrying cargo, and the use of later 220 hp Wright J-5 engines. The TA-1s were put into service in 1927 and 1928 with the US Marines in Nicaragua. The T-for-Transport designation conflicted with that of torpedo planes, so the TA-1s were soon redesignated RA-1, the letter R being adopted as the identifying symbol for multi-engine Navy transport and cargo aircraft. The three RA-1s became RA-3s following a change to 300 hp Wright J-6-9 engines.

Following delivery of the TA-1s, the Navy ordered three later versions similar to the Army C-2A as TA-2 (later RA-2). The principal difference involved was the use of a longer wing. When the RA-2s were also refitted with J-6 engines, they too were redesignated RA-3, thereby giving the Navy RA-3s with two different wings. One RA-3 was bought as such.

The final Navy/Marine Fokker was the RA-4, a stock F-10A airliner as used by a number of American airlines in the 1929–31 period. Installation of equipment required to meet military requirements seriously upset the balance and control of the RA-4, and major alterations were made to try to meet performance specifications. These came to naught, and the Navy rejected the RA-4.

TECHNICAL DATA (TA-1)

Manufacturer: Atlantic Aircraft Corporation, Teterboro Airport, NJ.
Type: Personnel and cargo transport.
Accommodation: Two pilots and eight passengers.
Power plant: Three 220 hp Wright J-5s.
Dimensions: Span, 63 ft 4½ in; length, 49 ft 1 in; height, 13 ft 4 in; wing area, 630 sq ft.
Weights: Empty, 5,400 lb; gross, 9,000 lb.
Performance: Max speed, 116 mph at sea level; climb, 8·6 min to 5,000 ft; service ceiling, 12,050 ft; range, 460 st miles.
Serial numbers:

TA-1 (RA-1, RA-3): A7661–A7563. TA-3 (RA-3): A8157.
TA-2 (RA-2, RA-3): A8007–A8008, A8018. RA-4: A8841.

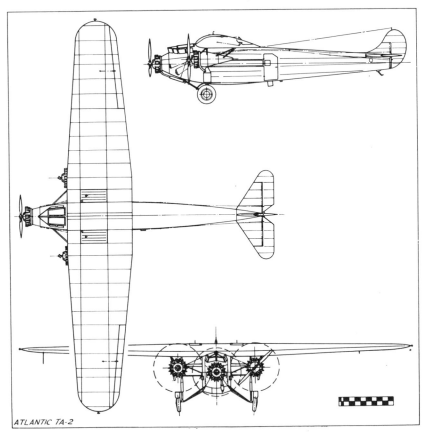

ATLANTIC TA-2

Beech JRB-1 serial 2543, the first Model 18 acquired by the Navy, fitted with a cupola to facilitate the control of drone aircraft. (*Beech photo*)

Beech JRB Expeditor, SNB Kansan, Navigator

The first five of more than 1,500 twin-engine Beech 18s used by the Navy and Marine Corps were ordered in 1940 with the designation JRB-1. Developed from the civil model C-18S and the USAAC F-2 variant of the Beech 18, they were intended primarily for photography. In addition to special fittings in the fuselage, they had a distinctive fairing over the cockpit to improve all-round visibility for the crew. Six more JRB-1s soon followed the first five, and at the same time the Navy ordered 15 JRB-2s for use as light transports.

For a short time, the five-seat JRB transports were known as Voyagers, but the official 'popular' name eventually become Expeditor and further quantities were obtained during World War II. These included 23 JRB-3s, which were similar to the USAAF's C-45B and were fitted for photography; and the 328 JRB-4s, which were the Naval equivalent of the standardized seven-seat UC-45F.

Beech SNB-1 gunnery, bombing and pilot trainer. (*Beech photo*)

41

Beech SNB-5P with camera hatches in bottom of fuselage. (*Beech photo*)

Following adoption of the Beech 18 by the USAAF in 1941 as a navigation and bombardier trainer, the Navy acquired a quantity of similar aircraft in various training roles. Equivalent to the AT-11, the SNB-1 Kansan in 1942 had a dorsal turret and a modified nose with a bomb-aimer's station and was used to train crews for Naval patrol aircraft. Outwardly similar to the JRB-2 and equivalent to the Army's AT-7, the SNB-2 Navigator also was ordered in 1942. Some were delivered with changed equipment as the SNB-2C (similar to AT-7C) and in 1945 others were modified for ambulance duties with the designation SNB-2H. The SNB-2P version carried cameras for training in photo-reconnaissance.

The Navy's Beech 18s remained in service in considerable numbers after the end of the war: some 1,200 were still in the inventory in 1950 and a sizeable quantity remained operational 20 years after V-J Day. As well as continuing in service as trainers and light transports they were adapted for special roles, such as the SNB-3Q trainer for electronic countermeasures techniques. A major modernization and remanufacturing scheme was undertaken in 1951, similar to that for USAF C-45s. This resulted in the

Beech JRB-4 in overall blue finish at Shanghai in November 1945. (*Peter M. Bowers*)

42

appearance of the JRB-6 transports and SNB-5 and SNB-5P trainers, the latter being modernized SNB-2Ps.

When the Defense Department introduced its unified designation system in 1962, surviving SNB-5s and -5Ps then became TC-45Js and RC-45Js respectively. Subsequently, the primary mission of the TC-45Js changed from training to utility transport, and these aircraft then became UC-45Js.

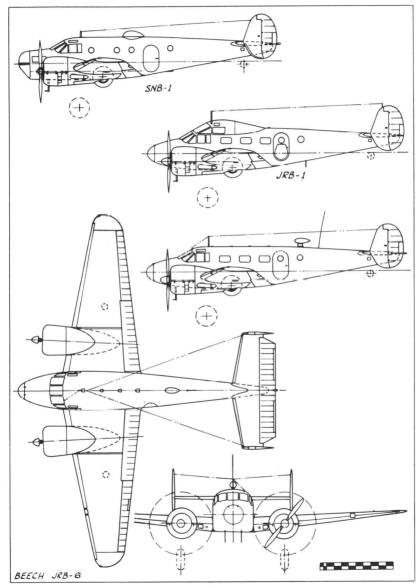

SNB-1

JRB-1

BEECH JRB-6

43

TECHNICAL DATA (JRB-2)

Manufacturer: Beech Aircraft Company, Wichita, Kansas.
Type: Light transport.
Accommodation: Crew of two and six passengers.
Power plant: Two 450 hp Pratt & Whitney R-985-50s or -AN-4s.
Dimensions: Span, 47 ft 8 in; length, 34 ft 3 in.; height, 9 ft 4 in; wing area, 349 sq ft.
Weights: Empty, 5,501 lb; gross, 7,850 lb.
Performance: Max speed, 225 mph at 6,770 ft; cruising speed, 129 mph; initial climb, 1,290 ft/min; service ceiling, 24,900 ft; range, 1,250 st miles.
Serial numbers:

JRB-1: 2543–2547; 4709–4710; 4726–4729; 09771.
JRB-2: 4711–4725; 90522–90523.
JRB-3: 76740–76759; 86294; 87752; 84032.

JRB-4: 44315; 44555–44684; 48246–48251; 66395–66471; 76760–76779; 85096–85135; 86293, 86295, 86296; 87753; 90532–90581.

TECHNICAL DATA (SNB-1)

Manufacturer: Beech Aircraft Company, Wichita, Kansas.
Type: Multi-engine, instrument and gunnery trainer.
Accommodation: Crew of four or five.
Power plant: Two 450 hp Pratt & Whitney R-985-AN-3s.
Dimensions: Span, 47 ft 8 in; length, 34 ft 3 in; height, 9 ft 4 in; wing area, 349 sq ft.
Weights: Empty, 6,203 lb; gross, 8,000 lb.
Performance: Max speed, 209 mph at sea level; cruising speed, 117 mph; initial climb, 1,620 ft/min; service ceiling, 21,500 ft; range, 780 st miles.
Armament: One 0·30-in gun in dorsal turret. Ten 100 lb bombs.
Serial numbers:

SNB-1: 39749–39998; 51025–51094.
SNB-2: 03553–03562; 12354–12389; 39192–39291; 51200–51293; 67100–67129; 67155–67383.

SNB-2C: 23757–23856; 29551–29664; 51095–51199; 51294–51349.

Beech SNB-5 in orange and white trainer colouring after redesignation as TC-45J in 1962.
(Peter M. Bowers)

44

Bell HTL-7 instrument trainer with fuselage side panels removed. (*Bell photo*)

Bell HTL, HUL

First flown on December 8, 1945, and granted the first-ever US type approval for a commercial helicopter in March 1946, the Bell Model 47 was still in production 20 years later for both civil and military users. A number of different versions were procured for Navy use between 1947 and 1958, starting with ten HTL-1s. These were part of a service evaluation quantity of Bell 47As with 178 hp Franklin O-335-1 engines. Designated as training helicopters, they were used to gain experience in rotary wing operations and were followed in 1949 by 12 HTL-2s, equivalent to the commercial Model 47D with its enclosed 'goldfish-bowl' cockpit. A change to a 200 hp Franklin engine in the Bell 47E distinguished the two-seat HTL-3. Nine were delivered in 1950 and 1951, plus three under MAP (Military Assistance Program) to Brazil.

First major Navy version of the Bell 47 was the HTL-4, equivalent of the commercial 47D-1 which dispensed with the fabric covering on the rear fuselage framework. Deliveries totalled 46 in 1950–1, and these were followed in 1951–2 by 36 similar HTL-5s with O-335-5 engines. To meet continuing requirements for training helicopters, the Navy next procured 48 HTL-6s in 1955–6; these were similar to the three-seat civil Model 47G incorporating a number of refinements, including a small elevator geared to operate in unison with rotor tilt for better stability.

The four-seat Model 47J, with its lengthened cabin, was also procured by the Navy, 28 being delivered as HUL-1s in 1955–6 for general utility operations by Squadron VU-2, including service aboard ice-breakers. To replace two HTL-1s which had passed into service with the US Coast Guard, two of the utility helicopters were transferred to the Coast Guard as HUL-1Gs. Two others were modified to have the Allison YT63-A-3

Bell HTL-1, Helicopter Training and Utility Squadron, Lakehurst, NJ. (*Howard Levy*)

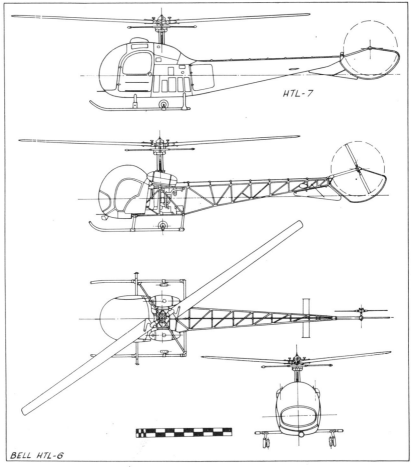

HTL-7

BELL HTL-6

46

turboshaft engine as HUL-1M (Bell 47L) and operated at gross weights up to 2,850 lb compared with the usual 2,800 lb of the HUL-1.

Final Navy version of the Bell helicopter was the HTL-7 (Bell 47K) which comprised an HUL-1 fuselage, engine and transmission with a new two-seat cabin incorporating full dual-control'and all-weather flight instrumentation. The 18 HTL-7s delivered in 1958 were assigned as instrument trainers. Their production brought total Navy/Marine procurement of the Model 47 to 209. As well as their primary use as trainers, some were used for general utility duties, and several HTL-4s were used in the Korean War by Marine units. Designations of versions still serving in 1962 were integrated in the Army/AF H-13 series as follows: HTL-4 to TH-13L, HTL-6 to TH-13M, HTL-7 to TH-13N, HUL-1 to UH-13P, HUL-1G to HH-13Q and HUL-1M to UH-13R.

TECHNICAL DATA (HTL-6)

Manufacturer: Bell Helicopter Company, Fort Worth, Texas.
Type: Training and general purpose helicopter.
Accommodation: Pilot and two passengers.
Power plant: One 200 hp Franklin O-335-5B.
Dimensions: Rotor diameter, 35 ft; length, 27 ft 4 in; height, 9 ft 6 in; disc area, 965 sq ft.
Weights: Empty, 1,435 lb; gross, 2,350 lb.
Performance: Max speed, 100 mph at sea level; cruising speed, 70 mph; initial climb, 780 ft/min; service ceiling, 10,900 ft; range, 250 st miles.
Serial numbers·

HTL-1: 122452–122461.
HTL-2: 122952–122963.
HTL-3: 124561–124569;
    144693–144695 (MAP).
HTL-4: 128621–128636;
    128887–128916.
HTL-5: 129942–129977.

HTL-6: 142373–142396;
    143148–143171.
HTL-7: 145837–145854.
HUL-1: 142363–142372;
    143134–143147;
    147578–147581; 148277.
HUL-1M: 149838–149839.

Bell HUL-1 with pontoons operating with squadron HU-1 in Antarctica. (*US Navy*)

47

Bell UH-1E utility transport helicopter of the US Marine Corps Headquarters and Maintenance Squadron 24. (*US Navy*)

Bell UH-1 Iroquois, AH-1 Sea Cobra

First Navy procurement of a version of the Bell UH-1 Iroquois helicopter was made on behalf of the Marine Corps, which raised a requirement for an assault combat helicopter in March 1962. The variant chosen by the USMC to fulfil this requirement was similar to the Army UH-1B (Bell Model 204) in most respects except instruments and avionics, and was designated UH-1E. The engine was a single Lycoming T53-L-11 rated at 1,100 shp and deliveries began on February 21, 1964, to Marine Air Group 26 at New River, NC. A total of 250 UH-1Es was procured, plus 20 TH-1Es for use as crew trainers.

In 1968, the Navy began procurement of two new versions of the Iroquois to meet its own needs for a rescue helicopter and a training helicopter respectively. The 27 HH-1Ks were similar to UH-1Es apart from having the Lycoming T53-L-13 engine, different avionics and overall orange finish to conform to international standards for rescue aircraft. Deliveries were made in 1970. The TH-1L was also similar to the UH-1E, but with special provision for the training role, and the T53-L-13 engine de-rated, as in the HH-1K, to 1,100 shp. The first of 90 TH-1Ls was delivered to NAS Pensacola on November 26, 1969, and the Navy also procured eight UH-1Ls for use in the utility role.

A major improvement of the basic Bell Model 204/205 series of helicopters was undertaken during 1968, based around the installation of the Pratt & Whitney (UACL) PT6T Twin Pac power plant. This comprised two of the Canadian company's PT6A turboshaft engines coupled to a common gear box and offered a significant increase in installed thrust, plus twin-engined reliability. This Bell Model 212 was designated UH-1N for US Air Force, Navy and Marine Corps service and it was made the subject of major procurement for the Navy and Marine Corps. The USN role of the UH-1N was primarily local base rescue at naval air stations and to support the annual *Operation Deep Freeze* in the Antarctic. Marine Corps roles included command and control mobility for troop commanders, and movement of troops, equipment and cargo in amphibious assault and shore-based operations; other missions included medical evacuation, combat reconnaissance support and various light utility missions.

Deliveries of the UH-1N began in 1971 and total procurement of 254 was planned, to equip six USMC squadrons and meet Navy needs. Annual purchases were continuing beyond 1976 to achieve this inventory total, including six VH-1N executive transports.

The Marine Corps also became interested in the potential of the armed version of the Iroquois developed as the AH-1G HueyCobra (Bell Model 209) for the Army. With the same transmission/rotor system/power plant as the Model 204, the AH-1G featured a new slim fuselage seating pilot and gunner in tandem, and carried a heavy load of ordnance for ground attack.

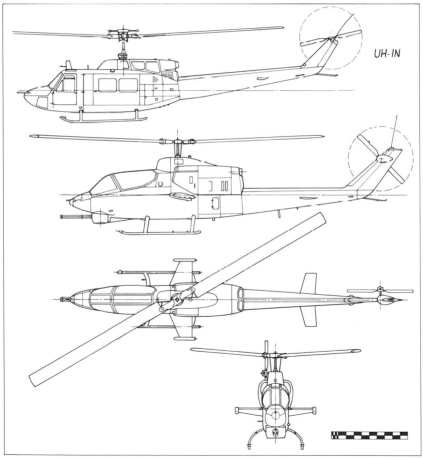

UH-1N

US Marine Corps Bell AH-1G Hueycobra. (*Bell photo*)

This armament included a tactical turret in the nose with two multi-barrel Mini-guns or two grenade launchers, plus four strong points on stub wings carrying up to 76 rockets or two gun pods.

To begin training and tactical deployment, the Marine Corps acquired 38 Army-standard AH-1Gs in 1969 and then switched to the AH-1J SeaCobra, which had the same Twin Pac power unit as the UH-1N. The basic nose armament of the AH-1J comprised a General Electric turret mounting an XM-197 three-barrel 20-mm weapon, and the four wing strong points could carry a variety of loads. Deliveries began in mid-1970 and the Marine Corps established a total requirement for 124. After production of 67 AH-1Js, however, an improved model was developed, with T400-CP-402 engine, increased weights and performance, provision for TOW missiles, advanced avionics, and a lengthened fuselage. Two prototypes were built for testing in 1976, after which the USMC planned to acquire 55 more of these improved versions as AH-1Ts.

TECHNICAL DATA (UH-1N)

Manufacturer: Bell Helicopter Company, Fort Worth, Texas, USA.
Type: Utility and transport helicopter.
Accommodation: Pilot plus up to 14 passengers; external load capability, 4,000 lb.
Power plant: One Pratt & Whitney (UACL) T400-CP-400 Twin Pac coupled turboshaft engine flat rated at 1,250 shp for take-off.
Dimensions: Rotor diameter, 48 ft 2½ in; overall length, 57 ft 0 in; height, 14 ft 4¾ in.
Weights: Typical empty weight, 6,000 lb; max take-off weight, 10,000 lb.
Performance: Max level speed, 127 mph at sea level; initial rate of climb, 1,745 ft/min; service ceiling, 17,300 ft; hovering ceiling, IGE, 13,900 ft; hovering ceiling, OGE, 9,300 ft; max range, 286 miles.
Armament: None.
Serial numbers:

UH-1E: 151266–151299; 151840–151887; 152416–152439; 153740–153767; 154750–154780; 154943–154969; 155337–155367; 157177–157203.	UH-1L: 157851–157858.
TH-1E: 154730–154749.	TH-1L: 157806–157850; 157859–157903.
	AH-1G: 157204–157241.
	AH-1J: 157757–157805.
	UH-1N: 158230–158291 (plus later contracts).

50

Boeing C-5 seaplane as tested by Navy before purchase of the Model Cs in quantity.
(*US Navy*)

Boeing C, C-1F

The sixth and seventh Boeing aeroplanes built, designated C-5 and C-6 by the manufacturer, were submitted to the Navy early in 1917 for test as primary trainers. The design was a slight modification of the C-4, incorporating an unusual degree of stagger and dihedral on the wings to achieve inherent stability. On the basis of the tests, the Navy ordered 50 production models that established the tiny Boeing Airplane Company as a 'major' manufacturer. The production versions were still called Model C by Boeing, and individual machines were identified by appending the Navy serial number to the designation, the whole lot thus being identified as C-650 through C-699. Although not a naval aircraft, an additional Model C built for Mr William E. Boeing in 1918 was called C-700 simply because it followed the last of the Navy machines through the factory.

An unusual feature of the Model C, apart from the amount of stagger to the wings, was the absence of a fixed horizontal tailplane. This last feature became a point of controversy between the pilots, who insisted there should be one, and the designer, who maintained that because of the stability imparted by the stagger a stabilizer was not necessary and performance was increased by the weight saving.

Otherwise a conventional twin-float seaplane, the Model C was not used at Naval training schools after delivery because of the extremely poor performance of the Hall-Scott A-7A engine. Most of the Cs were sold as surplus after the war, still in their original packing crates. The Army had similar experience with the Hall-Scott, and grounded its Standard Model J primary trainers, which used the same engine.

51

Recognizing the deficiencies of the Hall-Scott, the Navy ordered an additional Model C, powered with the 100 hp Curtiss OXX-6 engine, under the designation of C-1F, meaning a C with one main float instead of the usual two.

TECHNICAL DATA (Boeing Model C)

Manufacturer: Boeing Airplane Company, Seattle, Washington.
Type: Training seaplane.
Accommodation: Two pilots.
Power plant: One 100 hp Hall-Scott A-7A.
Dimensions: Span, 43 ft 10 in; length, 27 ft; height, 12 ft 7 in; wing area, 495 sq ft.
Weights: Empty, 1,898 lb; gross, 2,395 lb.
Performance: Max speed, 72·7 mph at sea level; cruising speed, 65 mph at sea level; climb, 23·5 min to 5,000 ft; service ceiling, 6,500 ft; range, 200 st miles.
Serial numbers:
   C: A147–A148, A650–A699.   C-1F: A4347.

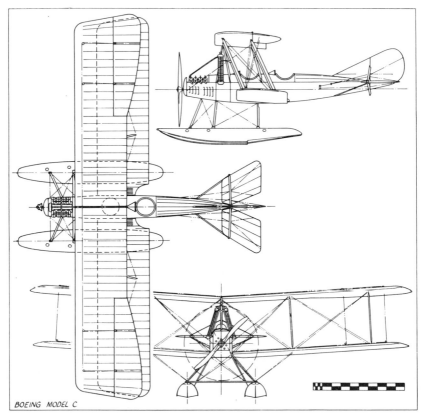

BOEING MODEL C

The first production Boeing NB-1 as a seaplane. (*Boeing photo*)

Boeing NB

Having successfully launched into the single-seat fighter business with the Model 15 in 1923 (see page 55), the Boeing Company turned its attention to primary trainers to meet Navy requirements. As the Model 21, the new type took shape as a conventional biplane powered by an un-cowled Lawrance J-1 engine, with two open cockpits in tandem. Unusual features were the split-axle undercarriage and N-struts between the wings to eliminate the wing incidence wires.

The Navy tested the first Model 21 under the designation VNB-1, in which the V, used by the Navy for heavier-than-air aircraft types, should not have been incorporated in the actual designation. Initial tests showed that the VNB-1 lacked the ability to be spun and the snappiness required in a trainer's handling characteristics: it was, in fact, too docile adequately to teach pupils how to cope with the more advanced biplanes of the day. Modifications were made to allow the aircraft to be spun, and the Navy then ordered 42 production NB-1s, including the modified prototype. Deliveries began on December 5, 1924.

Lawrance J-1, J-2 and J-4 engines were used in the NB-1s, some of which were later fitted with 220 hp Wright J-5s. A second batch was ordered with the 180 hp Wright-Hispano E-4 water-cooled inline engine, to use up surplus stocks. Thirty were built, taking the designation NB-2. Some operated as seaplanes with a single main float and wingtip stabilizers. Several NB-1s were used for experimental purposes, notably the two last production examples, which became NB-3 and NB-4 respectively and were used in trials to improve the spinning characteristics.

The NB-1s and NB-2s served as pilot and gunnery trainers, the rear cockpits having a Scarff ring for a single 0·30-in gun. The majority were based at Pensacola Naval Air Base where Squadrons VN-1D8 and VN-4D8 were using 32 at the end of 1927. Others were in use by VN-6D5 at Hampton Roads Naval Air Station and VN-7D11 at San Diego Naval Air Station. Some were also used by the Marines and Squadron VO-6M was using NB-2s modified for crop spraying in Puerto Rico in 1929.

TECHNICAL DATA (NB-1)

Manufacturer: Boeing Airplane Company, Seattle, Washington.

Type: Primary and gunnery trainer.

Accommodation: Instructor and pupil.

Power plant: One 200 hp Lawrance J-1.

Dimensions: Span, 36 ft 10 in; length (seaplane), 28 ft 9 in; height, 11 ft 8 in; wing area, 344 sq ft.

Weights: Empty, 2,136 lb; gross, 2,837 lb.

Performance: Max speed, 99·5 mph at sea level; cruising speed, 90 mph at sea level; climb, 9·8 min to 5,000 ft; service ceiling, 10,200 ft; range, 300 st miles.

Armament: One 0·30-in machine gun on Scarff ring in rear cockpit.

Serial numbers:

VNB-1: A6749.	NB-3: A6856.
NB-1: A6750–A6768; A6836–A6857.	NB-4: A5857.
NB-2: A6769–A6798.	

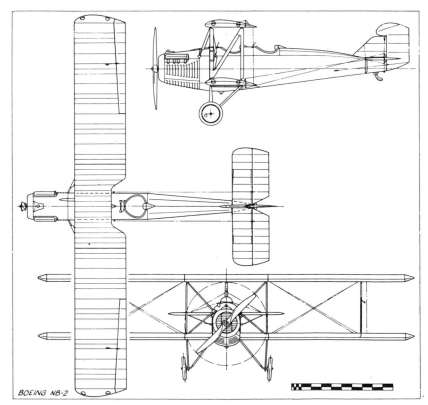

BOEING NB-2

One of ten Boeing FB-1 shore-based fighters assigned to US Marine Corps and used in China in 1927/28. (*US Navy*)

Boeing FB

In developing its first original fighter, Boeing was influenced by two streams of experience. The company had already won a contract to build 200 Thomas Morse MB-3As for the Army Air Corps, and this work brought intimate acquaintance with contemporary American practice in single-seat, high-performance fighter design; in addition, a comprehensive study had been made of European production techniques, leading Boeing engineers to conclude that the Fokker D.VII had a number of particularly meritorious features. A number of these features could be detected in the Boeing Model 15, which first flew on June 2, 1923, and was tested by the Army as the XPW-9. Its success launched Boeing into the production of a series of fighters which were to be its major activity for the next decade.

US Navy interest in the Model 15 followed closely upon Army production orders, and a contract for 16 was placed early in 1925. The first 10, delivered between December 1 and 22, 1925, were virtual duplicates of the Army PW-9s, with the 435 hp Curtiss D-12 engine and an armament of two 0·30-in (or one 0·30-in and one 0·50-in) machine guns. The FB-1s went to the Marine Corps, serving at first with Squadrons VF-1M, VF-2M and VF-3M; nine of the ten were assigned to the Expeditionary Force in China in 1927/28, operated by VF-10M (VF-3M redesignated) and the tenth was then assigned to VO-8M in San Diego.

The first Boeing fighters for naval use were not equipped for carrier operations, but two FB-2s (Boeing Model 53) were suitably modified, with strengthened fuselage structure and cross-axle undercarriage. Three FB-3s (Model 55) were similar, but had the FB-1 type split-axle undercarriage and provision for twin-float operation. They also had 510 hp Packard 1A-1500 engines in place of the original Curtiss power plant. Another experimental model was the FB-4 (Model 54), delivered to the Navy with a 450 hp Wright P-1 radial engine (and provision for floats) but converted later to the FB-6 with a 400 hp Pratt & Whitney Wasp.

Succeeding the FB-1 as the next production fighter for the Navy was the FB-5 (Model 67) first flown on October 7, 1926. This variant had a 520 hp Packard 2A-1500 engine, in a cowling similar to that originated on the FB-3, and increased wing stagger, the top wing being moved forward and the lower wing aft. Aerodynamically balanced rudders with increased area were also introduced on all but the first FB-5. Production totalled 27, and all were delivered on January 21, 1927; the delivery process was probably unique, the aircraft being loaded on barges at Boeing's water's-edge factory in Seattle and ferried to the carrier USS *Langley* in Seattle harbour, making their first flights from the carrier deck. They initially equipped Navy Squadrons VF-1B and VF-6B aboard the *Langley*, and later VF-3B flew FB-5s from the *Lexington* before a number were passed on to Marine units, including VF-6M at San Diego.

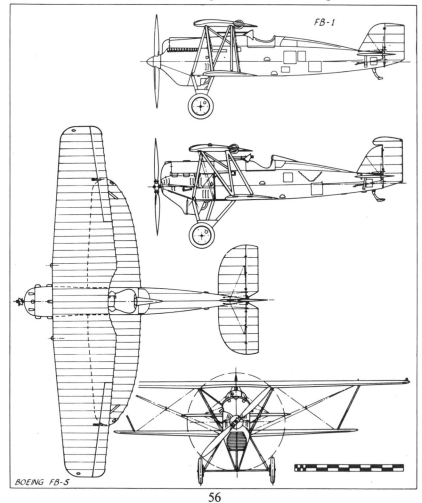

FB-1

BOEING FB-5

56

Plans to produce an FB-7 (Model 67A) with a Wasp engine were passed over after the Navy had tested the XF2B-1, an extensive redesign of the initial Boeing fighter to take advantage of the new engine (see page 58).

TECHNICAL DATA (FB-1)

Manufacturer: Boeing Airplane Company, Seattle, Washington.
Type: Shore-based fighter.
Accommodation: Pilot only.
Power plant: One 435 hp Curtiss D-12.
Dimensions: Span, 32 ft; length, 23 ft 5 in; height, 8 ft 2 in; wing area, 260 sq ft.
Weights: Empty, 1,936 lb; gross, 2,835 lb.
Performance: Max speed, 159 mph; cruising speed, 142 mph; initial climb, 1,630 ft/min; service ceiling, 18,925 ft; range, 390 st miles.
Armament: One fixed forward-firing 0·30-in and one 0·50-in guns.
Serial numbers:

FB-1: A6884–A6893. FB-4 (FB-6): A6896.
FB-2: A6894–A6895. FB-5: A7101–A7127.
FB-3: A6897, A7089, A7090.

Land-based Boeing FB-5 in markings of Squadron VF-6M. (*US Navy*)

Boeing FB-5s as entered in military aircraft racing events at the 1927 National Air Races. (*US Navy*)

Boeing F2B-1 of Squadron VF-1B, with auxiliary belly fuel tank. (*William T. Larkins*)

Boeing F2B, F3B

Pratt & Whitney's development of the radial air-cooled 400 hp Wasp—flown for the first time in the Wright F3W-1 Apache on May 5, 1926—offered important advantages over the liquid-cooled inline engines in use in the mid-twenties, and after the Navy had made a trial installation in the single FB-6 (see page 55), Boeing applied a similar installation to a new fighter, Model 69. Apart from the engine, Model 69 closely resembled the Model 66 which had been tested by the Army as the XP-8. Many features of the PW-9 and FB series were retained, but the difference between the spans of the upper and lower wings was lessened.

Built as a Boeing private venture, the Model 69 first flew on November 3, 1926, and was tested by the Navy with the designation XF2B-1, being one of the first prototypes to carry the X designation after its adoption by the Navy. Armament was the Navy standard of two 0·30-in or one 0·30 and one 0·50 guns in the forward fuselage, and provision was made for five 25-lb bombs to be carried under the wings and fuselage.

Successful testing of the XF2B-1, which had a top speed of 154 mph, led to a Navy order for 32 F2B-1s in March 1927. These aircraft differed from the prototype in two important respects: the large spinner, which had faired the propeller into the fuselage lines, was deleted and a balanced rudder was adopted in place of the unbalanced type.

Deliveries of the F2B-1 began on January 30, 1928, and these aircraft were assigned to two units aboard the USS *Saratoga*, VF-1B and VB-2B; the latter was a bomber squadron, despite the primary fighting role of the F2B-1. This was also the unit which formed the Three Sea Hawks aerobatic team with F2B-1s, famous for its tied-together formation flying.

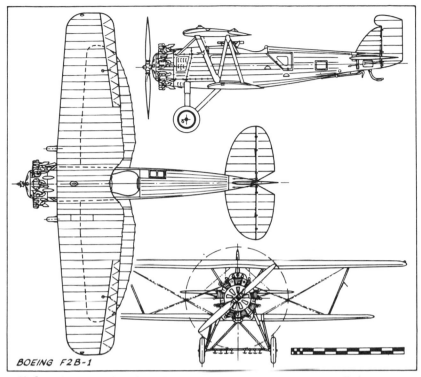

BOEING F2B-1

Only a few months after the Navy began to accept the F2B-1s, the first F3B-1 was delivered. Development of this third Boeing fighter design for the Navy began with the Model 74, first flown on March 2, 1927. This closely resembled the F2B-1 but had an FB-5 type undercarriage and provision for operation with a single main float and wingtip outriggers. Tested by the Navy in 1927 as XF3B-1, it was not adopted for production

Boeing F3B-1 as delivered to Navy in 1928. (*National Archives 72-AF-25821*)

A production Boeing F3B-1 with Townend anti-drag ring around the engine, a post-delivery modification by the Navy. (*Bowers Collection*)

but was converted by Boeing to Model 77 configuration with a lengthened nose and completely new wings, tail and landing gear. Unlike the earlier wing arrangements of the Boeing fighters, Model 77 had constant chord on upper and lower wings, and only the upper panels were swept back. Semi-monocoque all-metal construction was introduced for the first time, for the tail and ailerons.

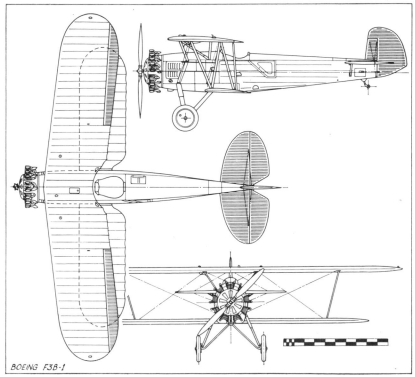

BOEING F3B-1

Boeing F3B-1 with Townend ring and wheel fairings, showing partial blue finish for an administrative and command aircraft for Captain's rank. (*John C. Mitchell*)

Tested as the F3B-1, this new design was first flown on February 3, 1928, and was accepted for production. The Navy ordered 74, including the prototype, and deliveries began in August 1928, the first examples going to VF-2B aboard USS *Langley*. By the end of the year, all were in service, the other units initially flying the F3B-1 being VB-2B (*Saratoga*) and VF-3B and VB-1B (*Lexington*). The designation of VB-2B reverted to its original VF-6B while it still flew F3B-1s. Most F3B-1s had been replaced in front-line squadrons by 1932, but they continued in use for several more years as staff and command transports.

TECHNICAL DATA (F2B-1)

Manufacturer: Boeing Airplane Company, Seattle, Washington.
Type: Carrier-based fighter.
Accommodation: Pilot only.
Power plant: One 425 hp Pratt & Whitney R-1340B.
Dimensions: Span, 30 ft 1 in; length, 22 ft 11 in; height, 9 ft 2¾ in; wing area, 243 sq ft.
Weights: Empty, 1,989 lb; gross, 2,805 lb.
Performance: Max speed, 158 mph at sea level; cruising speed, 132 mph; initial climb, 1,890 ft/min; service ceiling, 21,500 ft; range, 317 st miles.
Armament: One fixed forward-firing 0·30-in and one 0·50-in guns.
Serial numbers:
    XF2B-1: A7385.   F2B-1: A7424–A7455.

TECHNICAL DATA (F3B-1)

As F2B-1 except:
Power plant: One 425 hp Pratt & Whitney R-1340-80.
Dimensions: Span, 33 ft; length, 24 ft 10 in; height, 9 ft 2 in; wing area, 275 sq ft.
Weights: Empty, 2,179 lb; gross, 2,945 lb.
Performance: Max speed, 157 mph at sea level; cruising speed, 131 mph; initial climb, 2,020 ft/min; service ceiling, 21,500 ft; range, 340 st miles.
Serial numbers:
    XF3B-1: A7674.   F3B-1: A7675–A7691; A7708–A7763.

Boeing F4B-1 of VF-5B. (*J. M. F. Hasse*)

Boeing F4B

Development of the most famous of the Boeing biplane fighters, the F4B/P-12/Model 100 series, began as a company private venture intended to produce a replacement for the F2B/F3B Navy fighters and the PW-9 serving with the Army. Keynote of the new design was orthodoxy; there was nothing radically new, but all of the company's experience of fighter design and production went into refining the proven features. As a result, a considerable gain in performance was achieved while using the same engine, a Pratt & Whitney Wasp, as the F3B.

Two nearly identical prototypes were built by Boeing, as Model 83 and Model 89. There were differences in undercarriage design, and the 83 had arrester gear, while only the 89 had provision for carrying a 500-lb bomb under the fuselage because of its split-axle landing gear. First flights were made, respectively, on June 25, 1928, at Seattle, and August 7, 1928, at Anacostia, and both aeroplanes underwent Navy evaluation, during which time they became known as XF4B-1s. As a result of these tests, the Navy ordered 27 production model F4B-1s and also purchased the two prototypes after their conversion to production standard.

First flown on May 6, 1929, the F4B-1 (Boeing Model 99) had the split-axle undercarriage of the Model 89, plus the bomb provisions, and the arrester gear of the Model 83. The F4B-1s were powered by Wasp R-1340-8 engines and were delivered between June and August 1929 to equip, initially, VB-1B and VF-2B, aboard the USS *Lexington* and *Langley* respectively; VB-1B later became VF-5B while still flying the F4B-1s. During their period of service, the F4B-1s underwent several changes of configuration: the individual fairings behind each cylinder were removed

A Boeing F4B-2 of VF-5B landing aboard USS *Lexington*. (*National Archives 80-CF-54854-1*)

and, eventually, ring cowlings were fitted and F4B-4 type fins were substituted for the original type. One aircraft was modified for use as an executive aeroplane by David S. Ingalls, then the Assistant Secretary of the Navy, and was designated F4B-1A (serial A8133).

Design improvement of the original F4B-1 followed its adoption by the Army Air Corps as the P-12, and the next Navy model, ordered in June 1930, was equivalent to the P-12C. Designated F4B-2 (Boeing Model 223) it had a ring cowling, Frise ailerons, spreader-bar axle between the wheels like the original Model 83, and a tailwheel. Delivered between

Boeing F4B-3 assigned to NAS Anacostia, with special Alclad finish. (*Fred E. Bamberger, Jr.*)

63

January and May 1931, the 46 F4B-2s went initially to VF-6B (*Saratoga*) and VF-5B (*Lexington*). Like the F4B-1s, they eventually were fitted with F4B-4 fins.

After Boeing had developed a new variant of the design as Model 218 with a semi-monocoque metal fuselage and other changes, the Navy ordered 21 similar F4B-3s (Model 235) on April 23, 1931. They went to VF-1B on the *Saratoga* in December 1931 and January 1932, and were soon followed by the first of 92 F4B-4s, which had a larger fin and rudder but were otherwise identical. Delivery began in July 1932 and was completed in February 1933, the final 45 aircraft also having an enlarged head-rest. Wing bomb racks were fitted, for two 116-lb bombs. First Navy squadrons to fly the F4B-4 were VF-3B on the USS *Langley* and VF-6B on the *Saratoga*; the Marines received 21 of the total, at first equipping VF-10M and later VF-9M. A 22nd F4B-4 was produced for the Marines from spare parts supplied with the original order.

The F4B-4s remained in service with Navy squadrons aboard carriers

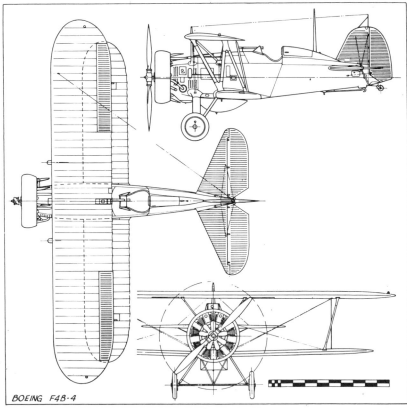

BOEING F4B-4

64

Boeing F4B-4 in the special markings of VF-9M, with belly fuel tank and underwing racks.
(William T. Larkins)

until 1937, but by the middle of 1938 they had been replaced by the faster Grumman biplanes. They remained in use for several more years, however, at shore stations for utility tasks. Also, in 1940, the Navy acquired 23 assorted P-12s from the Air Corps, designated F4B-4As. These and remaining F4Bs were eventually adapted for use as radio-controlled drones. The Boeing biplanes saw little action after the end of 1941, at which time 34 were still on US Navy charge.

TECHNICAL DATA (F4B-1)

Manufacturer: Boeing Airplane Company, Seattle, Washington.
Type: Carrier-borne fighter.
Accommodation: Pilot only.
Power plant: One 450 hp Pratt & Whitney R-1340-8.
Dimensions: Span, 30 ft; length, 20 ft 1 in; height, 9 ft 4 in; wing area, 227·5 sq ft.
Weights: Empty, 1,950 lb; gross, 2,750 lb.
Performance: Max speed, 176 mph at 6,000 ft; climb, 2·9 min to 5,000 ft; service ceiling, 27,700 ft; range, 371 st miles.
Armament: Two fixed forward-firing 0·30-in guns.
Serial numbers:
    XF4B-1: A8128–A8129.   F4B-1: A8130–A8156.

TECHNICAL DATA (F4B-4)

As F4B-1 except:
Power plant: One 550 hp Pratt & Whitney R-1340-16.
Weights: Empty, 2,354 lb; gross, 3,611 lb.
Performance: Max speed, 188 mph at 6,000 ft; initial climb, 2·7 min to 5,000 ft; service ceiling, 26,900 ft; range, 370 st miles.
Serial numbers:
    F4B-2: A8613–A8639; A8791–A8809.   F4B-4: 8912–8920; 9009–9053; 9226–
    F4B-3: 8891–8911.   9263; 9719.
                                      F4B-4A: 2489–2511.

Boeing PB-1W in overall blue finish. (*Gordon S. Williams*)

Boeing PB-1 Fortress

Both the US Navy and US Coast Guard operated small numbers of Boeing B-17 Fortress bombers acquired as surplus from the USAAF in 1945 and modified for new roles. Prior to the end of the War, two B-17s had been used by the Navy for assorted test projects, but by mid-1945 a need had arisen for large land-based monoplanes to serve operationally as radar-equipped anti-submarine and reconnaissance aircraft. This was a role which the B-17 was well able to fulfil.

On July 31, 1945, the Navy assigned the designation PB-1 to its original B-17F and B-17G, this being the second time the designation had been used for a Boeing patrol aeroplane, the first occasion being for a 1925 flying-boat prototype.

New B-17Gs, held by the USAAF in a supply pool, were picked up by the Navy and ferried to the NAMU (Naval Aircraft Modification Unit) at Johnsville, Penn., for modification. Primary change effected was the installation of the APS-20 search radar with its large scanner in a fairing beneath the centre fuselage. Operating primarily in the early-warning role, these aircraft were designated PB-1W and given an overall blue finish; the bomb-bays were sealed (no weapons being carried) and extra fuel tanks were installed; underwing tanks could also be carried. Initial deliveries to VPB-101 were made in the spring of 1946, and the PB-1Ws served operationally until replaced by Lockheed WV-2s. A few of the 31 PB-1Ws eventually were modified to have radomes above the fuselage instead of beneath, and one became a jet engine test-bed as XPB-1W.

Also during 1945 the B-17 was adopted for use by the US Coast Guard

Boeing PB-1G in US Coast Guard colours on its last mission in October 1959.
(*Peter M. Bowers*)

as a long-range air–sea rescue aircraft. Its use in this role followed successful development by the USAAF of the B-17H carrying a lifeboat externally. Designated PB-1Gs, these aircraft also undertook long-range iceberg reconnaissance flights, and one was modified for aerial survey work carrying a special nine-lens camera. This aircraft was the last of 17 PB-1Gs in service, making its last flight on October 14, 1959.

TECHNICAL DATA (PB-1W)

Manufacturer: Boeing Aircraft Company, Seattle, Washington (some also built by Lockheed (Vega) Aircraft Corporation, Burbank, California).
Type: Early warning patrol.
Accommodation: Crew of 13.
Power plant: Four 1,200 hp Pratt & Whitney R-1820-97s.
Dimensions: Span, 103 ft 9 in; length, 74 ft 4 in; height, 19 ft 1 in; wing area, 1,420 sq ft.
Weights: Empty, 36,135 lb; gross, 55,400 lb.
Performance: Max speed, 310 mph at 25,000 ft; cruising speed, 200 mph; initial climb, 7 min to 5,000 ft; service ceiling, 35,000 ft; range, over 2,500 st miles.
Serial numbers:
   PB-1W: 34106; 34114; 77137–77138; 77225–77244; 77258; 83992–83998.
   PB-1G: 77245–77257; 82855–82857 (plus one unknown).

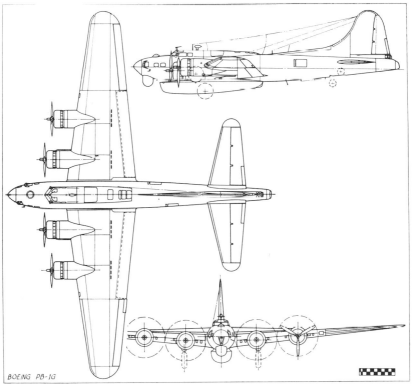

BOEING PB-1G

Boeing Vertol CH-46D Sea Knight of Navy Squadron HC-3. (*US Navy*)

Boeing Vertol H-46 Sea Knight

The Boeing Company's Vertol Division (formerly the independent Vertol Aircraft Corporation and now the Boeing Vertol Company) successfully competed in a 1960 design competition to select a new assault helicopter for the US Marine Corps with a military version of the Model 107-II tandem-rotor helicopter. Designated Model 107M by the company, the submission in the competition differed from the existing commercial 107 in several details, including up-rated engines, provision for powered folding of the rotor blades, and an integrated loading system. The submission was adjudged the winner on February 20, 1961, and an initial contract for 14 was placed.

The designation HRB-1 was at first allocated to the new Marine helicopter, but this was later changed to CH-46A, with the type name of Sea Knight. Its specified mission was to carry a 4,000-lb load (cargo or up to 17 fully equipped troops) over a combat radius of 115 miles at a speed of 150 mph. The folding rotor blades were required for stowage aboard aircraft carriers, and the combat mission required facilities for rapid loading and unloading of troops or cargo.

First flight of the CH-46A was made on October 16, 1962, and additional contracts in annual increments brought the total number of Sea Knights procured to 624 by the end of 1970. Official acceptance trials of the CH-46A were complete by November 1964, and 160 CH-46As were delivered, with five Marine squadrons equipped by mid-1965. A few of these helicopters modified for mine-hunting and sweeping operations were designated RH-46A. A new production standard introduced in mid-1966 was identified as the CH-46D, this version differing primarily from the CH-46A in having 1,400 shp General Electric T58-GE-10 engines in place of the earlier 1,250 shp T58-GE-8Bs or -8Fs. Production totalled 266, and a switch was then made to the CH-46F, which differed only in having added avionics and instrument panel changes. During 1975, the USMC began a programme to convert about 300 of its Sea Knights to CH-46E standard with 1,870 shp T58-GE-16 engines.

During 1964 the US Navy purchased a version of the Sea Knight for

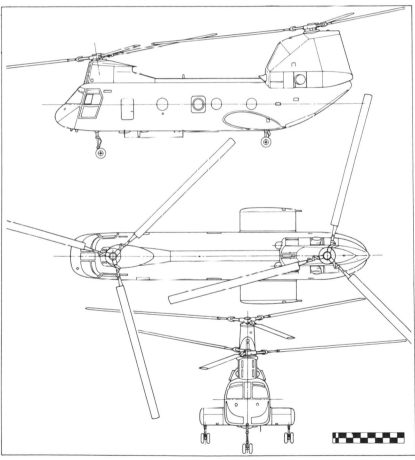

USMC Boeing Vertol CH-46A Sea Knight on carrier deck. (*Boeing Vertol photo*)

its Vertical Replenishment Program, providing for the transferring of supplies from AFS and AOE supply ships to combatant vessels at sea. The Navy designation was UH-46A, the first of these helicopters being delivered to VHU-1 in July 1964. Fourteen were delivered, followed by 10 UH-46Ds with the up-rated -10 engines.

TECHNICAL DATA (CH-46D)

Manufacturer: The Boeing Company, Vertol Division, Morton, Pennsylvania.
Type: Combat assault helicopter.
Accommodation: Crew of three and up to 17 assault troops or 15 stretchers with two attendants.
Power plant: Two 1,400 shp General Electric T58-GE-10 shaft turbines.
Dimensions: Rotor diàmeter, each 51 ft; length of fuselage, 44 ft 10 in; height, 16 ft 8½ in; disc area, total, 4,086 sq ft.
Weights: Empty, 13,065 lb; gross, 23,000 lb.
Performance: Max speed, 166 mph at sea' level; cruising speed, 154 mph; initial climb, 1,715 ft/min; service ceiling, 14,000 ft; range, 230 st miles.
Serial numbers:

CH-46A: 150265–150278; 150933–150964; 151906–151961; 152496–152553.
CH-46D: 152554–152579; 153314–153403; 153951–154044; 154789–154844.
CH-46F: 154845–154862; 155301–155318; 156418–156477; 157649–157726.
UH-46A: 150965–150968; 151902–151905; 152490–152495.
UH-46D: 153404–153413.

70

Brewster F2A-2 in 1941. (*Peter M. Bowers*)

Brewster F2A Buffalo

A US Navy contract placed with the Brewster Aeronautical Corporation on June 22, 1936, initiated construction of the prototype of what was to become the first monoplane fighter equipping a Navy squadron. The Model 139 fighter, designated XF2A-1 by the Navy, bore a distinct family resemblance to the XSBA-1. The design was characterized by the short, stubby fuselage, stalky inward-retracting main gear with wheel wells in the belly, and mid-mounted wing of low aspect ratio. The power plant around which the original design was planned was the 950 hp Wright XR-1820-22, and the armament comprised two 0·50-in guns in the engine cowling plus two wing guns added later.

First flight of the XF2A-1 was made in December 1937, and it was delivered for Navy testing a year later. Meanwhile, the first production contract had been placed on June 11, 1938, for 54 F2A-1s (Model 239) and the first of these was rolled out in June 1939. They were powered by the 940 hp R-1820-34 engine and had a larger fin. In the course of the following year, 11 F2A-1s reached the Navy, and 10 of these were in service with Fighting Squadron VF-3 aboard the USS *Saratoga* by June 1940. The remaining aircraft on the original contract were released by the United States for sale to Finland early in 1940, to be replaced by 43 F2A-2s.

The F2A-2 had originated on March 22, 1939, when the USN ordered conversion of the prototype to XF2A-2 with a 1,200 hp R-1820-40 engine and redesigned rudder. Other changes were from hydraulic to electric propeller, and addition of a high-altitude carburettor system. Testing of the XF2A-2 began in July 1939, and production deliveries started in September 1940. Finally, the Navy ordered 108 F2A-3s on January 21, 1941, and took delivery of these between July and December the same year. Features of this model were increased armour protection for the pilot and for the fuel tank, and improved equipment, the weight of which adversely

71

affected the aircraft's performance. Navy Squadron VF-2 operated F2A-3s aboard the USS *Lexington*, and VS-201 took a small number aboard the escort carrier *Long Island*; one Marine squadron, VMF-221, also flew F2A-3s (and F2A-2s). The fighter achieved little operational success, its only major battle being that at Midway when the Marine squadron suffered major losses.

TECHNICAL DATA (F2A-3)

Manufacturer: Brewster Aeronautical Corporation, Long Island City, NY.
Type: Carrier-based fighter.
Accommodation: Pilot only.
Power plant: One 1,200 hp Wright R-1820-40.
Dimensions: Span, 35 ft; length, 26 ft 4 in; height, 12 ft; wing area, 209 sq ft.
Weights: Empty, 4,732 lb; gross, 7,159 lb.
Performance: Max speed, 321 mph at 16,500 ft; cruising speed, 161 mph; initial climb, 2,290 ft/min; service ceiling, 33,200 ft; range, 965 st miles.
Armament: Two fixed forward-firing 0·50-in guns in upper cowling; two 0·50-in guns in the wings.
Serial numbers:

XF2A-1, -2: 0451.	F2A-2: 1397–1439.
F2A-1: 1386–1396.	F2A-3: 01516–01623.

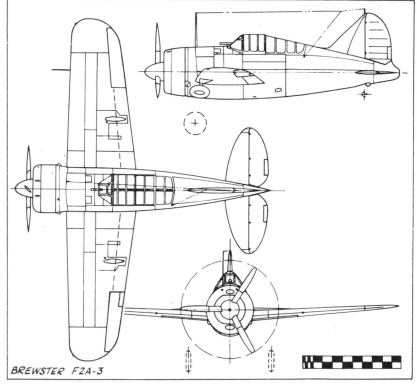

BREWSTER F2A-3

72

Brewster SB2A-4 photographed in November 1943. (*US Navy*)

Brewster SB2A Buccaneer

One of the least successful combat aircraft put into production in the USA during the period of the second world war, the SB2A was developed by the Brewster Company from its first aircraft for the US Navy, the XSBA-1 (see page 427). In overall configuration, the Brewster Model 340 was very similar to its predecessor, with a mid-wing, inward-retracting undercarriage and internal weapon bay. The shapes of the major components were also similar, but the overall dimensions were increased to double the internal bomb-load to 1,000 lb and to permit installation of the 1,700 hp Wright Cyclone R-2600 two-row radial.

Another innovation in the Model 340 design was the provision for a power-operated turret in the fuselage aft of the wing, taking the place of the open gun positions in the SBA-1. The armament comprised two 0·30-in guns in the turret, two in the forward fuselage and two in each wing.

The US Navy ordered a prototype of the new Brewster design on April 4, 1939, with the designation XSB2A-1, and this aircraft made its first flight on June 17, 1941, powered by a 1,700 hp Wright R-2600-8 engine. Large-scale production at Brewster's Johnsville factory in Pennsylvania had already started, as the company had succeeded in selling the Model 340 to the British Purchasing Commission which ordered 750 in July 1940 as Bermudas, and to the Dutch Government which ordered 162 for the Royal Netherlands Indies Army. On December 24, 1940, the US Navy ordered 140 aircraft, and during 1941, after passage of the Lend–Lease Act through Congress, the USAAF and USN jointly assumed responsibility for the aircraft which had been ordered by Britain. Production aircraft dispensed with the dorsal turret, which was carried in dummy form by the prototype.

During 1943 the USN began accepting 80 SB2A-2s, similar to the SB2A-1 but having two 0·50-in calibre nose guns and only two 0·30-in wing guns plus the two dorsal 0·30s. Final production variant was the SB2A-3, ordered in 1942 and delivered early in 1944; the 60 aircraft of this type had folding wings and arrester hooks, making them the only Buccaneers capable of carrier operations. Meanwhile, the 162 Dutch

73

aircraft had been repossessed by the USN before delivery and designated SB2A-4. Lacking some of the equipment of the SB2A-2, these aircraft were somewhat lighter and carried the original armament of eight 0·30s. They were used as trainers by Marine squadrons, particularly VMF(N)-531, the first Marine night fighting unit. There are no records to show that any of 771 Buccaneers and Bermudas were ever involved in combat.

TECHNICAL DATA (SB2A-2)

Manufacturer: Brewster Aeronautical Corporation, Johnsville, Bucks County, Pennsylvania.

Type: Land-based or carrier-based scout-bomber.

Accommodation: Pilot and observer/gunner.

Power plant: One 1,700 hp Wright R-2600-8.

Dimensions: Span, 47 ft; length, 39 ft 2 in; height, 15 ft 5 in; wing area, 379 sq ft.

Weights: Empty, 9,924 lb; gross, 14,289 lb.

Performance: Max speed, 274 mph at 12,000 ft; cruising speed, 161 mph; initial climb, 2,080 ft/min; service ceiling, 24,900 ft; range, 1,675 st miles (scout role, no bomb).

Armament: Two fixed forward-firing 0·50-in guns in fuselage; two 0·30-in. guns in wings and two in dorsal mounts; 1,000 lb internal bomb load.

Serial numbers:

XSB2A-1: 1632, 01005.
SB2A-2: 00803–00882.

SB2A-3: 00883–00942 (00890 not delivered).
SB2A-4: 29214–29375.

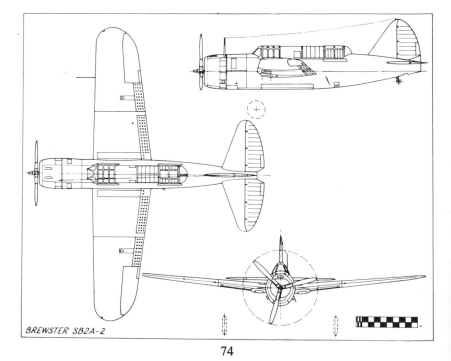

BREWSTER SB2A-2

74

Consolidated NY-2 seaplane serving with VN-15. (*Gordon S. Williams*)

Consolidated NY

Large numbers of these biplane trainers were used by the US Navy to give primary flight instruction and training in seaplane techniques and in gunnery. The basic design originated in 1922 as the Dayton-Wright TW-3, a side-by-side primary trainer built under US Army Air Service contract as a successor for the Curtiss Jenny. The Dayton-Wright Company, of Dayton, Ohio, having ceased to exist, a production contract for 20 TW-3s went to the Consolidated Aircraft Corporation, founded in 1923 by former Dayton-Wright personnel.

Consolidated's first original design, the PT-1 Army trainer, was derived from the TW-3 with a redesigned fuselage offering tandem seating in place of the unpopular side-by-side layout, and a new tail-unit. Features of the TW-3 which were retained included its wooden wings with thick Clark Y aerofoil, steel-tube fuselage structure and 180 hp Wright-Hispano E (war surplus) engine.

When the PT-1 was ordered into large-scale production in 1924 (the Army bought 221) the Navy indicated its interest in buying a similar aircraft subject to certain modifications being made. These comprised, primarily, substitution of the 200 hp Wright J-4 (R-790) radial engine and modification of the structure to permit installation, when required, of a single large float under the fuselage and small stabilizing floats under the wingtips. The fin and rudder were also redesigned to increase the area, primarily for the seaplane configuration. This variant of the PT-1 was adopted by the Navy, which ordered a total of 76 under the designation NY-1; these were the first Consolidated aircraft to carry a Navy designa-

Consolidated NY-2 seaplane. (*National Archives*)

tion. In service, a number of these trainers were fitted with armament for use as gunnery trainers, and were redesignated NY-1As.

As a seaplane, the NY-1 proved to be rather heavily loaded, and Consolidated developed long-span wings for the second production variant, the NY-2. The span increased from 34 ft 6 in to 40 ft, and the 220 hp Wright J-5 (R-790) engine was fitted. The Navy purchased 186 of this model and had 108 in active use in 1929, with 35 more assigned to reserves. A further 25 were bought as NY-2As with armament for gunnery training. As a result of the considerable improvement bestowed by the larger wings, these and the J-5 engines were later fitted to a number of the NY-1s, which were then designated NY-1B.

Final development of the design to serve with the Navy was the NY-3, 20 examples of which were purchased and assigned to Navy and Marine reserve squadrons. They had 240 hp Wright R-760-94 engines but were

Consolidated NY-2 landplane. (*James C. Mathiesen*)

otherwise similar to the NY-2. The last example did not disappear from Naval service until 1939. The single XN3Y-1 (A7273) tested in 1929 was a development of the NY-2 with a Wright R-790-A engine.

TECHNICAL DATA (NY-2)

Manufacturer: Consolidated Aircraft Corporation, Buffalo, NY.
Type: Primary trainer.
Accommodation: Pupil and instructor in tandem open cockpits.
Power plant: One 220 hp Wright R-790-8.
Dimensions: Span, 40 ft; length, 27 ft 10¾ in (31 ft 4½ in as seaplane); height, 9 ft 11 in (11 ft 10 in as seaplane); wing area, 370 sq ft.
Weights: Empty, 1,801 lb; gross, 2,627 lb.
Performance: Max speed, 98 mph at sea level; climb, 7·8 min to 5,000 ft; service ceiling, 15,200 ft; range, 300 st miles.
Serial numbers:

NY-1:	A7163–A7202;	A7205–A7220;	NY-2A: A8158–A8172; A8401–A8410.
	A7351–A7360;	A8173–A8182.	NY-3: A8487–A8506.
NY-2:	A7456–A7525;	A7693–A7707;	
	A7764–A7795;	A7970–A7977;	
	A8013–A8017;	A8183–A8192;	
	A8360–A8400.		

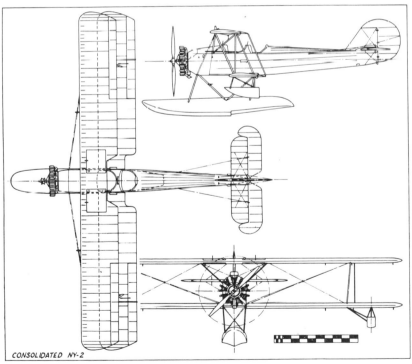

CONSOLIDATED NY-2

Consolidated P2Y-2, a modified P2Y-1 in service with VP-15. (*Gordon S. Williams*)

Consolidated P2Y

While several companies were busy in 1928 producing for the US Navy a series of biplane flying-boats based on the PN boats designed at the Naval Aircraft Factory, the first steps were taken to give the Navy its first monoplane patrol boats. The Bureau of Aeronautics drew up plans for a parasol monoplane with a fabric-covered, 100-ft span wing, two engines and a slender, single-step hull. Construction of the prototype was entrusted to Consolidated Aircraft Company, on a contract dated February 28, 1928, but a subsequent production order went to Glenn L. Martin (see page 316).

Although Consolidated lost the production order for the first patrol monoplane flying-boat to Martin, a contract was placed by the Navy on May 26, 1931, for a new prototype of a developed version of the original XPY-1. Designated XP2Y-1, this new prototype had the same 100-ft parasol wing, but became a sesquiplane with a smaller wing mounted lower, at the top of the hull, replacing the booms that had supported the stabilizing pontoons. Two engines, 575 hp Wright R-1820Es, were located close beneath the top wing and had narrow-chord cowlings; a third similar engine was strut-mounted on the centreline above the wing when the XP2Y-1 was first tested at NAS Anacostia in April 1932, but was removed the following month.

The Navy had ordered 23 P2Y-1s from Consolidated on July 7, 1931, and these were all delivered in the twin-engine configuration. They were serving by mid-1933 with VP-10F at home and VP-5F at Coco Solo.

After a prototype modification of a P2Y-1 to the XP2Y-2, the Navy ordered 23 similar production models as P2Y-3s on December 27, 1933. In this model, the engines were raised to the wing leading edge, reducing the drag of the installation, and more power was provided by the R-1820-88s in the XP2Y-2 and the R-1820-90s in the P2Y-3s, the latter rated at 750 hp for take-off and 700 hp at 4,000 ft. At least 21 of the P2Y-1s were modified to P2Y-2s with the raised engines in 1936, remaining with VP-5F and VP-10F until 1938, when they were transferred to VP-14 (later VP-52) and VP-15. The first P2Y-3s reached VP-7F in 1935, and this version was subsequently flown by VP-4F at Pearl Harbor and in 1939 was in operation with VP-19, VP-20 and VP-21 (these three squadrons being redesignated VP-43, VP-44 and VP-14 respectively). By the end of 1941 all the P2Y-2s and P2Y-3s had been withdrawn from operational use and were at Pensacola Naval Air Station.

TECHNICAL DATA (P2Y-3)

Manufacturer: Consolidated Aircraft Company, Buffalo, NY.
Type: Patrol flying-boat.
Accommodation: Crew of five.
Power plant: Two 750 hp Wright R-1820-90s.
Dimensions: Span, 100 ft; length, 61 ft 9 in; height, 19 ft 1 in; wing area, 1,514 sq ft.
Weights: Empty, 12,769 lb; gross, 25,266 lb.
Performance: Max speed, 139 mph at 4,000 ft; cruising speed, 117 mph; initial climb, 650 ft/min; service ceiling, 16,100 ft; range, 1,180 st miles with 2,000 lb bomb-load.
Armament: One flexible 0·30-in Browning in bows; two dorsal gun hatches behind wings.
Serial numbers:

XP2Y-1: A8939.	P2Y-1: A8986–A9007.
XP2Y-2: A9008.	P2Y-3: A9551–A9571; A9618–A9619.

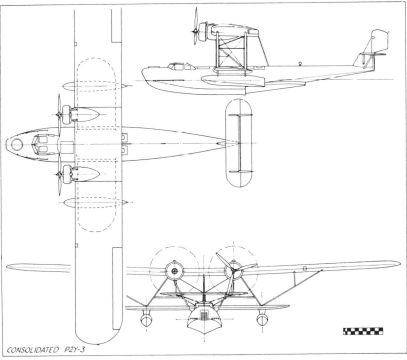

CONSOLIDATED P2Y-3

Consolidated PBY-3 showing the double bar tail marking. (*Gordon S. Williams*)

Consolidated PBY Catalina, PBN Nomad

To take advantage of new structural techniques, as well as experience gained with the first Navy monoplane patrol boats (see page 78), prototypes were ordered during 1933 from Consolidated and Douglas of new flying-boats which could replace the P2Y and P3M. The Douglas prototype, XP3D-1, was not destined to serve in quantity, but the Consolidated aircraft, XP3Y-1, became the progenitor of the most successful flying-boat operated by US forces in World War II and the most-produced boat of all time.

.Ordered by the Navy on October 28, 1933, the Consolidated Model 28 was designed by Isaac M. Laddon and had several distinctive features. Like the P2Y it was to succeed, it had a parasol-mounted wing, but internal bracing made it possible to dispense with external struts and, save for two small members between the hull and centre wing each side, the wing was a true cantilever. Adding to its aerodynamic cleanness was the use of unique retractable stabilizing floats which folded upwards to become the wingtips in flight. Two 825 hp Pratt & Whitney R-1830-58 engines were mounted on the wing leading edge. The hull itself, and arrangement of the crew accommodation, was similar to that in the P2Y, but a single cruciform tail was used on the newer type. Up to 2,000 lb of bombs could be carried in addition to the four 0·30-in machine guns.

The prototype XP3Y-1 was completed in Consolidated's Buffalo factory, and the first flight was made on March 28, 1935. Early trials confirmed the Navy's optimism for the type, and plans for an initial production batch of 60 were formalized on June 29, 1935. These were designated PBY-1, indicating that the ability to carry a substantial bomb load had brought the aircraft into the patrol bomber category. While work on these was put in hand at the company's new factory in San Diego, the XP3Y-1 was returned to Consolidated in October 1935 for modification up to production standard. Fitted with more powerful R-1830-64 engines and a redesigned, less angular fin and rudder, it flew again on May 19, 1936, as the XPBY-1, and was delivered to Patrol Squadron VP-11F in October; this unit was the first equipped with PBY-1s, which began to reach the squadron about the same time as the prototype. VP-12 equipped with PBY-1s in 1937, while VP-11 received the first PBY-2s.

Continuation contracts were placed with Consolidated for 50 PBY-2s on July 25, 1936, 66 PBY-3s on November 27, 1936, and 33 PBY-4s on December 18, 1937. All were generally similar in appearance. The PBY-3s had 1,000 hp R-1830-66 engines, and the PBY-4s had 1,050 hp R-1830-72s; all but one of the latter had bulbous transparent covers over the beam gunners' positions, previously covered by sliding hatches.

Introduction of these aircraft proceeded rapidly during 1937 and 1938; by the middle of the latter year, some 14 squadrons were flying the type, including five based at Pearl Harbor and three at Coco Solo. With the outbreak of war in Europe, the demand for Consolidated's flying-boat increased rapidly. Britain, Canada, Australia and the Dutch East Indies all ordered the type in quantity, and on December 20, 1939, the US Navy ordered another 200 examples, primarily to equip the Neutrality Patrol. These were of the PBY-5 type, first deliveries of which were made on September 18, 1940; they had 1,200 hp R-1830-92 engines, waist blisters like the PBY-4 and a further change in fin design. By the end of 1941 sixteen Navy patrol squadrons were flying PBY-5s, which had replaced most of the earlier models; three squadrons remained operational with the PBY-3 and two were using the PBY-4.

On November 22, 1939, Consolidated flew the first amphibious version of the design—actually the last PBY-4 converted to the XPBY-5A. The retractable tricycle undercarriage added greatly to the PBY's utility while detracting little from its performance, and the Navy ordered the final 33 PBY-5s to be completed to this standard and ordered 134 more PBY-5As on November 25, 1940. Deliveries began at the end of 1941, the first units being VP-83 and VP-91; by this time, the name Catalina, proposed by the company and used by the RAF, had been adopted also by the US Navy. When America entered the war in December 1941, the Catalina was thus the principal patrol bomber flying-boat in service with the US Navy, and it consequently played a prominent part in operations in the Pacific area right

Consolidated PBY-5A amphibian in all-white finish, including wheels and tyres.
(*A. U. Schmidt*)

81

Consolidated PB2B-2 built by Boeing in Canada. (*Courtesy Gordon S. Williams*)

from the start. In RAF service, it had already been in action over the English Channel, the Mediterranean and the Atlantic; the USN later joined the RAF in flying anti-submarine patrols over the Atlantic and in other theatres.

During 1941 and 1942, further contracts were placed with Consolidated by the USN for 586 more PBY-5s, 627 more PBY-5As and 225 PBY-5B flying-boats to be supplied under lend–lease arrangements to Britain. Other Catalinas of various types were diverted to the RAF from USN contracts and from RCAF orders. The total offensive load of these later PBYs was increased to a maximum of 4,000 lb of bombs, two torpedoes or four 325 lb depth charges under the wings. On later PBY-5As, a second 0·30-in gun was added in the bows, in addition to two 0·50-in beam guns and one 0·30-in tunnel gun in the hull.

Additional production of Catalinas began in Canada in 1941 at the Canadian Vickers Ltd. plant in Cartierville and the Boeing Aircraft of Canada factory in Vancouver. The Canadian Vickers aircraft were designated PBV-1A by the Navy, but all 230 built (67832–68061) in fact went to the USAAF as OA-10A amphibians; another 149 built by Canadian Vickers went to the RCAF under the service name Canso. Boeing in Canada built a total of 362 Catalinas and Cansos, the first of which flew on May 12, 1943. This total included 240 flying-boats similar to the PBY-5, designated PB2B-1, plus 50 PB2B-2s similar to the PBN-1 described below,

Consolidated PBY-6A amphibian with tall tail, in Navy Reserve colours.
(*Edgar Wischnowski*)

82

built on lend–lease contracts for the RAF, RAAF and RNZAF; and 17 Catalinas and 55 Canso amphibians for the RCAF.

Also in 1941 the Naval Aircraft Factory at Philadelphia received a contract to build 156 examples of the Consolidated flying-boat. Designated PBN-1 and renamed Nomad, these aircraft introduced a number of improvements, especially in hull design and wingtip float design. The fuel capacity was increased, the tail was heightened and the operating weights went up. A 0·50-in gun replaced the smaller calibre weapon in the nose. Deliveries began in February 1943, and 138 of the total went to Russia under lend–lease. A version similar to the PBN-1 but with two 0·50-in nose guns, search radar in a radome above the cockpit and amphibious undercarriage was ordered on July 9, 1943, from a new Consolidated factory at New Orleans. These were designated PBY-6A; 900 were ordered, but production ended after delivery of 48 to Russia, 75 to the USAAF (as OA-10B) and 112 to the USN. Production at New Orleans

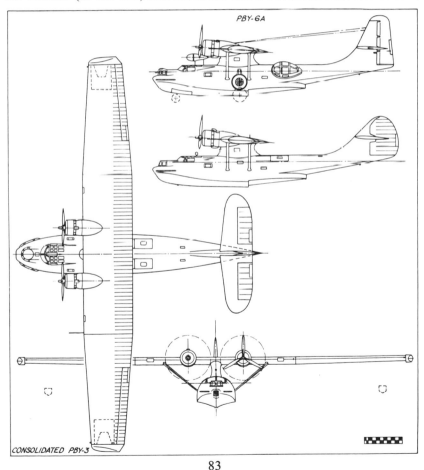

PBY-6A

CONSOLIDATED PBY-3

83

ended in April 1945 and brought the total of PBY types built by the parent company to 2,398. A further 892 examples were built by the NAF and in Canada, plus an unspecified quantity in Russia with the designation GST.

The flying-boat versions of the Catalina were quickly retired by the Navy after the end of the war, but amphibious PBY-5A and -6A variants continued in use for a few years, primarily for air–sea rescue. In this role, they often carried an air-dropped lifeboat beneath one wing.

TECHNICAL DATA (PBY)

Manufacturer: Consolidated Aircraft Corporation, San Diego, California; Canadian Vickers Ltd, Montreal; Boeing Aircraft of Canada, Ltd, Vancouver; and Naval Aircraft Factory, Philadelphia, Pennsylvania.
Type: Patrol bomber flying-boat and amphibian.
Accommodation: Crew of seven–nine.

	PBY-2	PBY-5	PBY-5A	PBN-1
Power plant:	2 × 900 hp R-1830-64	2 × 1,200 hp R-1830-92	2 × 1,200 hp R-1830-92	2 × 1,200 hp R-1830-92
Dimensions:				
Span, ft, in	104 0	104 0	104 0	104 3
Length, ft, in	65 10	63 10	63 10	64 8
Height, ft, in	18 6	18 6	20 2	21 3
Wing area, sq ft	1,400	1,400	1,400	1,400
Weights:				
Empty, lb	14,668	17,526	20,910	19,288
Gross, lb	28,400	34,000	35,420	38,000
Performance:				
Max speed, mph/ft	178/8,000	189/7,000	175/7,000	186/6,700
Cruising speed, mph	103	115	113	111
Climb, ft/min	830	690	620	—
Service ceiling, ft	20,800	18,100	13,000	15,100
Range, miles	2,110	2,990	2,350	2,590
Armament:				
Guns	2 × 0·30-in 2 × 0·50-in	2 × 0·30-in 2 × 0·50-in	3 × 0·30-in 2 × 0·50-in	3 × 0·50-in 1 × 0·30-in
Bombs	4 × 1,000 lb	4 × 1,000 lb	4 × 1,000 lb	4 × 1,000 lb

Serial numbers:
XP3Y-1 (XPBY-1): 9459.
PBY-1: 0102–0161.
PBY-2: 0454–0503.
PBY-3: 0842–0907.
PBY-4: 1213–1244.
PBY-5: 2289–2455;   04425–04514;   08124–08549; 63992.
XPBY-5A: 1245.

PBY-5A: 2456–2488; 7243–7302; 02948–02977;   04399–04420;   04972–05045;   08030–08123;   21232; 33960–34059;   46450–46638; 48252–48451.
PBY-6A: 46639–46698;   46724;   63993–64099; 64101–64107.
PBN-1: 02791–02946.
PB2B-1: 44188–44227; 72992–73116.
PB2B-2: 44228–44277.
PBV-1A: 67832–68061.

84

Consolidated PB2Y-5R, with turrets removed for transport role. (*Consolidated Vultee photo*)

Consolidated PB2Y Coronado

Within three months of the first flight of the Consolidated XP3Y-1 prototype Catalina (see page 80), the US Navy was making plans for the development of much larger flying-boats with better operational performance. Prototypes were ordered from Sikorsky and Consolidated, respectively in June 1935 and July 1936, of large four-engine flying-boats in the patrol bomber category. The Consolidated design, Model 29, made use of retractable wingtip floats similar to those on the Catalina, but in all other respects it was a wholly new design with a high-mounted wing and a capacious hull with accommodation for a crew of ten.

First flown on December 17, 1937, the XPB2Y-1, as Consolidated's boat was designated by the Navy, was powered by four 1,050 hp Pratt & Whitney XR-1830-72 Twin Wasp engines and had an armament of two 0·50-in guns (nose and tail) and three 0·30-in guns (two waist and one tunnel). As first tested, it had a tall single fin and rudder, but a twin tail unit was later fitted and the under hull redesigned.

With almost all Navy flying-boat procurement limited to PBYs, orders for the big new aircraft were delayed until March 31, 1939, when Consolidated received a contract for six PB2Y-2s, each costing as much as three PBYs. They had R-1830-78 engines with two-stage superchargers, increased armament of six 0·50-in guns and a redesigned, deeper hull. Despite higher operating weights, the PB2Y-2s achieved 255 mph at 19,000 ft. The first to be delivered reached VP-13 on December 31, 1940. A further production contract placed on November 19, 1940, established the PB2Y-3 in production, with R-1830-88 engines, self-sealing fuel tanks and eight 0·50-in guns. These Coronados, as the type was named, often carried ASV radar in a fairing just behind the cockpit. In all, 210 PB2Y-3s were built.

Little operational use was made of the PB2Y-3, but 10 -3Bs supplied to Britain under lend–lease as Coronados were operated on freight services across the Atlantic to and from the United Kingdom, and subsequently 31 for the USN were converted to PB2Y-3R transports with R-1830-88 engines, faired-over turrets and other modifications. Some of the other PB2Y-3s became PB2Y-5s and PB2Y-5Rs when fitted with low-altitude

R-1830-92 engines and increased fuel capacity, or PB2Y-5Hs with the same modifications plus cabin arrangements for 25 stretchers. A single example with Wright R-2600 Cyclones was designated XPB2Y-4. The Coronados were withdrawn from active service before the end of 1945.

TECHNICAL DATA (PB2Y-3)

Manufacturer: Consolidated Aircraft Corporation, San Diego, California.
Type: Patrol bomber flying-boat.
Accommodation: Crew of ten.
Power plant: Four 1,200 hp Pratt & Whitney R-1830-88s.
Dimensions: Span, 115 ft; length, 79 ft 3 in; height, 27 ft 6 in; wing area, 1,780 sq ft.
Weights: Empty, 40,935 lb; gross, 68,000 lb.
Performance: Max speed, 213 mph at 20,000 ft; cruising speed, 141 mph at 1,500 ft; initial climb, 440 ft/min; service ceiling, 20,100 ft; range, 1,490 st miles.
Armament: Eight 0·50-in guns, all flexibly mounted. Up to four 1,000 lb bombs external and eight 1,000 lb bombs internal.
Serial numbers:

XPB2Y-1: 0453.	XPB2Y-3; 1638.
PB2Y-2: 1633-1638.	XPB2Y-4; 1636.
PB2Y-3: 7043–7242; 02737–02746.	

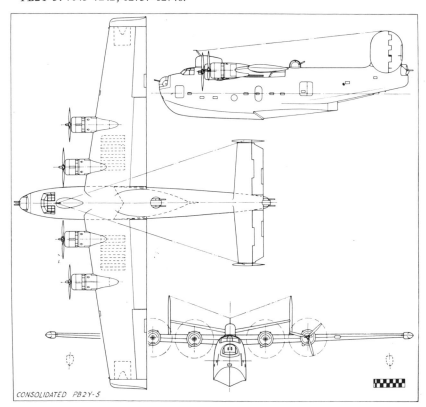

CONSOLIDATED PB2Y-5

Consolidated PB4Y-1 in anti-submarine colours with gull-grey top and white sides and undersurfaces. (*US Navy*)

Consolidated PB4Y Liberator and Privateer

Early war-time experience with flying-boat patrol bombers demonstrated the limitations as well as the advantages of this category of aircraft. Consequently, the US Navy began pressing early in 1942 for a force of B-24 Liberator land-based bombers to fly long-range overwater patrols against shipping and submarines. The RAF was already using the Liberator in this role with outstanding success, and, despite the pressing USAAF need for heavy bombers in 1942, an agreement on July 7 that year allowed the Navy a share of B-24 production.

Designated PB4Y-1, versions of the B-24D began to reach the Navy in August, some going to the TTSP(PAC), a transitional training squadron, and others to the first operational unit, based in Iceland, which scored its first success against a U-boat on November 5, 1942. Deliveries built up slowly at first, but in August 1943 the USAAF agreed to hand over its anti-submarine B-24 squadrons to the Navy, and accordingly disbanded the Anti-Submarine Command at the end of that month. All the USAAF's ASV-equipped B-24s were transferred to the Navy as PB4Y-1s in exchange for an equal number of unmodified Liberators already in production for the Navy. Most PB4Y-1s carried an armament of eight 0·50-in machine guns and up to 8,000 lb of bombs; some, equivalent to the B-24J model, had Erco nose turrets.

Two PB4Y-1s of Navy Squadron VB-110, based in the UK, were used

Convair PB4Y-2 with individual crew insignia on nose, Shanghai 1945. (*Peter M. Bowers*)

87

for a unique operation during 1944, being launched against targets in occupied Europe as pilotless flying bombs. Under the code name of Project *Anvil*, the scheme was devised at short notice to attempt to destroy a German V-2 installation in occupied France. A PB4Y-1 was fitted with remote control gear, a forward-looking television with a transmitter to send the picture to the control aircraft and 25,000 lb of high explosive. As there was no time to develop equipment to permit a remotely-controlled take-off, the aircraft was to be flown to 2,000 ft by a squadron pilot, who would then switch it to automatic flight and bail out.

The operation was mounted on August 12, 1944, with two pilots in the Liberator—Lt Joseph P. Kennedy, Jr, brother of John F. Kennedy (later the US President) and Lt Wilford J. Willy—accompanied by a Lockheed PV-1 control aircraft and a USAAF B-17 to monitor the TV transmission. Twenty minutes after take-off, the PB4Y-1 exploded, killing both pilots, apparently as a result of a malfunction in the fuse system.

A second PB4Y-1 was under test before the end of August and after several successful test flights was launched against an airfield target in Heligoland on 3 September, 1944, the original V-2 site in France having by this time been overrun by advancing Allied forces. Lt Ralph Spaulding made the take-off and parachuted to safety, leaving the Liberator under the control of the PV-1, and the flight across the North Sea was made at about 300 ft. Control was effected without difficulty, and an engineer from VB-110 flying in the B-17 had a clear view, via the television transmitter, of the Dutch countryside on the approach to the target. The TV camera was put out of action by flak just before the PB4Y-1 struck its target, but control was not impaired and the observers in the other aircraft saw the massive explosion as the Liberator struck. This was the only recorded operational use of a guided bomb by Allied forces in Europe during World War II.

In addition to a total of 977 PB4Y-1s, the Navy received a number of

A Coast Guard Convair PB4Y-1G, showing observer's station in modified nose.
(*William T. Larkins*)

transport versions of the Liberator, designated RY-1 and RY-2, respectively equivalent to the USAAF C-87A and C-87. A number of the Navy Liberators were modified for reconnaissance duties as PB4Y-1Ps and served until 1950 with Navy Squadrons VP-61 and VP-62; they were redesignated P4Y-1Ps in 1951 shortly before being withdrawn completely.

Development of a B-24 variant more specifically suited to Navy requirements began on May 3, 1943, when Convair was instructed to allocate three B-24Ds for conversion into XPB4Y-2s. This new model, named Privateer, retained the same wing and landing gear, but had a lengthened fuselage, a tall single fin and rudder and different engine nacelles. The engines were R-1830-94s, no turbo-superchargers being fitted since the aircraft spent most of their time at low altitude.

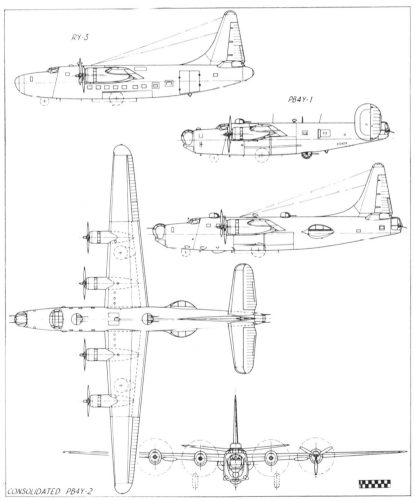

The first of the three prototypes made its initial flight on September 20, 1943, and a production contract for 660 PB4Y-2s was placed in October, followed by a second for 710 a year later. Deliveries began in March 1944 and ended in October 1945, cancellations reducing the number built to 736. A transport version of the Privateer was also put into production as the RY-3.

Little operational use was made of the PB4Y-2 during the war, although one squadron, VP-24, was equipped with PB4Y-2Bs carrying beneath each wing an ASM-N-2 Bat anti-shipping glide-bomb with radar homing. At least five other Navy squadrons flew the Privateer for several years after the war, some being PB4Y-2S with added anti-submarine radar, and in 1951 the aircraft in service were redesignated P4Y-2, P4Y-2B and P4Y-2S. A few were also used by the US Coast Guard for air–sea rescue and weather reconnaissance duties, designated P4Y-2G as well as earlier PB4Y-1Gs.

Final version of the Liberator operational with the Navy was a target drone, designated P4Y-2K. In the unified designation system of 1962 this version became the QP-4B; no other Liberator variants were redesignated.

TECHNICAL DATA (PB4Y-1)

Manufacturer: Consolidated (later Consolidated Vultee) Aircraft Corporation, San Diego, California.
Type: Land-based patrol bomber.
Accommodation: Crew of nine to ten.
Power plant: Four 1,200 hp Pratt & Whitney R-1830-43s or -65s.
Dimensions: Span, 110 ft; length, 67 ft 3 in; height, 17 ft 11 in; wing area, 1,048 sq ft.
Weights: Empty, 36,950 lb; gross, 60,000 lb.
Performance: Max speed, 279 mph at 26,500 ft; cruising speed, 148 mph at 1,500 ft; initial climb, 830 ft/min; service ceiling, 31,800 ft; range, 2,960 st miles.
Armament: Eight flexible 0·50-in guns in nose, dorsal and tail turrets and waist mounts. Up to eight 1,600 lb bombs internal.
Serial numbers:

PB4Y-1: 31936–32287;	32288–32335;	RY-1: 67797–67799.
38733–38979;	46725–46737;	RY-2: 39013–39017.
63915–63959;	62587–65396;	
90132–90271;	90462–90483.	

TECHNICAL DATA (PB4Y-2)

Manufacturer: Consolidated Vultee Aircraft Corporation, San Diego, California.
Type: Land-based patrol bomber.
Accommodation: Crew of 11.
Power plant: Four 1,350 hp Pratt & Whitney R-1830-94s.
Dimensions: Span, 110 ft; length, 74 ft 7 in; height, 30 ft 1 in; wing area, 1,048 sq ft.
Weights: Empty, 37,485 lb; gross, 65,000 lb.
Performance: Max speed, 237 mph at 13,750 ft; cruising speed, 140 mph; initial climb, 1,090 ft/min; service ceiling, 20,700 ft; range, 2,800 st miles.
Armament: Twelve flexible 0·50-in guns in turrets and waist mounts. Up to eight 1,600 lb bombs.
Serial numbers:

XPB4Y-2: 32086; 32095–32096.	RY-3: 90020–90021; 90023–90059.
PB4Y-2: 59350–60009; 66245–66324.	

Curtiss A-1 Triad resting on its wheels in shallow water. Both sets of elevators are 'down'. (*Curtiss photo*)

Early Curtiss Pushers

Identifying the various Curtiss pushers supplied to the US Navy from 1911 through 1913 by either a factory or Navy model number is impossible. Curtiss had no firm designation system at the time and the Navy system was very general, applying originally only to manufacturer and sequence of procurement. This was soon changed to a basic type, as acroplane, and to subtype, as hydroplane or flying-boat. The first Navy aeroplane was a Curtiss pusher seaplane designated A-1. This was a waterborne version of the basic Curtiss pusher that had been in production since 1909 and which had been developed into the world's first consistently successful seaplane. After several unsuccessful float configurations were tried, Glenn Curtiss made his first seaplane flight at San Diego, California, on January 26, 1911. It was this ability to operate from water, plus a visit by Curtiss in his seaplane to the battleship USS *Pennsylvania* anchored off San Diego from a shore station on February 17, 1911, that crystallized the Navy's existing

Naval Aviator No 1, Lt T. G. Ellyson, in a Curtiss Pusher at San Diego, January 1911, prior to purchase of Navy's first aeroplane. (*A. U. Schmidt Collection*)

91

interest in aeroplanes and led to the purchase of the A-1 in July 1911. Earlier, the Navy co-operated with Curtiss in the operation of Curtiss-owned aeroplanes from ships of the fleet. On November 14, 1910, the Curtiss pilot Eugene Ely flew a pusher from a platform rigged on the aft deck of the cruiser USS *Birmingham* anchored in Hampton Roads, Virginia. Later (on January 18, 1911) he landed aboard and took-off from a similar platform built on the battleship USS *Pennsylvania* anchored in San Francisco Bay.

Lt Theodore G. Ellyson, who had been an official naval observer at Curtiss' first seaplane flight, was sent to the Curtiss Flying School at Hammondsport, NY, to receive flight instruction that qualified him as Naval Aviator No. 1. After the A-1 was test hopped by Curtiss on July 1, 1911, Curtiss took Ellyson up on a familiarization flight after which Ellyson made two more flights alone. After acceptance by the Navy, the A-1 figured prominently in early Naval aircraft developments. One experiment consisted of converting the A-1 to an amphibian, called the Triad, by adding retractable wheels to the float.

Launching methods were given high priority. On September 7, 1911, Ellyson was able to take-off in the A-1 from an inclined wire. Directional control was maintained by a groove in the float through which the wire ran. Later, experiments were conducted in which the A-1 was to be launched by a compressed-air catapult based on the successful torpedo launchers then in use by the Navy. The initial trials, undertaken on a dock at Annapolis, Maryland, ended in failure due to the A-1 not being properly restrained. It lifted off from the carriage without control as soon as it gained forward speed. The first successful catapult launch was made by Ellyson in the A-3 at the Washington Navy Yard. The A-1, meanwhile, being the only aeroplane that the Navy had, set an impressive number of records merely by exceeding its previous performance. With Lt John H. Towers as passenger, Ellyson flew 112 miles from Annapolis, Maryland, to Milford Haven, Virginia, in 122 minutes. The A-1 was the first Navy

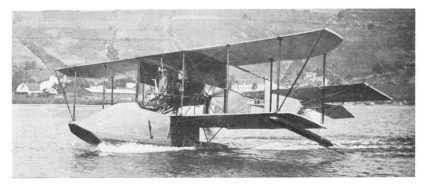

Curtiss A-2 with built-up superstructure above pontoon. This aircraft later became the E-1 (later AX–1) when wheels were added. (*Bowers-Williams Collection*)

aeroplane to carry a radio, although without success, and established a seaplane altitude record of 900 ft on June 21, 1912.

The A-2 was delivered to the Navy on July 13, 1911, in landplane configuration. As a seaplane, it remained in the air for 6 hr 10 min on October 6, 1912. By October 1912, the A-2 had been converted to a flying-boat by the simple expedient of building a superstructure from the pontoon deck upward to enclose the crew. Retractable tricycle landing gear similar to that of the Triad was fitted, and the designation of the modified machine was changed to E-1. This was also called the OWL for Over Water and Land and was briefly known as the AX-1.

The next two Navy aeroplanes, designated A-3 and A-4, were also single-float Curtiss pushers with minor differences in detail. A-3 achieved a degree of fame on June 13, 1913, by establishing an American seaplane altitude record of 6,200 ft. By the time the Navy was ready to order additional Curtiss pusher seaplanes, the designation system had been changed and the aeroplanes were known as AH for Airplane, Hydro,

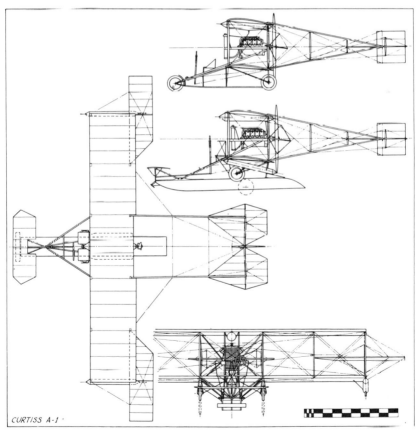

CURTISS A-1

Curtiss AH-13 two-seater bearing the legend 'U.S.N. AH-13' on the rudder. (*US Navy*)

followed by a sequence number. The first two procured under the new system were AH-8 and AH-9, both of which were referred to as Type AH-8 in the manner of naval ships, where the first of a new class or design gave its name to the entire group. The AH-8s were followed by five more, starting with AH-11. These were followed by a further three that Navy records refer to as AH-8 type.

Further designation changes took place within the procurement period of the Curtiss pushers, the last 11 being ordered under a system whereby each Navy aeroplane was identified by a consecutive serial number, which was prefixed by the letter A-for-Airplane. The Navy had wisely decided to designate its aeroplanes by the manufacturer's own name and model number, but since the Curtiss aircraft had no firm number at the factory, the remaining 11 of the basic design (A60–A62, A83–A90) were again designated as Type AH-8. A83 was the AH-9 rebuilt, and A84-A90 may have been similarly reassigned.

The original AH-8, meanwhile, had been turned over to the US Army. It survived World War I, still in Army hands, and was put in storage. It was resurrected in 1928 and flown briefly by Capt Holden C. Richardson, Chief of Design and Materiel Division, Bureau of Aeronautics, and Naval Aviator No. 13, on February 10.

The Curtiss A-1 had a 75 hp Curtiss V-8 engine for most of its 285 flights. The A-2, flown 575 times before becoming the E-1, had an 80 hp engine eventually, earlier units rated at 60 hp and 75 hp being fitted. Span, 37 ft 1 in; length, 27 ft 2 in; gross weight, 1,547 lb; speed 60 mph.

TECHNICAL DATA (A-1)

Manufacturer: Curtiss Aeroplane and Motor Co, Inc, Hammondsport, NY.
Type: Pusher biplane, landplane or seaplane.
Accommodation: Pilot and passenger.
Power plant: One 75 hp Curtiss V-8.
Dimensions: Span, 37 ft; length, 28 ft 7$\frac{1}{8}$ in; height, 8 ft 10 in; wing area, 286 sq ft.
Weights: Empty 925 lb; gross, 1,575 lb.
Performance: Max speed, 60 mph.
Serial numbers: None.

Curtiss C-1 flying-boat (later AB-1), with Cmdr T. G. Ellyson in the cockpit. (*US Navy*)

Curtiss F-Boat

The Curtiss Model F was an early development of the original Curtiss pusher flying-boat of 1912, the world's first successful flying boat. Designation of the US Navy models is confusing because procurement of the basic design bridged the changeover from the original Navy system of designating aircraft by code letters for the manufacturer and type to the use of the manufacturer's own model designation. Also, there was enough change between early consecutive production examples of a single basic design to make the rigid application of an all-inclusive model number somewhat unrealistic. The designation is confused further by the later adoption of the British F (for Felixstowe, see page 114) symbol for later developments of the twin-engine Curtiss H series, and the practice of using F as a type letter applied to small single-engine pusher flying-boats developed by other manufacturers.

The major production version of the Curtiss F used a hull built up of laminated wood veneer strips which were shaped and glued up in a jig before being applied to the wooden hull frame. The student and his instructor sat side by side in a single cockpit ahead of the wings. Wing shape varied during the production life of the F, which continued into 1918. Some had equal-span wings with ailerons mounted between the panels, and some had the ailerons built into the upper wing. Overhang was added to the upper wing to increase lifting area on some models. Early models had fabric stretched between front and rear struts just outboard of the engine to serve as anti-skid vanes; later models had fabric

95

applied between the kingposts that braced the upper wing overhang to serve the same purpose.

The first five Navy F-boats, all differing in detail, were procured as Navy models C-1 through C-5. In March 1914 these machines were re-designated AB-1 through AB-5, the letter A designating Curtiss as the first manufacturer of aircraft for the Navy and the letter B identifying the type of the machine as a flying-boat. As ABs, these machines contributed much to early Navy aeronautical development. In December 1912 one boat, believed to have been the C-1 at the time, was launched from a catapult mounted on a dock at the Washington Navy Yard. AB-2 was later catapulted from a barge moored at the Pensacola Naval Air Station in Florida and on November 5, 1915, Lt Cdr H. C. Mustin, Naval Aviator No. 11, made the first Navy catapult launch from a ship when he success-fully left the battleship USS *North Carolina* in the AB-2. The ship was at anchor at the time and the catapult was directed straight astern.

Several of the ABs accompanied the Navy to Vera Cruz, Mexico, in April 1914. Carried aboard the USS *Mississippi*, AB-3 was flown from the water by Lt (Jg) P. N. L. Bellinger in the first operation of US military air-craft against another country. The first flight, made on April 25, was for observation purposes and to look for mines in the harbour. The second, made on April 28, was a photographic mission. Upon its return to the United States, AB-3 had its wings shortened and finished its career as a non-flying 'penguin' trainer.

At least 144 additional trainer flying-boats were ordered from Curtiss under the F designation. More show up in Navy serial number listings, but these are mixed among the later Curtiss MF models that began to replace the Fs in 1918 and Fs built to Curtiss' or their own design by other manu-facturers. The Burgess Company of Marblehead, Mass., was to have built

Curtiss F flying-boat, showing the overhanging top wing and inter-wing ailerons.
(*US Navy*)

Curtiss Fs under licence but was switched to the production of Curtiss N-9s, and produced only one Curtiss F (A2281).

TECHNICAL DATA (F-Boat)

Manufacturer: Curtiss Aeroplane and Motor Co, Inc, Hammondsport, and Buffalo, NY.
Type: Flying-boat trainer.
Accommodation: Pilot and instructor side by side.
Power plant: One 100 hp Curtiss OXX.
Dimensions: Span, 45 ft 1$\frac{3}{8}$ in; length, 27 ft 9$\frac{3}{4}$ in; height, 11 ft 2$\frac{3}{4}$ in; wing area, 387 sq ft.
Weights: Empty, 1,860 lb; gross, 2,460 lb.
Performance: Max speed, 69 mph at sea level; initial climb, 10 min to 2,300 ft; service ceiling, 4,500 ft; endurance, 5·5 hrs.
Serial numbers (*known*):
   A386; A387; A390–A393; A408; A2279–A2281; A2295–A2344; A4079–A4108; A5258.

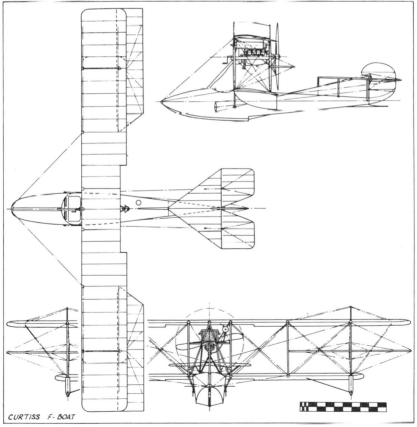

CURTISS F- BOAT

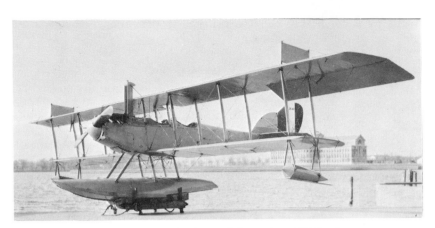

Curtiss N-9H with Hispano-Suiza engine. (*US Navy*)

Curtiss N-9

The Curtiss N-9, originally powered with a 100 hp Curtiss OXX-6 engine, was the standard Navy primary and advanced seaplane trainer of World War I. It was developed as a private venture by Curtiss as a seaplane version of the JN-4B landplane trainer then in production. The N-8 landplane model was almost identical to the JN-4, having only minor differences in the aerofoil section and the control system, and evolution into the N-9 was simple. Intended from the beginning as a seaplane, the N-9 was faced with the weight problem of the heavy central float and the stabilizing tip floats. The extra 10 hp of the OXX-6 engine over the standard OX-5 was not enough, so additional wing area was obtained by increasing the span of each wing by 10 ft. Instead of building longer panels, a wider centre section was built for the upper wing and an extra 5-ft panel was fitted between the fuselage and standard-size lower panels.

N-9s entered service with the Navy even before the United States entered the war in April 1917. The Army bought 14 N-9s at this time, too, since it conducted relatively extensive seaplane operations. In the primary training role, the 100 hp N-9 was satisfactory, but more performance was required for such advanced operations as bombing and gunnery training. To satisfy this requirement, Curtiss replaced the OXX-6 with the 150 hp Hispano-Suiza Model A then being manufactured in the United States by the Simplex Division of the Wright-Martin Company. This improved model was designated N-9H.

Five hundred and sixty N-9s were built for the Navy during World War I and the type remained in service as late as 1926. Of this total, only 100 were built by Curtiss, the majority being produced under licence by the Burgess Company of Marblehead, Mass. An additional 50 N-9Hs were

98

created by the Navy in the post-war years by the practice of assembling available spare parts and engines into entirely new airframes.

TECHNICAL DATA (N-9H)

Manufacturer: Curtiss Aeroplane and Motor Co, Inc, Garden City, LI, and Buffalo, NY; and the Burgess Company, Marblehead, Mass.

Type: Training seaplane.

Accommodation: Two in tandem.

Power plant: One 150 hp Hispano-Suiza A.

Dimensions: Span, 53 ft 3¾ in; length, 30 ft 10 in; height, 10 ft 8½ in; wing area, 496 sq ft.

Weights: Empty, 2,140 lb; gross, 2,765 lb.

Performance: Max speed, 80 mph at sea level; climb, 10 min to 4,450 ft; service ceiling, 9,850 ft; range, 179 st miles.

Serial numbers:

N-9/N-9H (Curtiss): A60–A65; A85–A90; A96–A125; A201–A234; A294–A301; A342–A373; A2285–A2290.

N-9/N-9H (Burgess): A409–A438; A999–A1028; A2351–A2572; A2574–A2650.

N-9H (NAS Pensacola): A6528–A6542; A6618–A6632; A6733–A6742; A7091–A7100

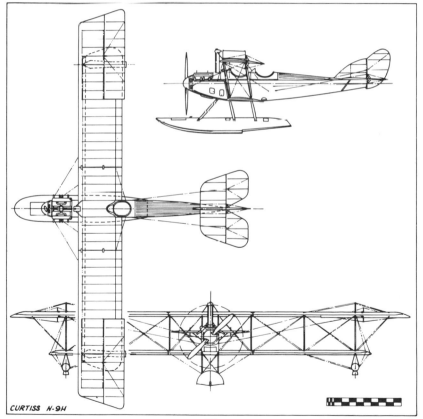

CURTISS N-9H

99

Curtiss JN-4H serving at NAS Pensacola. (*US Navy*)

Curtiss JN Series

The Curtiss JNs, particularly the JN-4 model, are widely known throughout the world as the Jenny, a logical expression of the model designation JN, which covered the result of combining the best features of the Curtiss Models J and N. In addition to being the most widely used trainers of the US Army and the Royal Canadian Air Force in World War I, the Jenny and its Canadian equivalent, the Canuck, embarked upon an entirely new career in the post-war years when cheap war-surplus models came into the hands of private owners. While over 4,000 aircraft in the JN series were built, with most going to the US Army, the RCAF and the RAF, a respectable number, 261, were used by the US Navy from 1916 into the early 1920s.

The design originated in England. B. Douglas Thomas, who had been an engineer with Avro and later with Sopwith, was engaged by Glenn Curtiss while still in England to develop a new tractor-type trainer to replace the Curtiss pushers that were then finding great disfavour with both the US Army and Navy training schools. Since Europe had the lead in tractor design at the time, Curtiss sought to save valuable time by hiring an engineer already experienced in this layout, which was as yet unfamiliar to American practice. Thomas's design became the Model J, an equal-span biplane built in the Curtiss plant at Hammondsport, New York. Initial flights were made with the fuselage uncovered. This design was tried both as a landplane and a single-float seaplane. A very similar Model N, differing mainly in the aerofoil used, followed. The J design was discontinued upon development of the JN, but the N model remained in production and was developed to the N-9 by war's end (see page 98). Some of the Navy JN-4s were obtained on direct purchase from Curtiss, but others were obtained by exchanges of aircraft with the US Army, which had occasion to use aircraft developed originally for the Navy.

The first Navy JN was an oddity, compared with the rest of the line, in that it was a twin-engine design using major JN components. Rather than being given an entirely new model designation by the manufacturer, it was simply called Twin JN; the Navy serial number was A93. This was evaluated as a landplane and as a twin-float seaplane, but was not ordered into production for the Navy, although the Army used a total of ten.

The first genuine Navy Jennies were two JN-1Ws (A149, A150), single-float seaplane versions of the Army JN-1. In spite of the relatively modern lines of this model compared with the open-air Curtiss pushers that it replaced, the old shoulder-yoke type of aileron control was retained. This survived in the contemporary N series through the N-8. One additional JN-1 (A198), fitted out as a gunnery trainer, was obtained later.

Subsequent procurement of Navy Jennies was not in strict sequence of model development, due partly to the exchanges with the Army. Three

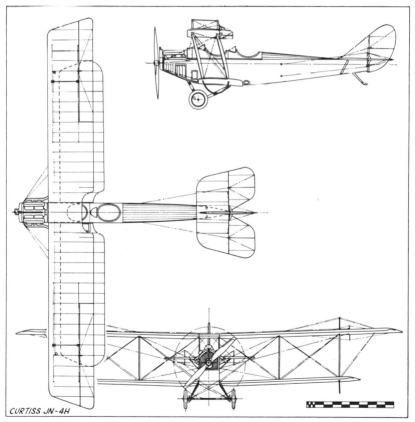

CURTISS JN-4H

Curtiss JN-1S seaplane, first Navy single-engined Jenny. (*Bowers-Williams Collection*)

JN-4Bs, late 1916 versions of the JN-1 but fitted with improved vertical tail surfaces and the wheel-type Deperdussin control, were obtained ahead of five JN-4As. These were followed by six additional JN-4Bs in 1918.

A major design change took place with the JN-4H, an advanced trainer fitted with the 150 hp Wright-Hispano engine, popularly called the Hisso. The letter H in the designation identified the engine and was not a reflection of model development. Thirty of the Hs were procured for advanced pilot training in 1918 and were followed by 90 gunnery trainers designated JN-4HG. Further minor changes resulted in the JN-6, which could be distinguished from the JN-4H mainly in being fitted with ailerons on both wings. A total of ten was procured, some of which were designated JN-6HG-I to identify them as gunnery trainers powered with the 150 hp Wright-Hispano Model I engine. Frequently this latter designation is misquoted as JN-6HG-1.

Procurement continued into the early post-war years, an additional 113 JN-4H landplane trainers being used by the Navy and Marines.

TECHNICAL DATA (JN-4H)

Manufacturer: Curtiss Aeroplane and Motor Co, Inc, Garden City, LI, and Buffalo, NY.
Type: Trainer.
Accommodation: Two in tandem.
Power plant: One 150 hp Wright-Hispano.
Dimensions: Span, 43 ft 7¾ in; length, 27 ft 4 in; height, 9 ft 10½ in; wing area, 352·6 sq ft.
Weights: Empty, 1,467 lb; gross, 2,017 lb.
Performance: Max speed, 93 mph at sea level; climb, 10 min to 4,350 ft; service ceiling, 10,525 ft; range, 268 st miles.
Serial numbers:

JN-4A: A388; A389; A995–A997.	JN-4HG: A4128–A4217.
JN-4B: A157–A159; A4112–A4117.	JN-6H: A5470–A5471; A5581–A5586;
JN-4H: A3205–A3234; A6193–A6247;	A5859.
A6271–A6288.	

102

Curtiss R-6L after modification for torpedo dropping. (*US Navy*)

Curtiss R-3, R-6, R-9

The Curtiss R series of 1915–18 was widely used by the US Army and Navy and the Royal Naval Air Service for scouting, observation and training. The Navy models, as well as a few Army, were twin-float seaplanes originally powered with the 150 hp Curtiss V-X engine. As was common practice at the time, the pilot occupied the rear of the two cockpits while the observer sat in front, although his vision was consequently handicapped by the wings. The basic design was merely an enlargement of the J and N models that had become standard Army observation and training models. The Army R-4 landplane model had a wing span of 48 ft 4 in, with two bays, but the Navy R-3 model, of which two were built, had the span increased to 57 ft 1 in in order to carry the weight of the floats. This was accomplished by building a wider centre section for the upper wing, in the manner of the N-9, and fitting an additional section between the standard-sized bottom wings and the fuselage.

The R-6 was an improved R-3 with a 200 hp Curtiss V-X-X engine and dihedral on the wings. A few of the Curtiss-powered R-6s were converted to R-9s, the main change being relocation of the pilot to the front cockpit. The last 40 of the Curtiss-powered R-6s were converted to R-6L in 1918 by the installation of the 360–400 hp Liberty V-12 engine, and an

Curtiss R-3 floatplane with early Navy anchor markings on rudder and wings and original AH-62 serial, later changed to A-66. (*US Navy*)

additional 40 R-9s were ordered as such. R-6s were the first US built aircraft to serve overseas with US armed forces in World War I, a squadron being based at Ponta Delgada in the Azores for anti-submarine patrols from January 17, 1918.

After the Armistice, R-6Ls were modified to carry naval torpedoes. The Navy had tried this on August 14, 1917, but the experiment was not successful. Such late adoption of the torpedo-carrier was rather ironic for the US Navy, as it had been an American admiral, Bradley A. Fiske, who had proposed such an aircraft before the outbreak of World War I. British

Curtiss R-6 landplane with Curtiss V-X-X engine. (*US Navy*)

104

and German forces both used torpedo-carriers successfully during that war, but the overall weight of the weapon cut down on the amount of explosive to such an extent that the torpedo was a less effective weapon than the standard aerial bomb.

TECHNICAL DATA (R-6L)

Manufacturer: Curtiss Aeroplane and Motor Co, Inc, Buffalo, NY.
Type: Observation, scouting and training.
Accommodation: Two in tandem.
Power plant: One 400 hp Liberty V-12.
Dimensions: Span, 57 ft 1¼ in; length, 33 ft 5 in; height, 14 ft 2 in; wing area, 613 sq ft.
Weights: Empty, 3,325 lb; gross, 4,500 lb.
Performance: Max speed, 100 mph at sea level; climb, 10 min to 6,000 ft; service ceiling, 12,200 ft; range, 565 st miles.
Serial numbers:
  R-3: A66–A67.   R-9: A302–A341.
  R-6: A162–A197; A873–A994.

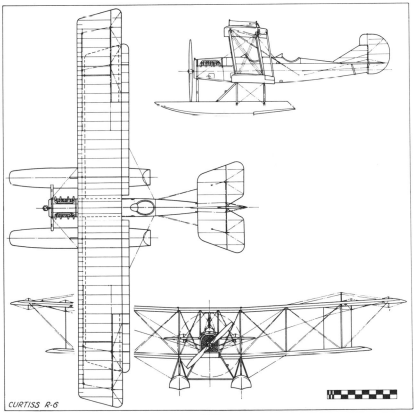

CURTISS R-6

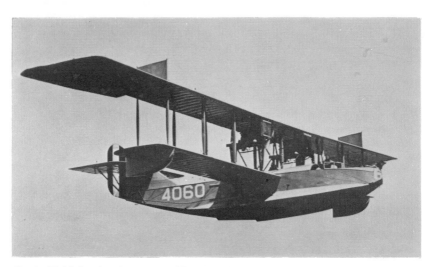

Curtiss H-16 showing the swept trailing edge of ailerons and pilot's cabin which distinguish this model from the F-5L. (*US Navy*)

Curtiss H-12, H-16

In 1914 Curtiss developed a then giant flying-boat to the special order of Mr Rodman Wanamaker, who planned to use it for a transatlantic flight. The outbreak of war cancelled these plans and the aeroplane, which had been named *America*, was sold to the Royal Naval Air Service. Under the factory designation Model H, sister ships with improved 150 hp engines were also sent to England, where as a class they were called Americas.

The original *America* was a daring design concept at the time, and left a permanent mark on subsequent large flying-boat development. Its effects could be seen on biplane designs that remained in production right up to World War II. The *America*, fitted with two 90 hp Curtiss OX engines, did not have the power to carry enough fuel for the trip, so a third engine was mounted above the wing. Earlier, flotation difficulties had been encountered when power was applied to the engines for take-off; because of their high location, they exerted a considerable downward push on the nose that tended to drive it under water. This was corrected by adding more flotation volume to the nose in the form of auxiliary structures called sponsons that were built on to the lower portion of the hull from the bow to the step. This was to remain a feature of many flying-boats built into the 1930s.

In 1916 the Navy ordered an improved version of the *America* under the

Curtiss H-12L in overall grey finish with 1918 roundels on wings. (*US Navy*)

designation of H-12. This featured the laminated wood veneer hull of the prototype, longer wings, and 200 hp Curtiss V-X-X engines. Equivalent models sold to Britain were called Large Americas. The initial Navy order was followed by another for 19 production versions. In 1918 some of these were converted to H-12L by the installation of the new Liberty engine.

A further improved model was introduced early in 1917 as the H-16, still powered with the 200 hp V-X-X. Many were sold to Britain in knockdown condition, still as Large Americas. They were then assembled and test flown in England, fitted with British engines. Commander Porte of the Royal Navy, who had assisted in the design of the original *America*, developed an improved hull design for the H-16, and the British versions were built at RNAS Felixstowe as F.2, F.3 and F.5.

By the time the US Navy became interested in production of the H-16, the Liberty engine was in the offing and was specified for the Navy's H-16s. However, in spite of the engine change there was no need to designate the production version as H-16L because there were no Curtiss-powered Navy models to require distinction. Because of the commitment of most of its production facilities to other war-time models, Curtiss could not meet the Navy's demand for H-16s, so the Navy undertook H-16 manufacture on its own at the Naval Aircraft Factory. This version was originally designated Navy Model C, as the third design built by Navy shops. This was the first aeroplane built by the new Naval Aircraft Factory, and the first example was completed on March 27, 1918. The original Curtiss designation was finally used.

In continuing attempts to improve the design, Curtiss built one H-16 with the engines turned around to drive pusher propellers. Because the engines had to be moved aft to get the propellers behind the wing, it became necessary to sweep the wings back slightly to relocate the centre of lift to match the new centre of gravity position. The Navy built 150 H-16s, and Curtiss built 124 for the Navy, some of which remained in service until 1928.

TECHNICAL DATA (H-16)

Manufacturer: Curtiss Aeroplane and Motor Co, Inc, Garden City, LI, and Buffalo, NY; and Naval Aircraft Factory, Philadelphia, Pennsylvania.

Type: Patrol/bomber flying-boat.

Accommodation: Crew of four.

Power plant: Two 400 hp Liberty 12s.

Dimensions: Span, 95 ft 0¾ in; length, 46 ft 1½ in; height, 17 ft 8⅝ in; wing area, 1,164 sq ft.

Weights: Empty, 7,400 lb; gross, 10,900 lb.

Performance: Max speed, 95 mph at sea level; initial climb, 10 min to 4,700 ft; service ceiling, 9,950 ft; range, 378 st miles.

Armament: Five–six flexible 0·30-in Lewis machine guns. Four 230 lb bombs.

Serial numbers:

  H-12: A152; A765–A783.

  H-16 (Curtiss): A784–A799; A818–A867; A1030–A1048; A4039–A4078.

  H-16 (NAF): A1049–A1098; A3459–A3558.

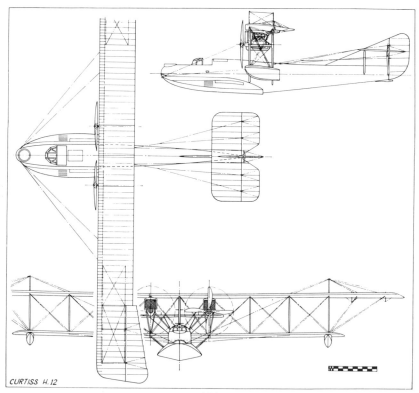

CURTISS H.12

Curtiss HS-1L patrol flying-boat. (*US Navy*)

Curtiss HS Series

The Curtiss HS-1 was a single-engine pusher flying-boat that was essentially a scaled-down version of the earlier twin-engine H models. In fact, the designation letters stood for 'H, Single engine'. Actually, the new H-boat stood about half-way between the smaller F (see page 95) and the larger H-12 (see page 107).

The original power plant of the HS-1, which was introduced early in 1917, was the 200 hp Curtiss V-X-X, a watercooled V-8. On October 21, 1917, the HS-1 was used as the test-bed for the first flight of the new twelve-cylinder Liberty engine, which in its original form developed 375 hp and was destined to become the major American aeronautical contribution to World War I and one of the world's great aircraft engines. In its later versions, this engine produced 420 hp, but was generally referred to as a 400 hp power plant.

Following the successful marriage of the HS-1 airframe and the Liberty engine, the Navy ordered the HS-1 into large-scale production as the Navy's standard coastal patrol flying-boat. The numbers required were beyond the capacity of the Curtiss plants, so additional manufacturers were asked to produce the boats under licence from Curtiss. The original Curtiss order was for 664 machines. The Standard Aircraft Corporation was given an order for 250, of which the last 50 were cancelled; Lowe,

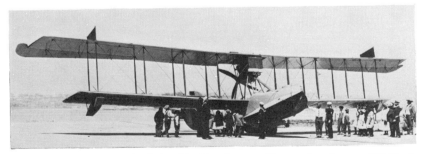

Boeing-built Curtiss HS-2L, distinguished by absence of lower-wing ailerons. (*US Navy*)

Willard and Fowler was given an order for a total of 200, but 50 were cancelled. (This firm later rebuilt one of its own HS-boats, A1171, which was then given the new serial number A5630.) The Gallaudet Aircraft Corporation produced 60, the Boeing Airplane Company built 25 of an original order for 50, and the Loughead Aircraft Corp (known today as Lockheed) built two. The Boeing versions could be distinguished from all the others in that they were fitted with horn-balanced ailerons only on the upper wing. The others had ailerons on both wings.

After the HS-1s, by this time designated HS-1L to identify installation of the Liberty engine, had entered service, it had been found that the standard 180 lb depth-bomb was ineffective against submerged submarines. Since two of these were all that the HS-1L could carry, it was decided to increase the wing span so that heavier 230 lb bombs could be carried. The modification was quite simple, an additional 6-ft panel was fitted between the centre section and the regular outer wing panels, increasing the overall span from 62 to 74 ft. This modification, which resulted in a new designation of HS-2L, was made to the majority of aircraft still on order, and there is no distinction in serial numbers between HS-1Ls and the HS-2Ls.

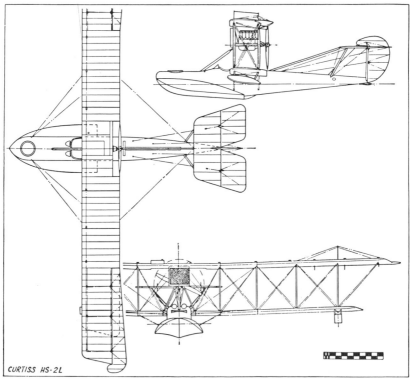

CURTISS HS-2L

HS-1s were the first American-built aircraft received by the US Naval forces in France, eight arriving by ship at the US Naval base at Pauillac on May 24, 1918. The first flight was made on June 13. Records indicate that 182 HS-1Ls and HS-2Ls were distributed among 10 of the 16 Naval Air Stations in France. Of the total, only 19 can be confirmed as HS-2Ls.

The HS-2L remained the standard single-engine patrol and training flying-boat in the post-war years, examples remaining in the inventory until 1926. An additional 24 HS-2Ls were obtained in the post-war years by assembling accumulated spare parts at various Naval Air Stations and assigning new serial numbers to the complete aircraft. The serials were assigned as follows:

A5564/5569	NAS Miami, Florida
A5615/5619	NAS Hampton Roads, Virginia
A5787	NAS Key West, Florida
A5808	NAS Anacostia, Maryland
A6506	NAS Coco Solo, Canal Zone
A6507/6513	Naval Aircraft Factory, Philadelphia, Pa.
A6553/6556	NAS San Diego, California

The HS-3 was an improved model with revised hull lines under development at war's end. Curtiss built five (A5459–A5462) and the Naval Aircraft Factory two (A5590–A5591). After the war, many surplus HS-2Ls were acquired by civil owners who used them for passenger carrying and even scheduled airline operations. A few were still in use as survey planes in Canada in the early 1930s.

TECHNICAL DATA (HS-2L)

Manufacturer: Curtiss Aeroplane and Motor Co, Inc, Garden City, LI, and Buffalo, NY; Standard Aircraft Corporation, Elizabeth, NJ; Lowe, Willard and Fowler, College Point, LI; Gallaudet Aircraft·Corporation, East Greenwich, Conn.; Boeing Airplane Company, Seattle, Wash; Loughead Aircraft Corporation, Santa Barbara, California.

Type: Patrol flying-boat.

Accommodation: Crew of two or three.

Power plant: One 350 hp Liberty 12.

Dimensions: Span, 74 ft 0½ in; length, 39 ft; height, 14 ft 7¼ in; wing area, 803 sq ft.

Weights: Empty, 4,300 lb; gross, 6,432 lb.

Performance: Max speed, 82·5 mph at sea level; initial climb, 10 min to 2,300 ft; service ceiling, 5,200 ft; range, 517 st miles.

Armament: One flexible 0·30-in Lewis gun. Two 230 lb bombs.

Serial numbers:

HS-1L/2L (Curtiss): A800–A815; A1549–A2207.

HS-2L (Standard): A1399–A1548 (50 cancelled).

HS-2L (LWF): A1099–A1398 (50 cancelled).

HS-2L (Gallaudet): A2217–A2276.

HS-2L (Loughead): A4228–A4229.

HS-2L (Boeing): A4231–A4255.

HS-2L (NAS): A5564–A5569; A5615–A5619; A5787; A5808; A6506–A6513; A6553–A6556.

HS-3 (Curtiss): A5459–A5462.

HS-3 (NAF): A5590–A5591.

Curtiss MF training flying-boat. (*Curtiss photo*)

Curtiss MF

The Curtiss MF was an improved 1918 model flying-boat' that was intended to replace the venerable F model that had been on hand since 1912. While the letters of the designation stood for Modified F there was no detail resemblance between the two other than the general configuration of a wooden-hulled pusher flying-boat. The MF drew upon the later design experience of the H-boats and the F-5Ls and used a flat-sided hull with additional forward buoyancy provided by sponsons added to the sides. The initial order to Curtiss was for six machines. This was followed by a production order for 47, but because of the Armistice, only the first 16 were delivered. An additional 80 were ordered into production at the Naval Aircraft Factory after the war.

Curtiss, meanwhile, quickly brought out a civil version of the MF under the name Seagull. Some of these, fitted with Wright-Hispano engines of 150 hp, or the new Curtiss K-6s of 150–160 hp were modified to carry as many as four people. Curtiss even tried an amphibious version in its quest to capture the sportsman-pilot market in the face of competition from cheap war-surplus machines, but the Seagull in its different versions was not a commercial success.

Navy interest in an amphibious version was reflected by a contract given to the Elias brothers aircraft company, Buffalo, NY, for the conversion of one of the NAF MFs (A5484) to an amphibian. As the MFs, both Curtiss and Navy-built, became surplus to the needs of the service, they were bought up by the Cox-Klemin Aircraft company, College Point, Long Island, and modified for civil use.

TECHNICAL DATA (MF-Boat)

Manufacturer: Curtiss Aeroplane and Motor Co, Inc, Garden City, LI, and Buffalo, NY; and Naval Aircraft Factory, Philadelphia, Pennsylvania.
Type: Flying-boat trainer.
Accommodation: Pilot and instructor side by side.
Power-plant: One 100 hp Curtiss OXX.
Dimensions: Span, 49 ft 9 in; length, 28 ft 10 in; height, 11 ft 7 in; wing area, 402 sq ft.
Weights: Empty, 1,850 lb; gross, 2,488 lb.
Performance: Max speed, 72 mph at sea level; initial climb, 10 min to 2,400 ft; service ceiling, 4,100 ft; range, 345 st miles.
Serial numbers:
 A2345–A2350; A4403–A4449 (last 31 cancelled); A5483–A5562 (by NAF)

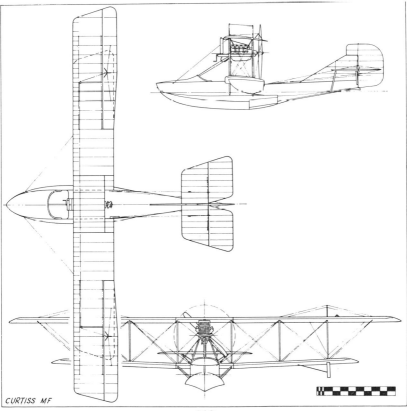

CURTISS MF

The first Curtiss F-5L built by the Naval Aircraft Factory. (*US Navy*)

Curtiss F-5L

Production of the F-5L in 1918 put Curtiss in the odd position of building an improved foreign version of one of its own designs. A number of Curtiss H-12s and H-16s, developed from the original *America* flying-boat of 1914, had been sold to Britain in 1915–16. These designs had been improved upon by the Royal Naval Air Station at Felixstowe, and were put into large-scale production as F (for Felixstowe) -2, -3 and -5, powered with British engines. The wings and tail were essentially Curtiss, but the major improvement was in hull design, which permitted quicker take-off under heavy load and stood up better on the surface of the rough North Sea.

Although Curtiss was producing later versions of the H-16, roughly equivalent to the British F-3, late in 1917, the Navy decided to adapt the F-5 to American manufacture and power it with the new Liberty engine. In addition to the Naval Aircraft Factory and Canadian Aeroplanes, Ltd, Curtiss was selected as a manufacturer and built 60. Canada built 30 and the NAF built 137, the last two of which were completed as improved versions, F-6L.

The principal feature distinguishing the F-5L from the Liberty-powered H-16 was the use of ailerons with parallel leading and trailing edges instead of the distinctively tapered trailing edges of those on the earlier boats. Balance area was also added to the F-5L rudder, but this area was set into the fin beneath the horizontal tail and was not noticeable. After

114

A 1924 photograph of a Curtiss F-5L with modified fin and rudder. (*US Navy*)

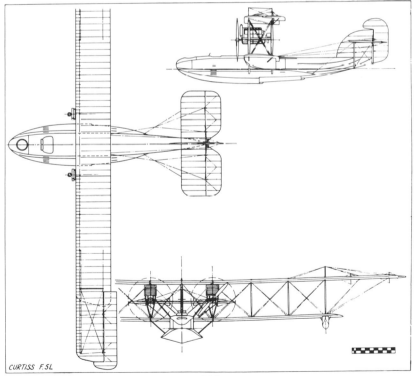

CURTISS F.5L

115

the war all the F-5Ls in service were fitted with much larger vertical tails of entirely new design, which the Navy had developed on the two F-6Ls.

The F-5L was considered to be a US Navy design, rather than a Curtiss, and when the new designating system of 1921 was adopted, the F-5L was assigned the designation PN-5 while the F-6Ls became PN-6. However, in practice the new designations were not used for designs in production before adoption of the new system, and the F-5Ls were called such until their retirement in 1928. Improved versions, in which newer wings and engines were fitted to the basic F-5L hull, did use the new designations, starting with the PN-7 (see page 334).

The wooden-hulled F-5Ls, along with the near-duplicate H-16s, remained the standard patrol boats of the Navy until replaced by production versions of the NAF PN-12 built by Douglas, Martin and Keystone in the late 1920s.

TECHNICAL DATA (F-5L)

Manufacturer: Curtiss Aeroplane and Motor Co, Inc, Garden City, LI, and Buffalo, NY; Canadian Aeroplanes Ltd, Toronto; Naval Aircraft Factory, Philadelphia, Pennsylvania.

Type: Patrol flying-boat.

Accommodation: Crew of four.

Power plant: Two 400 hp Liberty 12As.

Dimensions: Span, 103 ft 9¼ in; length, 49 ft 3¾ in; height, 18 ft 9¼ in; wing area 1,397 sq ft.

Weights: Empty, 8,720 lb; gross, 13,600 lb.

Performance: Max speed, 90 mph at sea level; initial climb, 2,200 ft in 10 min; service ceiling, 5,500 ft; range, 830 st miles.

Armament: Six to eight flexible 0·30-in machine guns. Four 230 lb bombs.

Serial numbers:
F-5L (Canadian): A3333–A3362; A3363–A3382 (cancelled).
F-5L (NAF): A3559–A4038 (of which 343 were cancelled).
F-5L (Curtiss): A4281–A4340.
F-6L (NAF): A4036–A4037.

Curtiss F-5L with post-war tail modification and checkerboard markings on hull for better visibility during fleet manoeuvres. (*US Navy*)

116

Curtiss 18-T1 serial A3326 as flown in the 1923 National Air Races. (*US Navy*)

Curtiss 18-T

Although the Navy had only two Curtiss 18-Ts, which the manufacturer named Wasp, they set numerous records and had a significant effect on subsequent designs. The two were ordered on March 30, 1918, and were intended as two-seat fighters. Unusually clean design was achieved in spite of triplane configuration by the fact that the fuselage was a beautifully streamlined structure built up of cross-laminated strips of wood veneer formed over a mould and then attached to the inner structure of longerons and formers. This method was well-established in Germany during the World War I years by such manufacturers as Roland and Pfalz, but had been used only on a limited scale in the United States, most notably by Lowe, Willard and Fowler (L.W.F.). Curtiss had previous experience with similar construction in the hulls of the Model F flying-boats and the early H-boats, and this feature was to reappear in several subsequent Curtiss designs, most notably the Pulitzer and Schneider racers of the early 1920s.

At a time when equivalent American military designs were being planned round the Liberty engine, the 18-T used an entirely new power plant, the Curtiss K-12, developed for Curtiss by the well-known engine designer Charles Kirkham. The aircraft and the engine were designed for each other, and because of this, the 18-T and its equivalent biplane version, the 18-B Hornet for the Army, were often referred to as Curtiss-Kirkhams. The compact power plant, which produced 400 hp, developed into the famous Curtiss D-12 (V-1150) of the 1920s and the larger V-1570 Conqueror of the 1930s, some examples of which remained in use until World War II. The K-12 enabled the 18-T to set an official world's speed record of 163 mph with full military load on August 19, 1918. On July 25, 1919, Curtiss test pilot Roland Rholfs established an American altitude record of

30,100 ft, later raising it to an official world record of 34,610 ft on September 19.

In its original form the 18-T, which first flew on July 5, 1918, had single-bay unswept wings of 31 ft 11 in span. Because of tail heaviness resulting from the heavy wood construction, five degrees of sweepback were built in to move the centre of lift aft. The Navy loaned the first modified 18-T, A3325, to the Army for test, after which the Army ordered two examples of its own for comparison with the biplane version then under construction. In order to improve the altitude capability of the design and accommodate the additional weight of floats, longer two-bay wings were built and installed, the short-wing version becoming 18-T1 and the long-wing 18-T2.

While never ordered into production as a service type, the performance of the Navy's two 18-Ts was such that they were kept on hand as racers. In the short-wing landplane configuration, both were entered in the 1920 Pulitzer Race, but dropped out before the finish with engine trouble. They were inactive as racers in 1921 because the Navy withdrew from racing that year, but reappeared as short-wing seaplanes for the 1922 Curtiss Marine Trophy Race, one of the events in the 1922 National Air Races held at Detroit, Michigan. Lt R. Irvine, flying A3325 which had been painted bright green for identification purposes, dropped out in the fifth lap. Lt L. H. Sanderson, flying the yellow-painted A3326, was leading the field in the last lap with an average speed of 125·3 mph round the 20-mile triangular course when he ran out of fuel and was forced to land short of the finishing line, damaging the plane in the process. A3326 was returned to the Naval Aircraft Factory for repair, along with its sister

The long-span Curtiss 18-T2 used for the altitude record flight in 1919. (*Curtiss photo*)

ship. Both were reconverted to landplanes and appeared at the 1923 National Air Races held in St Louis, Missouri. A3325 crashed during a preliminary trial flight. A3326, piloted by Lt L. G. Hughes, broke a crankshaft during the Liberty Engine Builders Trophy race and was destroyed in the resulting crash.

TECHNICAL DATA (Curtiss 18-T1)

Manufacturer: Curtiss Aeroplane and Motor Co, Inc, Garden City, LI, NY.
Type: Fighter.
Accommodation: Crew of two.
Power plant: One 400 hp Curtiss K-12.
Dimensions: Span, 31 ft 10 in; length, 23 ft 4 in; height, 9 ft 10¾ in; wing area, 309 sq ft.
Weights: Empty, 1,980 lb; gross, 3,050 lb.
Performance: Max speed, 160 mph at sea level; initial climb, 10 min to 12,500 ft; service
   ceiling, 23,000 ft; endurance, 5·9 hrs.
Armament: Two fixed forward-firing 0·30-in Marlin guns; two 0·30-in Lewis guns on
   flexible mount in rear cockpit.
Serial numbers:
   A3325–A3326.

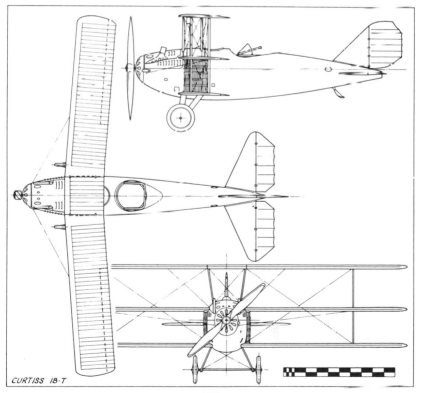

CURTISS 18-T

Curtiss Navy Racer No. 2 as used by Curtiss test pilot Bert Acosta to win the 1921 Pulitzer Trophy race. Note the external radiators. (*Curtiss photo*)

Curtiss CR

One of the outstanding US Navy designs of the 1921–4 period was the pair of Curtiss racers built in 1921. In the absence of an official designation system at the time, these were known merely as Curtiss Navy Racers. When an official system was adopted in 1922, they were initially classed as fighters with the designation of CF-1, but this was subsequently changed to CR-1 and CR-2 for Curtiss racer. Both were ordered on June 16, 1921, as the Navy's entries in the Pulitzer Trophy race to be held in conjunction with the National Air Races at Omaha, Nebraska, later in the year.

The racers drew heavily on Curtiss design experience with the 18-T triplanes (see page 118) and two privately financed racers that Curtiss had built for the 1920 Gordon Bennett Cup, the *Cactus Kitten* (a monoplane

Curtiss Navy Racer No. 2, designated CR-2, as modified for 1922 Pulitzer Trophy race with wing surface radiators and streamlined wheels. (*Curtiss photo*)

Curtiss CR-3, winner of the 1923 Schneider Trophy contest. (*Curtiss photo T2481*)

that was converted to a triplane to tame its unruly performance) and the *Texas Wildcat* (see page 425). These civil racers had each used a geared post-war development of the 1918 Curtiss-Kirkham engine, the C-12. This developed 435 hp at 2,250 rpm. The Navy racer fuselages were of the established Curtiss laminated wood veneer construction and were fitted with the Curtiss CD-12 engine, a direct-drive adaptation of the C-12 that developed 405 hp at 2,000 rpm. The radiators were a pair of pineapple-shaped French Lamblins mounted between the undercarriage V-struts. By the time the machines were ready to race in September, the Armed Services had decided not to participate in the National Air Races that year. However, the Navy loaned the new racers back to Curtiss, and factory test pilot Bert Acosta won the Pulitzer race in A6081 in the new closed-course record time of 176·7 mph. On November 3, Acosta used the same machine to set a world's absolute speed record of 197·8 mph over the Curtiss company's own field at Garden City, on Long Island, NY.

In 1922 the racers were again entered in the Pulitzer, this time by the Navy. New designations were assigned, A6080 with its original Lamblin radiators being CR-1, while A6081, which had been fitted with new wings incorporating surface radiators, was CR-2. Both had new CD-12 engines and slightly larger vertical tail surfaces plus other minor refinements. Piloted by Lt H. J. Brow, the CR-2 was placed third in the Pulitzer, held at Detroit, Michigan, at 193 mph for the closed course. Marine Corps Lt Alford J. Williams was placed fourth in the CR-1 at 187 mph. First and second places went to Army pilots flying Curtiss R-6s, which were 1922 developments of the CRs and used the same CD-12 engines and wing surface radiators.

In 1923 both CRs were converted to seaplanes as the US entries in the Schneider Trophy contest held at Cowes, Isle of Wight. Both machines, then designated CR-3, used the wing surface radiators and were fitted with larger vertical tail surfaces. The race was won by Lt David Rittenhouse in

A6081 at a speed of 177·38 mph around the closed course, Lt Rutledge Irvine coming in second in A6080 at 173·46 mph.

All European entries withdrew from the 1924 Schneider race, to have been held at Baltimore, Maryland, and, rather than win by default, the US cancelled that year's races. For prestige purposes, however, the Navy staged a massive assault on existing world's records in a public show that substituted for the race. A6081, fitted with a later D-12 engine and re-designated CR-4, was used by Lt G. T. Cuddihy to set a new closed-course seaplane speed record of 188·078 mph on October 25. After being used briefly in 1925 for high-speed research and the training of that year's Schneider racing team which consisted of both Army and Navy pilots, A6081 was scrapped.

TECHNICAL. DATA (CR-3)

Manufacturer: Curtiss Aeroplane and Motor Co, Inc, Garden City, LI, NY.
Type: Racing seaplane.
Accommodation: Pilot only.
Power plant: One 450 hp Curtiss D-12.
Dimensions: Span, 22 ft 8 in; length, 25 ft 0⅜ in; height, 10 ft 8¾ in; wing area, 168 sq ft.
Weights: Empty, 2,119 lb; gross, 2,746 lb.
Performance: Max speed, 194 mph at sea level; initial climb, 6 min to 5,000 ft; service
  ceiling, 19,200 ft; range, 522 st miles.
Serial numbers:
  CR-1, CR-3: A6080.   CR-2, CR-3, CR-4: A6081.

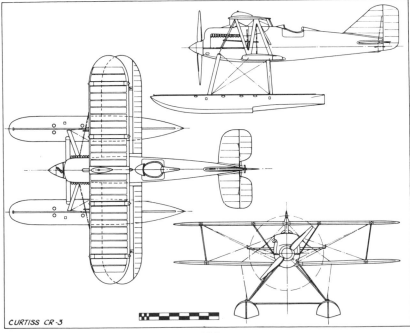

CURTISS CR-3

122

Martin SC-1 floatplane, production version of the Navy-designed Curtiss CS-1. (*US Navy*)

Curtiss CS (Martin SC)

Much effort was devoted to the development of torpedo-bombers for the US Navy in the years following World War I, and for some time these types followed a clearly defined pattern. Powered by inline, water-cooled engines, they were large biplanes with a crew of two or three, provision for torpedo stowage under the fuselage, and interchangeable wheel or float chassis. Conforming closely to this pattern, the Curtiss CS-1 appeared at the end of 1923 as the company's first torpedo-bomber; it was designed to serve also in the scout and bombing roles, and was unusual in having the span of the upper wing shorter than that of the lower.

Powered by a 525 hp Wright T-2 engine, the CS-1 was a three-seater. Six were built for the Navy, plus two CS-2s with Wright T-3 engines, more fuel capacity and an optional third float when used as a seaplane. Delivered in April 1924, they served with Squadron VT-1 with sufficient success to be chosen to replace the Douglas DT-2s, and a production contract was put out to tender in the aircraft industry.

Glenn Martin underbid Curtiss to win the contract for 35 aircraft identical with the CS-1 but designated SC-1. Delivered in the first half of 1925, the SC-1s were issued to VT-1, VT-2 and VS-1, being quickly followed into service with these and other units by the SC-2. Also built by Martin, the latter had the T-3 engine; 40 were built and all had been delivered by the end of 1925.

By mid-1927, only one Fleet unit, VT-2B, was flying SC-1s and SC-2s; 15 of each type were serving in VN-3D8 training squadron at Pensacola, and the remainder had been withdrawn.

TECHNICAL DATA (SC-2 landplane)

Manufacturer: Glenn L. Martin Company, Cleveland, Ohio (SC-1, SC-2); Curtiss Aeroplane and Motor Co, Inc, Buffalo, NY (CS-1, CS-2).
Type: Land-based or seaplane torpedo-bomber.

123

Accommodation: Pilot, torpedo-man and gunner.

Power plant: One 585 hp Wright T-3.

Dimensions: Span, 56 ft 6⅞ in; length, 37 ft 8¾ in; height, 14 ft 8 in; wing area, 856 sq ft.

Weights: Empty, 5,007 lb; gross, 8,422 lb.

Performance: Max speed, 103 mph at sea level; initial climb, 2,000 ft in 10 min; service ceiling, 8,000 ft; range, 1,018 st miles.

Armament: One 1,618 lb torpedo, externally carried.

Serial numbers:

CS-1: A6500–A6505. SC-1: A6801–A6835.

CS-2: A6731–A6732. SC-2: A6928–A6967.

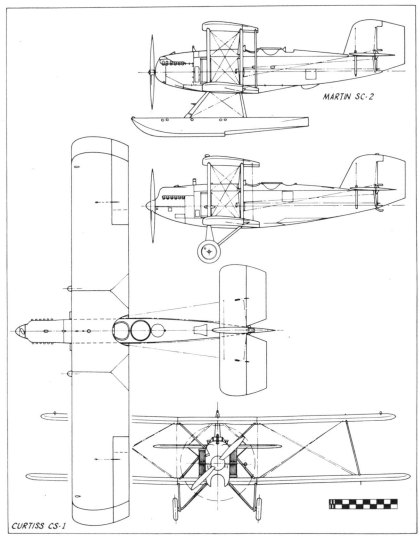

MARTIN SC-2

CURTISS CS-1

Curtiss R2C-1 serial A6692, winner of the 1923 Pulitzer Trophy race. (*Walter J. Addems*)

Curtiss R2C, R3C

The R2C was a logical design development of the CR-1 and the following Army R-6s, and two examples were rushed to completion as Navy entries in the Pulitzer event of the 1923 National Air Races at St Louis on October 6. First place was taken by Lt Alford J. Williams in A6692 at 243·68 mph, while second place went to Lt H. J. Brow, with 241·77 mph in A6691. The R2Cs certainly looked racy, with the upper wing resting on the top of the fuselage, I-type interplane struts and a single-strut undercarriage braced by streamlined wires. The now famous Curtiss skin radiators were installed on both wings. Power was supplied by the Curtiss D-12A engine boosted to 488 hp.

In 1924 A6692 was converted to a seaplane as R2C-2, but was not raced as such because of cancellation of the Schneider Trophy race. A6691 was sold to the US Army Air Service for $1.00 and was redesignated R-8. It crashed shortly before the 1924 Pulitzer, and A6692, being used as a trainer for the 1926 US Schneider cup team, crashed at Anacostia on September 13, 1926.

Following the R-8 crash, the Army and Navy together planned entries for the 1925 races and jointly ordered three more racers from Curtiss under the designation R3C-1, two to be flown by the Navy and one by the Army. Outwardly, the R3Cs could be distinguished from the R2Cs only by the filleted lower ends of the interplane struts. However, power was increased to 610 hp with an enlarged 1,400 cu in engine, the V-1400 which had developed out of the stock D-12. Performance was further improved by the use of a new aerofoil section. The 1925 Pulitzer race, held on October 10, was not run in the same manner as previous races, and the two R3Cs were the only pure racers entered; the other four entries were stock pursuit planes, so the race was run in two heats. Lt Cyrus Bettis of the Army flew A6979 and Lt A. J. Williams flew A6978 in races against the clock rather than each other, making separate starts two minutes apart. This system, plus the fact that diving starts were now prohibited since an Army R-6 lost its wings at the start of the 1924 race, made it a dull show for the spectators. Bettis won at a speed of 248·99 mph for the four laps of the 50-kilometre triangle. The four pursuits flew the second heat as a proper race.

In 1925 the Schneider Trophy race followed close on the heels of the Pulitzer on October 26, so all three R3Cs were put on twin floats and re-designated R3C-2. Again the race was an Army victory, with Lt James Doolittle turning in a winning speed of 232·573 mph in the same machine that had won the Pulitzer. Lt George Cuddihy, until then in second place, was forced out just short of the finishing line when A6979 ran out of oil and the engine overheated. Lt Ralph Ofstie, flying in third place in A7054, was forced out in the sixth lap with engine trouble.

For the 1926 Schneider Trophy, in which they were flown by Navy pilots, two of the R3C-2s were modified by the installation of larger engines and larger floats that curved downward noticeably at the tops of the bows. A7054 became R3C-3 with a larger 700 hp Packard 2A-1500 engine. This installation was made at the Naval Aircraft Factory. A6978 was refitted by Curtiss with an improved 700 hp Curtiss V-1550 engine and became R3C-4. The unmodified R3C-2 that had won the 1925 race was used as a training machine and a spare racer. The R3C-3 capsized on landing after a practice flight and sank, and the R3C-2 and R3C-4 started the race. Lt Cuddihy was in second place on the last lap when he was

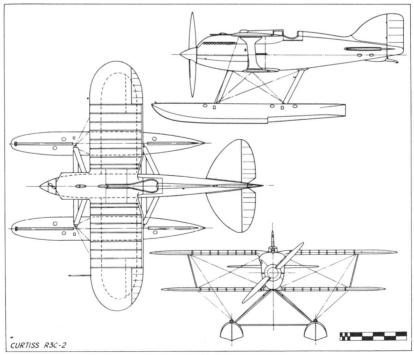

CURTISS R3C-2

126

Curtiss R2C-2 serial A6692, used as a trainer for the 1926 Schneider Trophy team. (*US Navy*)

forced out with fuel pump trouble. Lt C. F. Schilt then came second in the R3C-2. Because of its impressive racing record, the R3C-2, in its 1925 Pulitzer race markings, was placed in the National Air Museum of the Smithsonian Institution.

Although flown as racers under racing plane designations, the R2Cs and R3Cs were also given paper designations as fighters, F2C-1 and F3C-1. Other racers ordered at the same time as the R2C/F2C were raced under fighter designations.

TECHNICAL DATA (R2C-1)

Manufacturer: Curtiss Aeroplane and Motor Co, Inc, Garden City, LI, NY.
Type: Racing seaplane.
Accommodation: Pilot only.
Power plant: One 488 hp Curtiss D-12A.
Dimensions: Span, 22 ft; length, 19 ft 8½ in; height, 8 ft 1 in; wing area, 144 sq ft.
Weights: Empty, 1,677 lb; gross, 2,150 lb.
Performance: Max speed, 266 mph at sea level; climb, 3·6 min to 10,000 ft; service ceiling, 31,800 ft; range, 430 st miles.
Serial numbers:
    R2C-1, R-8: A6691.   R2C-1, R2C-2: A6692.

TECHNICAL DATA (R3C-1)

Manufacturer: Curtiss Aeroplane and Motor Co, Inc, Garden City, LI, NY.
Type: Racing seaplane.
Accommodation: Pilot only.
Power plant: One 610 hp Curtiss V-1400.
Dimensions: Span, 22 ft; length, 19 ft 8½ in; height, 8 ft 8 in; wing area, 149 sq ft.
Weights: Gross, 2,150 lb.
Performance: Max speed, 265 mph at sea level; initial climb, 3·2 min. to 10,000 ft; service ceiling, 26,400 ft; range, over 250 miles.
Serial numbers:
    R3C-1, R3C-2: A6979.   R3C-1, R3C-2, R3C-3: A7054.
    R3C-1, R3C-2, R3C-4: A6978.

Curtiss F6C-3 landplane. (*US Navy*)

Curtiss F6C Hawk

Paralleling the purchase by the Navy of versions of the Boeing biplane fighter evolved for the Army Air Corps, Curtiss also succeeded in selling its Army fighters to the Navy in the mid twenties. The first of the famous line of Curtiss Hawk fighters was the P-1, developed from and closely resembling the PW-8; and in March 1925 the Navy ordered nine examples of the P-1 with only minor modifications, including provision for float operation. Seven of these aircraft were delivered as F6C-1s at the end of 1925, with no arrester gear, and three were issued to Squadron VF-2; the remainder were used primarily for test and development, but three went to the Marine Corps in 1927 and were used by VF-9M with the East Coast Expeditionary Force in Quantico.

The F6C-1 had a 400 hp Curtiss D-12 engine and the standard two-gun armament. Two other aircraft on the initial contract were delivered

Curtiss F6C-3 on Macchi floats, as flown in 1928 by VB-1B. (*US Navy*)

as F6C-2s with arrester hooks and strengthened landing gear for trial operation by VF-2 from carriers. These same features were adopted for the 35 F6C-3s ordered by the Navy in 1927 and put into service with Squadron VF-5S. This unit, famous as the Red Rippers and responsible for some spectacular aerobatic demonstrations, was temporarily redesignated VB-1B in July 1928, and flew its Hawks as twin-float seaplanes for a time in the same year. A number of F6C-3s also went into service with Marine Squadron VF-8M at Quantico in 1928.

Two F6C-1s were converted to F6C-3 standard, and another was converted to F6C-4 to evaluate installation of the Pratt & Whitney R-1340 engine. A production order for 31 of this version followed. Although

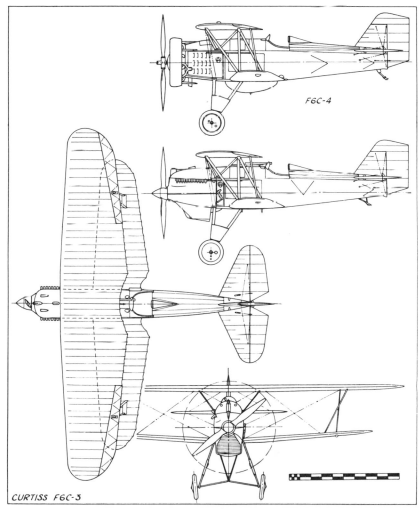

F6C-4

CURTISS F6C-3

Curtiss F6C-4 of VF-10M at San Diego in August 1931, with unusual location of national insignia under top wing. (*US Navy*)

lighter and more manoeuvrable than the Curtiss-engined versions, the F6C-4s were becoming outmoded by the time they were delivered, and they equipped only VF-2B aboard USS *Langley* until the beginning of 1930. Others were issued to the Navy's advanced fighter-training squadron VN-4D8, at Pensacola. After their replacement as front-line equipment with the Navy, some F6C-4s were passed to the Marine Corps.

Other designations in the F6C series applied primarily to experimental versions. They included the XF6C-5, which was the prototype F6C-4 with a 525 hp Pratt & Whitney R-1690 Hornet engine, an F6C-6 modified from an F6C-3 for the 1929 National Air Races, the XF6C-6 monoplane conversion of an F6C-3 for the 1930 Thompson Trophy races, and the XF6C-7, which was an F6C-4 with a Ranger SGV-770 inverted-V engine.

TECHNICAL DATA (F6C-4)

Manufacturer: Curtiss Aeroplane and Motor Co, Inc, Buffalo, NY.
Type: Carrier-based fighter.
Accommodation: Pilot only
Power plant: One 410 hp Pratt & Whitney R-1340.
Dimensions: Span, 37 ft 6 in; length, 22 ft 6 in; height, 10 ft 11 in; wing area, 252 sq ft.
Weights: Empty, 1,980 lb; gross, 3,171 lb.
Performance: Max speed, 155 mph at sea level; initial climb, 2·5 min to 5,000 ft; service ceiling, 22,900 ft; range, 360 st miles.
Armament: Two fixed forward-firing 0·30-in guns.
Serial numbers:

F6C-1: A6968–A6972; A6975–A6976.	XF6C-5: A6968.
F6C-2: A6973–A6974.	XF6C-6: A7147.
F6C-3: A6970; A6972; A7128–A7162.	F6C-6: A7144.
F6C-4: A6968; A7393–A7423.	XF6C-7: A7403.

Curtiss OC-2 in Marine markings, attached to NAS San Diego. (*US Army Air Corps*)

Curtiss F8C, OC, Falcon

Scarcely less famous than the Curtiss Hawk family of single-seat fighters, the Falcons were two-seaters with a distinctive wing arrangement featuring sweepback on the outer panels of the upper wings and straight, parallel-chord lower wings. The design originated in 1924 for an Army Air Corps competition to develop a new observation aircraft, and the prototype had a Liberty engine. A year later it was revised with a Packard 1A-1500 engine, but production models had the Liberty or Curtiss D-12 engine.

The Falcon had a riveted, tubular aluminium fuselage structure with fabric covering. Armament included one or two 0·30-in Lewis guns on a Scarff ring round the rear cockpit plus fixed forward-firing guns in the lower wings. The Navy preference for air-cooled radial engines led to selection of the Pratt & Whitney Wasp when the Falcon was picked for Naval use in June 1927. The first two Falcons were designated XF8C-1s and were delivered with four F8C-1s to the Marine Corps early in 1928, other examples going into service in Nicaragua and China as well as the USA. They were redesignated OC-1s as observation aircraft almost

Curtiss OC-2 showing Scarff ring. (*Curtiss photo*)

131

immediately after going into service and operated in an all-purpose role including fighting, dive-bombing, observation and air evacuation. A prototype XF8C-3 and 21 similar production models were delivered later in 1928, to serve, after redesignation as OC-2s, with the Marine Squadrons VO-8M and VO-10M. One of the XF8C-1s became the XOC-3 with an experimental Curtiss H-1640 Chieftain engine.

TECHNICAL DATA (OC-2)

Manufacturer: Curtiss Aeroplane and Motor Co, Inc, Buffalo, NY.
Type: Fighter-bomber and observation biplane.
Accommodation: Pilot and observer/gunner.
Power plant: One 432 hp Pratt & Whitney R-1340.
Dimensions: Span, 38 ft; length, 27 ft 11 in; height, 11 ft 8 in; wing area, 350 sq ft.
Weights: Empty, 2,515 lb; gross, 4,191 lb.
Performance: Max speed, 144 mph at sea level; initial climb, 1,010 ft/min; service ceiling, 16,130 ft; range, 650 st miles.
Armament: Two fixed forward-firing 0·30-in guns; one 0·30-in gun on Scarff ring in rear cockpit.
Serial numbers:

XF8C-1: A7671–A7672.	F8C-1 (OC-1): A7945–A7948.
XOC-3: A7672.	F8C-3 (OC-2): A7949–A7969.

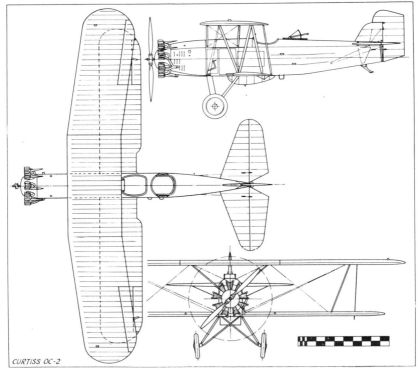

CURTISS OC-2

Curtiss N2C-1 trainer in markings of VN-13. (*Frank Shertzer Collection*)

Curtiss N2C Fledgling

The Curtiss XN2C-1 Fledgling, of which three examples were built, was the winner of a 1928 competition for a new Navy primary training plane. In view of the rapidly advancing state of the art of aeroplane design, which was featuring refined lines and fewer struts and wires, the XN2C-1 seemed to be a long step backward; with its maze of struts and wires, it looked like a throwback to an earlier age. Developed in parallel with commercial versions for the nation-wide schools of the Curtiss-Wright Flying Service, the Navy versions differed from the civil models principally in the details required by the military specification and the use of the 220 hp Wright J-5 Whirlwind (R-760-8) engine in place of the 165–186 hp Curtiss Challenger.

The prototypes were evaluated as both seaplanes and landplanes in competition with the Keystone XNK-1s and the Boeing XN2B-1. Upon winning the competition, Curtiss was given an order for 31 production aircraft. The first of these was tried with single-bay wings of shorter span,

Curtiss N2C-2 modified for use as radio-controlled drone, with nosewheel gear and wing flaps. (*US Navy*)

133

but the other machines were not so modified. The Wright J-5 engine went out of production soon after delivery of the N2C-1s, so the 20 N2C-2s of a subsequent order were powered with the new 240 hp Wright R-760-94.

The Fledglings spent the majority of their careers assigned to Naval reserve training squadrons throughout the country. In the later 1930s, a few surviving N2C-2s were fitted with tricycle undercarriages and were converted to radio-controlled anti-aircraft targets.

TECHNICAL DATA (N2C-2)

Manufacturer: Curtiss Aeroplane and Motor Co, Inc, Buffalo, NY, and Bristol, Pa.
Type: Primary trainer.
Accommodation: Instructor and pupil.
Power plant: One 240 hp Wright R-760-94.
Dimensions: Span, 39 ft 5 in; length, 27 ft 4¼ in; height, 10 ft 8½ in; wing area, 368 sq ft.
Weights: Empty, 2,138 lb; gross, 2,860 lb.
Performance: Max speed, 116 mph at sea level; initial climb, 7·5 min to 5,000 ft; service ceiling, 17,800 ft; range, 384 st miles.
Serial numbers:

XN2C-1: A7650–A7652.　　　　N2C-2: A8526–A8545.
N2C-1: A8020–A8050.

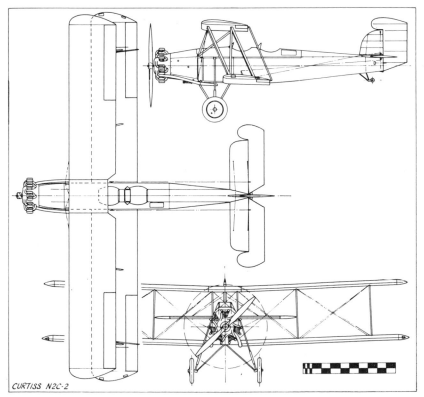

CURTISS N2C-2

134

Curtiss O2C-1, built as an F8C-5, after passing into service with the Marine Corps in 1934. (*David C. Cooke*)

Curtiss F8C, O2C Helldiver

Despite its origin as a fighter for the USAAC, the Curtiss Falcon evolved in Navy service into a multi-purpose type operating as a fighter, dive-bomber and on observation duties (see page 131). Experience with the early Falcons, including operational deployment in Nicaragua and China, suggested that further development of the type was possible.

To take full advantage of the Falcon's ability as a bomber and to match the Marine Corps' development of dive-bombing techniques, Curtiss developed a new aeroplane in 1929, with the ability to carry a 500 lb bomb under the fuselage or two 116 lb bombs under the wings. The forward firing guns were moved from the lower to the upper wing, the Wasp engine was cowled (on the second prototype) and the upper wing was reduced in span and area. Many other detail changes included a revision to the rudder shape. Because of the primary dive-bombing mission of this type, it was given the popular name of Helldiver, but was designated by the Navy in the same series as the Falcons. The two prototypes were XF8C-2 and XF8C-4 respectively, and 25 F8C-4s were built for the Navy, serving primarily with VF-1B aboard the USS *Saratoga*, and later with the Marine Corps. They passed from the active fleet to the Reserve in 1931.

A further batch of 63 Helldivers built primarily for observation duties were designated F8C-5 initially but served as O2C-1s, all going to Marine Corps squadrons from 1931 onwards. Like the F8C-4s, they were passed on to Reserve units, which received 55 in 1934.

Two F8C-5s temporarily fitted with wing flaps and leading-edge slots

135

were designated XF8C-6 until reconverted to original standard. One special model with a Wright Cyclone R-1820 engine was used as a VIP transport designated XO2C-2 (XF8C-7) and two more Helldivers with this same engine were O2C-2s (XF8C-8). One other O2C-2 was built with a Curtiss R-1510 engine, but later had the Cyclone engine fitted and became the XF10C-1 and, for a time, the XS3C-1.

TECHNICAL DATA (O2C-1)

Manufacturer: Curtiss Aeroplane and Motor Co, Inc, Buffalo, NY.
Type: Observation scout.
Accommodation: Pilot and observer/gunner.
Power plant: One 450 hp Pratt & Whitney R-1340-4.
Dimensions: Span, 32 ft; length, 25 ft $7\frac{7}{8}$ in; height, 10 ft 3 in; wing area, 308 sq ft.
Weights: Empty, 2,520 lb; gross, 4,020 lb.
Performance: Max speed, 146 mph at sea level; cruising speed, 110 mph; initial climb, 5·5 min to 5,000 ft; service ceiling, 16,250 ft; range, 720 st miles.
Armament: Two fixed and one flexible 0·30-in guns. One 500 lb or two 116 lb bombs.
Serial numbers:

XF8C-2: A7673.
XF8C-4: A8314.
F8C-4: A8421–A8445.
F8C-5 (O2C-1): A8446–A8456; A8589–A8597; A8748–A8790.

XO2C-2 (XF8C-7): A8845.
O2C-2 (XF8C-8): A8848–A8849.
O2C-2: A8847.

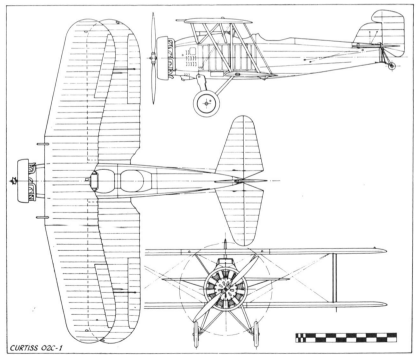

CURTISS O2C-1

Curtiss F9C-2 of the unique HTA Squadron attached to the airships *Akron* and *Macon*.
(*National Archives 80-A-441982*)

Curtiss F9C Sparrowhawk

The eight aeroplanes bearing the Sparrowhawk name occupy a unique place in the annals of US Naval aviation, being the only operational fighters to serve aboard the airships USS *Akron* and *Macon*. Contracts for the construction of these two airships, ZRS-4 and ZRS-5, were signed on October 6, 1928, and it was intended from the start that these airships should incorporate hangar space for four aeroplanes which could be launched and retrieved by means of a trapeze from which the aircraft would hang by a skyhook. While the airships were being built, however, plans for a suitable fighter to be carried by them lagged, and the Sparrowhawk, eventually adopted, began life as a conventional carrier-based fighter to Bureau of Aeronautics specification Design No. 96, issued on May 10, 1930. The requirement was for a very small aeroplane—so small, in fact, that none of the three prototypes tested by the Navy in the Design 96 competition (the Curtiss XF9C-1, Berliner-Joyce XFJ-1 and General XFA-1) proved acceptable for carrier operation.

The XF9C-1, however, had one special attribute so far as the airships were concerned—it was small enough to pass through the arbitrarily designed hangar door. Consequently, following conventional testing between March 31 and June 30, 1931, at Anacostia, it was fitted with a skyhook and transferred to Lakehurst, making the first hook-on to the experimental trapeze on the USS *Los Angeles* on October 27, 1931. Prior to this event the BuAer drew up specifications for a possible improved version of the F9C, and the Curtiss company set about construction of a second prototype as a private venture. Major changes were made, particularly by raising the top wing some four inches and giving it the distinctive gull configuration of the Sparrowhawk. The 421 hp Wright R-975-C engine in the XF9C-1 was replaced by a 438 hp Wright R-975-E3, and the landing gear and tail unit were redesigned.

Bearing a civil registration (NX 986M) this prototype was demonstrated to the Navy in October 1931 and later the same month six production examples were ordered, to be designated F9C-2. The prototype was purchased subsequently as the XF9C-2; both this and the XF9C-1 remained spare aircraft for the airship operation. The first F9C-2 flew on April 14, 1932, and made its first hook-on to *Akron* on June 29, 1932. All six were delivered to the Navy in September 1932 and began operating as scouts from the *Akron*. The latter was lost in 1933 with no Sparrowhawks

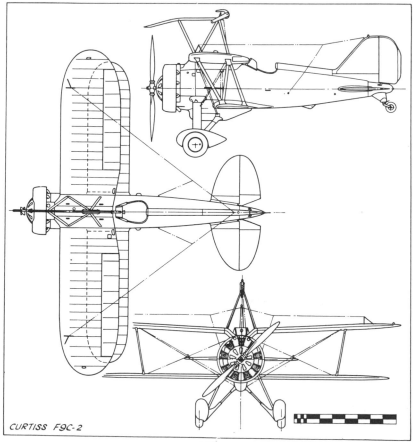

CURTISS F9C-2

138

Curtiss F9C-2 hooking-on to the USS *Macon*; after contact, the arm on the trapeze was swung down to steady the fuselage prior to aircraft being raised into airship hull. (*US Navy*)

on board, and flying continued from *Macon* until she, too, was lost in 1935, with four F9C-2s on board. In the course of the trials, the HTA (Heavier-than-Air) unit of F9Cs dispensed with the aircraft landing gear once they were aboard the airship, depending wholly upon their skyhook for launch and recovery. After the loss of the *Macon*, the two surviving F9C-2s and the XF9C-2 remained in Navy service for a year or two as utility aircraft, all with the XF9C-2 designation, and one is now in the National Air Museum.

TECHNICAL DATA (F9C-2)

Manufacturer: Curtiss Aeroplane and Motor Co, Inc, Buffalo, NY.
Type: Airship-based fighter/scout.
Accommodation: Pilot only.
Power plant: One 438 hp Wright R-975-E3.
Dimensions: Span, 25 ft 5 in; length, 20 ft 7 in; height, 10 ft 11½ in; wing area, 173 sq ft.
Weights: Empty, 2,089 lb; gross, 2,770 lb.
Performance: Max speed, 176·5 mph at 4,000 ft; initial climb, 1,700 ft/min; service ceiling, 19,200 ft; range, 350 st miles.
Armament: Two fixed forward-firing 0·30-in guns.
Serial numbers:

XF9C-1: A8731.	XF9C-2: 9264.
F9C-2: A9056-A9061.	

Curtiss BFC-2 of VB-2B in 1937; a camera gun is on the top wing centre section.

Curtiss F11C, BFC, BF2C, Goshawk

The success of the Curtiss and Wright companies in developing two-seat dive-bombers from designs originating as fighters led the US Navy to specify a similar capability for its new single-seat fighter under development in the early thirties. Among the aircraft affected by this decision was the Curtiss Goshawk, a fighter which was destined to achieve considerably more publicity than its short production run appeared to justify.

Two prototypes had been ordered from Curtiss on April 16, 1932, as the XF11C-1 and XF11C-2; the former was powered by a two-row 600 hp Wright R-1510-98 engine, eventually with a three-blade propeller, and the latter had a single-row 700 hp Wright SR-1820-78 Cyclone with two-blade propeller. In general design, the XF11C was typical of its era, with angular lines, an open cockpit, and faired cantilever mainwheel legs. The single bomb, of up to 500 lb capacity, was carried beneath the fuselage in a special crutch which swung down for release to prevent the bomb hitting the airscrew as it fell. In place of the fuselage bomb, four 112 lb bombs could be carried under the wings, and the fixed armament comprised two 0·50-in guns. A 50 gallon tank could be carried under the fuselage.

The XF11C-2 began its trials in June 1932 and was joined later in the year by the XF11C-1. A production order dated October 18, 1932, called for 28 F11C-2s with R-1820-78 engine and a somewhat modified cockpit. Deliveries began in February 1933, the only unit to operate the type being the Navy's famous High Hat Squadron, VF-1B, aboard the USS *Saratoga*. In March 1934 the aircraft were redesignated BFC-2s to indicate their fighter-bomber role and were modified to have a higher rear fuselage and a half-canopy over the cockpit. The Squadron was later renumbered VB-2B and then VB-3B, retaining its BFC-2s until February 1938. A few aircraft

140

then served briefly with VB-6, assigned to the new USS *Enterprise*, but never went aboard this carrier.

Delivery of 27 F11C-2s had been completed by May 27, 1933, when the remaining aircraft on the contract appeared as the XF11C-3 with an R-1820-80 engine and a hand-operated retractable landing gear. It was redesignated XBF2C-1, while the original XF11C-1 became the XBFC-1. A contract dated February 26, 1934, ordered 27 of the new fighter-bombers, the production models having R-1820-04 engines and a semi-enclosed cockpit. Delivered from October 1934 onwards, by which time the designation had been changed from F11C-3 to BF2C-1, they were assigned to VB-5 aboard the USS *Ranger*, but served only a few months before

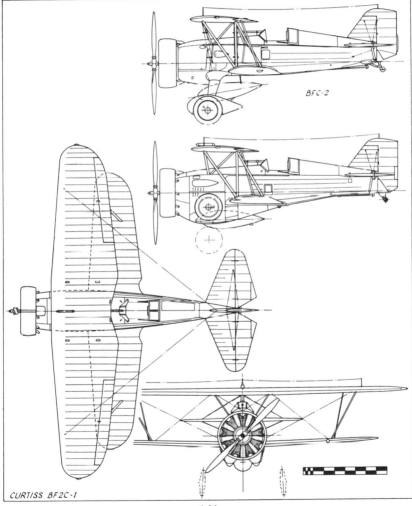

CURTISS BF2C-1

141

difficulties with the landing gear led to their withdrawal. They were the last Curtiss fighters accepted for service with the US Navy.

TECHNICAL DATA (BFC-2)

Manufacturer: Curtiss-Wright Corporation, Curtiss Aeroplane Division, Buffalo, NY.
Type: Carrier-borne fighter-bomber.
Accommodation: Pilot only.
Power plant: One 700 hp Wright R-1820-78.
Dimensions: Span, 31 ft 6 in; length, 25 ft; height, 10 ft 7¼ in; wing area, 262 sq ft.
Weights: Empty, 3,037 lb; gross, 4,638 lb.
Performance: Max speed, 205 mph; initial climb, 2·6 min to 5,000 ft; service ceiling, 24,300 ft; range, 560 st miles.
Armament: Two fixed forward-firing 0·30-in guns.
Serial numbers:
    XF11C-1 (XBFC-1): 9219.   XF11C-2 (XBFC-2): 9213.
    BFC-2: 9265–9268; 9270–9282; 9331–9340.

TECHNICAL DATA (BF2C-1)

Manufacturer: Curtiss-Wright Corporation, Curtiss Aeroplane Division, Buffalo, NY.
Type: Carrier-borne fighter-bomber.
Accommodation: Pilot only.
Power plant: One 700 hp Wright R-1820-04.
Dimensions: Span, 31 ft 6 in; length, 23 ft; height, 10 ft 10 in; wing area, 262 sq ft.
Weights: Empty, 3,329 lb; gross, 5,086 lb.
Performance: Max speed, 225 mph at 8,000 ft; initial climb, 2·6 min to 5,000 ft; service ceiling, 27,000 ft; range, 797 st miles.
Armament: Two fixed forward-firing 0·30-in Browning guns. Four 116 lb or one 474 lb bombs.
Serial numbers:
    XF11C-3 (XBF2C-1): 9269.   BF2C-1: 9586–9612.

Curtiss BF2C-1 of VB-5B. (*Howard Levy*)

Curtiss SOC-1 in grey finish adopted in 1941, attached to the cruiser USS *Astoria*.
(*Peter M. Bowers*)

Curtiss SOC Seagull

A thoroughly conventional biplane capable of operating on floats or wheels, the SOC Seagull was the last of the Curtiss biplanes in operational service with the US Navy. Designed in the early thirties, it was so successful in its appointed task that it survived until the end of World War II on first-line operations, actually outlasting two types produced to replace it.

The primary mission for the SOC, ordered in prototype form on June 19, 1933, as the XO3C-1, was scout-observation from battleships and cruisers. This required the aircraft to be a floatplane, catapulted for take-off and recovered by winch after alighting on the sea alongside the parent ship. The XO3C-1 had folding wings with full-span slots and flaps on the top wing, and a fully enclosed cockpit for the pilot and gunner. A single 0·30-in gun in the rear cockpit was given a clear field of fire by retracting the fuselage turtledeck behind the cockpit. As first flown in April 1934, the XO3C-1 had an amphibious landing gear, with twin wheels incorporated in the central main float. This feature was later abandoned, and the production model was a plain seaplane with alternative wheel undercarriage for land operation.

After successful evaluation in competition with the Douglas XO2D-1 and Vought XO5U-1, the Curtiss biplane was put into production as the SOC-1. The change in designation followed the combination in 1934 of the scouting and observation roles; previously, observation types had been deployed on battleships, and scouting types were attached to cruisers. The SOC-1 had an R-1340-18 engine, and deliveries began on November 12, 1935, principal Squadrons being VS-5B, VS-6B, VS-9S, VS-10S, VS-11S and VS-12S. Production totalled 135, followed by 40 land-based SOC-2s with minor improvements and R-1340-22 engines, and 83 SOC-3s

143

Curtiss SOC-3 with blue fuselage and tail indicating a command aircraft.
(*William T. Larkins*)

which were like the SOC-2s but had the interchangeable alighting gear. Aircraft equipped with arrester gear for carrier operations were designated SOC-2A and SOC-3A after modification in 1942.

To comply with a Navy policy to manufacture 10 per cent of its own Service aircraft, 64 Seagulls were ordered from the Naval Aircraft Factory. Equivalent to the SOC-3, they were designated SON-1, and SON-1A with arrester gear. Three final Seagulls built by Curtiss were designated SOC-4 to serve with the US Coast Guard; these were acquired by the Navy in 1942 and modified to SOC-3A standard.

Production of the SOC Seagull series ended in the spring of 1938, by which time work had begun on a monoplane scout intended as a replacement (see page 148). However, the newer aircraft, the Curtiss SO3C, had a somewhat unsatisfactory operational career and was withdrawn from service early in 1944. All remaining SOC biplanes were hastily restored to operational status and continued in service until the end of the war.

Curtiss SOC-4 serving with US Coast Guard in 1941. (*George H. Williams*)

When they were finally retired, they ended an era of Naval aviation, for no other types were produced to replace them aboard the battleships and cruisers of the fleet.

TECHNICAL DATA (SOC-1 seaplane)

Manufacturer: Curtiss-Wright Corporation, Curtiss Aeroplane Division, Buffalo, NY; and Naval Aircraft Factory, Philadelphia, Penn.
Type: Scouting and observation seaplane (or landplane).
Accommodation: Pilot and observer in tandem.
Power plant: One 600 hp Pratt & Whitney R-1340-18.
Dimensions: Span, 36 ft; length, 31 ft 5 in; height, 14 ft 9 in; wing area, 342 sq ft.
Weights: Empty, 3,788 lb; gross, 5,437 lb.
Performance: Max speed, 165 mph at 5,000 ft; cruising speed, 133 mph; initial climb, 880 ft/min; service ceiling, 14,900 ft; range, 675 st miles.
Armament: One fixed forward-firing 0·30-in. gun in wing; one 0·30-in gun flexibly mounted in rear cockpit. Two 325 lb bombs.
Serial numbers:

XO3C-1: 9413.	SOC-3: 1064–1146.
SOC-1: 9856–9990.	SOC-4: V171–V173 (48243–48245).
SOC-2: 0386–0425.	SON-1: 1147–1190.

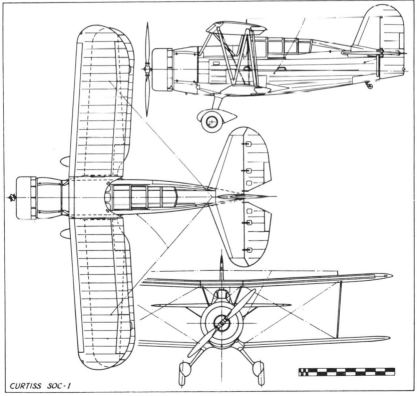

CURTISS SOC-1

145

Curtiss SBC-4 in 1940 Reserve Squadron markings. (*Courtesy William T. Larkins*)

Curtiss SBC Helldiver

Destined to become the last combat biplane produced in America, the Curtiss SBC Helldiver paradoxically began life as a monoplane. A Navy contract, dated June 30, 1932, with the Curtiss company produced the XF12C-1 prototype, a two-seat parasol-wing carrier-based fighter with retractable undercarriage. First flown in 1933, it was redesignated in the scout category in December that year as the XS4C-1, but one month later its role changed again to scout-bomber as the XSBC-1. Fitted originally with a 625 hp Wright R-1510-92 engine, this prototype later had an R-1820-80 and demonstrated a high speed of 217 mph, but the parasol wing proved unsuitable for the dive-bombing role and failed during tests in September 1934. A new prototype ordered in April 1935 with the same Navy serial number made its first flight on December 9 of the same year, as the XSBC-2 biplane. It reverted to the 700 hp XR-1510-12 engine and was an almost wholly new design.

Re-engined with a Pratt & Whitney R-1535-82 Wasp engine in March 1936, the prototype became the XSBC-3 and on August 29, the same year, the Navy contracted for 83 production examples with -94 engines. Deliveries began on July 17, 1937, to Navy Squadron VS-5.

Better performance, and the ability to carry a 1,000 lb bomb, was obtained in the SBC-4. The final aircraft on the SBC-3 production contract was completed as the XSBC-4 with a single-row Wright R-1820-22, and the 174 production aircraft of this series had the -34 engine. The first production contract was placed on January 5, 1938, and deliveries began in March 1939. In 1940 the Navy diverted 50 of its SBC-4s to France, which had already placed an order for 90 of these dive-bombers; these aircraft were replaced in 1941 by the final 50 of the biplanes off the Curtiss line, which had been laid down as French aircraft. They differed from the original Navy SBC-4s in having self-sealing fuel tanks.

146

At the time of the Japanese attack on Pearl Harbor, in December 1941, the Navy had 69 SBC-3s and 117 SBC-4s on strength, two squadrons of the latter, VB-8 and VS-8, being on board the USS *Hornet*. One Marine unit, VMO-151, was also flying SBC-4s, but the SBC-3s, which had served with Scouting Squadrons VS-3, VS-5 and VS-6, were already obsolescent.

TECHNICAL DATA (SBC-4)

Manufacturer: Curtiss-Wright Corporation, Curtiss Aeroplane Division, Buffalo NY.
Type: Carrier-based scout-bomber.
Accommodation: Pilot and observer/gunner.
Power plant: One 950 hp Wright R-1820-34.
Dimensions: Span, 34 ft; length, 28 ft 4 in; height, 12 ft 7 in; wing area, 317 sq ft.
Weights: Empty, 4,841 lb; gross, 7,632 lb.
Performance: Max speed, 237 mph at 15,200 ft; cruising speed, 127 mph; initial climb, 1,860 ft/min; service ceiling, 27,300 ft; range, 590 st miles with 500 lb bomb.
Armament: One fixed forward-firing 0·30-in gun and one flexible rear 0·30-in gun. One 500 lb or 1,000 lb bomb under fuselage.
Serial numbers:

XF12C-1 (XSBC-1, -2, -3): 9225. SBC-4: 1269–1325; 1474–1504; 1809–
SBC-3: 0507–0581; 0583 0589. 1843; 4199–4248.
XSBC-4: 0582; 1268.

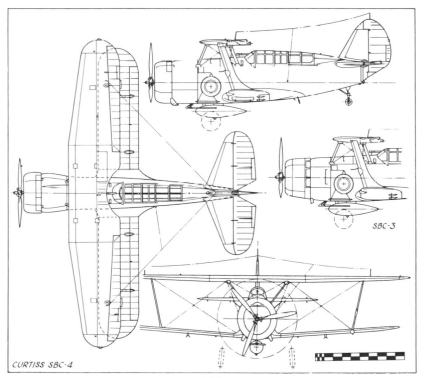

CURTISS SBC-4

SBC-3

Curtiss SO3C-1 floatplane, showing how front portion of dorsal fin slid forward with rear cockpit canopy. (*US Navy*)

Curtiss SO3C Seamew

To replace the Curtiss SOC Seagull biplanes aboard the fleet's battleships and cruisers, the US Navy drew up requirements in 1937 for a high-speed scouting monoplane. Both the Curtiss company and Chance Vought, long versed in Navy observation needs, were awarded prototype contracts for aircraft meeting this requirement and produced similar types.

The Curtiss monoplane was designated XSO3C-1 and was at first named Seagull like the SOC, although the British name of Seamew was subsequently adopted. The prototype contract was placed with Curtiss on May 9, 1938, and the XSO3C-1 made its first flight on October 6, 1939. It was a slender, mid-wing monoplane with a low aspect ratio wing and, in its basic configuration, a large central float and underwing stabilizers. Test flights in this form and, initially, as a landplane with fixed tail-down landing gear showed serious stability and control problems, and upturned wingtips and enlarged tail unit were introduced.

Deliveries of SO3C-1s from the Curtiss Buffalo factory began in July 1942, with the first operational aircraft assigned to the USS *Cleveland*. Powered by the Ranger V-770-6 engine, the SO3C-1 had underwing racks for two 100 lb bombs or two 325 lb depth charges. Following production of 141 of these aircraft, Curtiss switched to the SO3C-2 version, with arrester gear and provision for carrier operations, plus a bomb rack under the fuselage (in the landplane version) with 500 lb capacity. Production totalled 459, including 250 assigned to Britain under lend–lease. The British version was originally intended to be the SO3C-1B, but those delivered were actually SO3C-2Cs, this variant having a 24-volt electric system, V-770-8 engine, improved radio and hydraulic brakes.

The unsatisfactory operational record of the early SO3C-1s led to many of these aircraft being converted for use as radio-controlled targets,

30 of these also being assigned to the Royal Navy as Queen Seamews. The SO3C-3, introduced by Curtiss late in 1943, had the more powerful V-770-8 engine and a small reduction in weight in an attempt to improve overall performance, but production ended in January 1944 with 39 built. All Seamews were withdrawn from service early in 1944 and plans to produce the SO3C-4, similar to the -3 with arrester gear, and the Ryan SOR-1, similar to the SO3C-1, were abandoned.

TECHNICAL DATA (SO3C-2C)

Manufacturer: Curtiss-Wright Corporation, Curtiss Aeroplane Division, Buffalo, NY.
Type: Scouting and observation seaplane (or landplane).
Accommodation: Pilot and observer in tandem.
Power plant: One 600 hp Ranger V-770-8.
Dimensions: Span, 38 ft; length, 35 ft 8 in; height, 14 ft 2 in; wing area, 293 sq ft.
Weights: Empty, 4,800 lb; gross, 7,000 lb.
Performance: Max speed, 172 mph at 8,100 ft; cruising speed, 125 mph; initial climb, 720 ft/min; service ceiling, 15,800 ft; range, 1,150 st miles.
Armament: One fixed forward-firing 0·30-in gun, and one 0·50-in gun flexibly mounted in rear cockpit. Two 100 lb bombs or 325 lb depth-charges under wings; 100 lb or 500 lb bomb, or 325 lb weapon under fuselage.
Serial numbers (allocations):

XSO3C-1: 1385.	SO3C-2C: 4784–4792; 22007–22256.
SO3C-1: 4730–4783; 4793–4879.	SO3C-3: 04199–04348.
SO3C-2: 4880–5029; 04149–04198.	

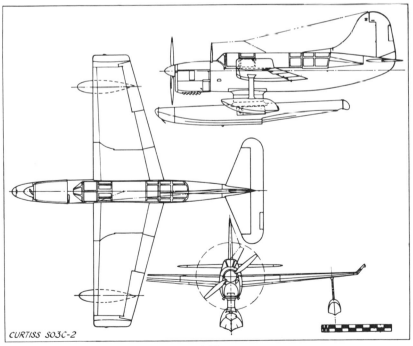

CURTISS SO3C-2

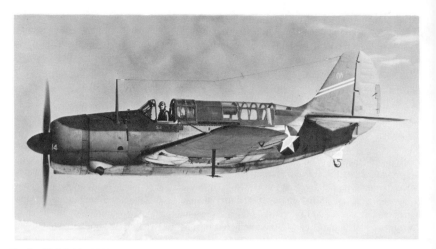

Curtiss SB2C-1 in 1943 camouflage; white stripes on the fin port side helped the landing signal officer on the carrier deck to estimate the angle of approach. (*Curtiss-Wright photo*)

Curtiss SB2C Helldiver

The long series of Curtiss combat aeroplanes built for the US Navy and Marine Corps between the two world wars was brought to a highly successful conclusion with the company's first monoplane bomber and the last to carry the name Helldiver. Ordered on May 15, 1939, while the earlier Helldiver biplane was still in quantity production (see page 146), the new type was a low-wing monoplane with the same general layout as the Brewster SB2A Buccaneer, with which it was in competition. It was a two-seat scout-bomber powered by the big Wright R-2600 Double Cyclone engine, and had an internal bomb-bay in the fuselage.

The prototype XSB2C-1, with a 1,700 hp R-2600-8, made its first flight on December 18, 1940, but was destroyed a few days later. Large-scale production had already been ordered on November 29, 1940, but a large number of modifications were specified for the production model. The size of the fin and rudder was enlarged, fuel capacity was increased and self-sealing added, and the fixed armament was doubled to four 0·50-in guns in the wings, compared with the prototype's two cowling guns. Curtiss established a new factory for SB2C production at Columbus, Ohio, and the first production model did not fly until June 1942. After the first 200 SB2C-1s, fixed armament was again changed to two 20-mm cannon, in the SB2C-1C version; in addition the Helldiver had two 0·30-in guns in the rear cockpit, and an internal bomb load of 1,000 lb.

Production at the new Columbus factory was protracted, and, although deliveries to VS-9 began in December 1942, 11 more months elapsed before the type had been brought up to operational effectiveness and was ready

150

for action. The first operational sortie was made on November 11, 1943, when VB-17 attacked Rabaul. Production of the SB2C-1 totalled 978; one of these was converted to the single XSB2C-2 floatplane, another became the XSB2C-5 and two became XSB2C-6s. The SB2C-3, which began to appear in 1944, had the R-2600-20 engine with a four-blade propeller, while the SB2C-4 had wing fittings for eight 5-in rockets or up to 1,000 lb of bombs and, in the SB2C-4E version, carried a small radar set. Production by Curtiss totalled 1,112 SB2C-3s and 2,045 SB2C-4s. Finally came 970 SB2C-5s, starting in February 1945, with increased fuel capacity. Two XSB2C-6 prototypes had R-2800-28 engines and longer fuselages.

Added capacity for Helldiver production was provided at two Canadian factories; Fairchild produced a total of 300, designated XSBF-1, SBF-1 SBF-3 and SBF-4E, while Canadian Car and Foundry built 894 designated SBW-1, SBW-3, SBW-4, SBW-4E and SBW-5, these models being respectively equivalent to their Curtiss-built counterparts.

Throughout 1944 the Navy's new Helldivers mounted an ever-growing offensive against Japanese targets in the Pacific, taking over from Douglas SBDs. So vital did they prove in this task that all production of the type

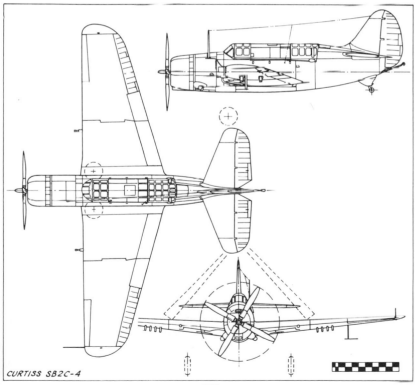

CURTISS SB2C-4

was retained by the US Navy with the exception of 26 Canadian-built aircraft supplied to Britain under lend–lease arrangements and designated SBW-1B for this purpose. The Marine Corps took on strength a large portion of the 900 Helldivers built by Curtiss for the USAAF as A-25As, and these were designated SB2C-1A after transfer.

Late versions of the Helldiver remained in US Navy service several years after the end of World War II, and others were supplied to foreign countries.

TECHNICAL DATA (SB2C-4)

Manufacturer: Curtiss-Wright Corporation, Airplane Division, Columbus, Ohio; Canadian Car & Foundry Co, Ltd, Montreal; and Fairchild Aircraft Ltd, Longueuil, PQ, Canada.

Type: Carrier-based scout-bomber.

Accommodation: Pilot and observer.

Power plant: One 1,900 hp Wright R-2600-20.

Dimensions: Span, 49 ft 9 in; length, 36 ft 8 in; height, 13 ft 2 in; wing area, 422 sq ft.

Weights: Empty, 10,547 lb; gross, 16,616 lb.

Performance: Max speed, 295 mph at 16,700 ft; cruising speed, 158 mph; initial climb, 1,800 ft/min; service ceiling, 29,100 ft; range, 1,165 st miles with 1,000 lb bomb-load.

Armament: Two fixed forward-firing 20 mm cannon in wings; two 0·30-in machine guns in rear cockpit. Up to 1,000 lb bombs internal and 1,000 lb external.

Serial numbers:

XSB2C-1: 1758.	SB2C-5: 83128–83751; 89120–89465.
SB2C-1: 00001–00200.	XSB2C-6: 18620–18621.
SB2C-1A: 75218–75588; 76780–76818.	SBF-1: 31636–31685.
SB2C-1C: 00201–00370; 01008–01208;	SBF-3: 31686–31835.
18192–18598.	SBF-4E: 31836–31935.
XSB2C-2: 00005.	SBW-1: 21192–21231; 60010–60035.
SB2C-3: 18599–19710.	SBW-3: 21233–21645.
SB2C-4: 19711–21191; 64993–65286;	SBW-4E: 21646–21741; 60036–60209.
82858–83127.	SBW-5: 60210–60295.
XSB2C-5: 18308.	

Curtiss SBW-4E, built by Canadian Car & Foundry, with Naval Reserve markings in 1946.
(Gordon S. Williams)

Curtiss SC-1 floatplane. (*US Navy*)

Curtiss SC Seahawk

Last of the long line of Navy scouting aeroplanes designed to serve aboard battleships as well as carriers and from land bases, the Curtiss SC-1 originated to a specification issued to industry in June 1942. The requirement was for a convertible land or floatplane with a much improved performance over the observation/scouts then in service and with provision for catapult launching.

The Curtiss design proposal in response to the specification was quickly adopted by the Navy, which issued a letter of intent on October 30, 1942, and a contract for two prototypes on March 31, 1943, with the designation XSC-1. A production order for 500 SC-1s followed in June 1943, and the first XSC-1 made its first flight on February 16, 1944.

The SC-1 was a stubby low-wing monoplane, designed to be as simple as possible to speed production. The pilot was seated beneath a 360 degree vision canopy, and provision was made in the rear fuselage for a stretcher to be stowed. All SC-1s were delivered as landplanes with fixed landing gear; an Edo float installation, with central float and cantilever wingtip floats, was purchased separately by the Navy and the aircraft were converted as required.

Deliveries of the SC-1 began in October 1944, the first operational aircraft being assigned to the USS *Guam*. A second contract for 450 was placed, but only 66 of these were completed before V-J day brought cancellation of the programme. Meanwhile, a further contract had been placed for the SC-2 (evolved by way of the XSC-1A) with 1,425 hp R-1820-76 engine, clear-view canopy, reprofiled fin and rudder and provision of a jump seat behind the pilot. Only 10 SC-2s had been delivered before cancellation of the contracts.

TECHNICAL DATA (SC-1)

Manufacturer: Curtiss-Wright Corporation, Curtiss Airplane Division, Columbus, Ohio

Type: Scout and anti-submarine patrol monoplane.

Accommodation: Pilot only.

Power plant: One 1,350 hp Wright R-1820-62.

Data for seaplane:

Dimensions: Span, 41 ft; length, 36 ft 4½ in; height (on beaching gear, tail down), 16 ft; wing area, 280 sq ft.

Weights: Empty, 6,320 lb; max gross, 9,000 lb.

Performance: Max speed, 313 mph at 28,600 ft; cruising speed, 125 mph; initial climb 2,500 ft/min; service ceiling, 37,300 ft; range, 625 st miles.

Armament: Two fixed forward-firing 0·50-in guns. Two underwing bomb racks capacity 325 lb each; two bomb-cells in central float optional.

Serial numbers:

XSC-1: 35298–35300.   SC-1: 35301–35797; 93302–93367.

SC-2: 119529–119538.

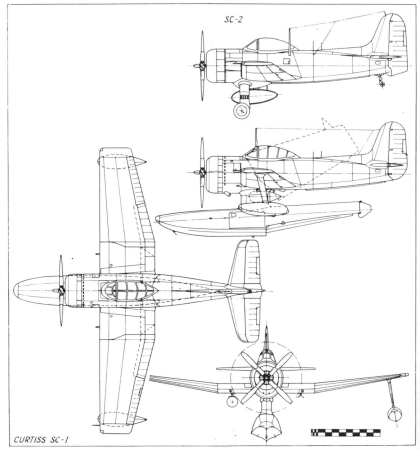

SC-2

CURTISS SC-1

154

De Havilland DH-4 in France, 1918, with US Marine Corps insignia. (*US Navy*)

De Havilland DH-4 Series

The two-seat De Havilland 4 'Liberty Plane' was the only American-made landplane to see service with US Naval forces in France in World War I. An American adaptation of the original British D.H.4 day bomber of 1916, the Liberty Plane was the major war-time product of the United States military aircraft programme. In spite of the really amazing rate of production achieved after the initial redesign and manufacturing problems were overcome, few DH-4s reached France, and their effectiveness was practically nil.

All of the scandals of politics in procurement, deficiencies in design and armament, suicidal war missions, and even the 'Billion Dollar Bonfire' in which most DHs overseas were piled and burned rather than being returned to the States after the Armistice, were directed at the Aircraft Production Board and the War Department. Historians have virtually ignored the fact that 51 of 145 Liberty Planes built by Dayton-Wright, and transferred from the Army to the Navy, served the US Navy and Marines in France. Most of these were with the 9th and 10th Marine Squadrons of the Northern Bombing Wing based at Dunkirk. Independent American operations with DH-4s were initiated against German installations in Belgium on October 14, 1918.

The original Liberty Plane had many shortcomings, and an improved version, the DH-4B, which borrowed many features of the later British D.H.9, notably the relocation of the pilot's cockpit and the main fuel tank

Boeing-built DH-4Ms delivered to the US Marine Corps as O2B-1s in 1925. (*US Navy*)

to lessen the chance of pilot fatality in even a minor crash, was in production by the war's end. The fuselage, mostly fabric-covered on the Liberty Plane, was completely covered with plywood on the DH-4B. Forty-two DH-4Bs were transferred to the Navy from the War Department and an additional 80 were rebuilt as DH-4Bs from surplus Liberty Planes by the Naval Aircraft Factory. A further improved version, the DH-4B-1, had the fuel capacity increased from 96 to 118 US gal and other minor refinements. Fifty of this model were also transferred from the War Department to the Navy.

In 1923 the Army instigated the use of welded steel-tube fuselage construction for more rebuilt DHs to be known as the DH-4M, for DH-4 Modernized. The initial work under this programme was accomplished by the Boeing Airplane Company, which used a new arc-welding process that it had developed. Thirty of these DH-4M1s, as they were known to distinguish them from the later Atlantic-Fokker DH-4M2s with gas-welded fuselages, were released from the Army contract and made available to the Navy, which bought them for the US Marine Corps. Although the Navy normally identified older aircraft by the designations in use before the adoption of standardized Naval designations in 1922, the DH-4M1s were given a new designation, O2B-1, to identify them as observation types built by Boeing. An earlier OB-1 amphibian based on a BuAer design had been cancelled, but the designation was not reassigned. The O2B-1s delivered in 1925, were indistinguishable from the DH-4Bs except by a return to fabric covering on the fuselage and a more forward location of the landing wheels.

The postwar Navy-Marine DH-4Bs and Ms were initially used in the observation and day bomber roles and were gradually down-graded to training and utility work. A degree of late fame was achieved by these aircraft in the US Marine Corps action against the Nicaraguan bandits in 1927. A few remained in service with the Marines into 1929. The last Army models were not retired until 1932.

156

TECHNICAL DATA (DH-4B-1)

Manufacturer: Dayton-Wright Company, Dayton, Ohio; Standard Aircraft Corporation, Patterson, NJ; Fisher Body Division of General Motors, Cleveland, Ohio; Naval Aircraft Factory, Philadelphia, Penn; Boeing Airplane Company, Seattle, Washington (O2B).

Type: Observation, day bomber and general purpose biplane.

Accommodation: Pilot and observer in tandem.

Power plant: One 400 hp Liberty.

Dimensions: Span, 42 ft 5½ in; length, 30 ft 1¼ in; height, 10 ft 6 in; wing area, 440 sq ft.

Weights: Empty, 2,647 lb; gross, 4,214 lb.

Performance: Max speed, 122·5 mph at sea level; initial climb, 6·8 min to 5,000 ft; service ceiling, 14,000 ft; range, 550 st miles.

Armament: Two fixed forward-firing 0·30-in guns; two flexible 0·30-in guns on Scarff ring.

Serial numbers:

DH-4B (War Department): A5809–A5814; A5834–A5839; A5870–A5884; A5982–A6001.

DH-4B (NAF): A6113–A6192; A6514.
DH-4B-1: A6352–A6401.
O2B-1: A6898–A6927.

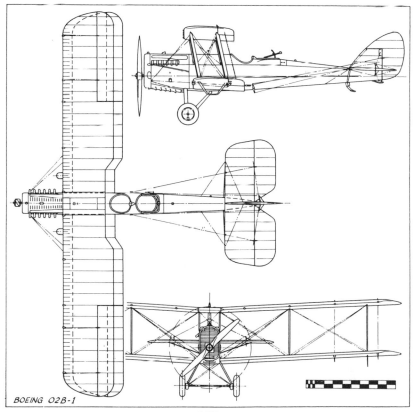

BOEING O2B-1

157

Douglas DT-2 floatplane of VT-2 in 1923. (*US Navy*)

Douglas DT

As the first military aeroplane produced by the Douglas Aircraft Company, the DT-1 is deserving of a place in history quite apart from its importance as one of the Navy's first successful torpedo-bombers. In fact the DT series of biplanes, derived in some measure from the Douglas Cloudster (first product of the embryonic Douglas organization in Los Angeles), gave the Navy notable service for a number of years.

Design of the DT-1 was completed in 1921 under the personal supervision of Donald W. Douglas, who had just set up his own company after a spell as chief engineer for Glenn L. Martin. The DT-1 was a chunky single-seat biplane with a 400 hp Liberty engine, folding wings and interchangeable wheel/float chassis. Three were ordered by the Navy in 1921, but only one was delivered as such, the others becoming DT-2s. The DT-1

Douglas DT-2 landplane built by L.W.F. (*US Navy*)

could carry a 1,835 lb torpedo under the forward fuselage and was distinguished by side radiators on the engine cowling.

Successful experience with the DT-1 led to further Navy orders for an eventual total of 38 two-seat DT-2s plus two of the original DT-1s, with conventional nose radiators but otherwise similar to the DT-1. Deliveries to the Naval Air Station at San Diego began in 1922 and continued until 1924, and these aircraft began to replace Curtiss R-6Ls and NAF PTs in torpedo squadrons. The first unit to receive the Douglas torpedo-bomber was VT-2; by 1925 they were in service at all major Navy bases, both in the United States and overseas. Production of the DT-2 was swelled by the Naval Aircraft Factory, which built six and by L.W.F. which built 20;

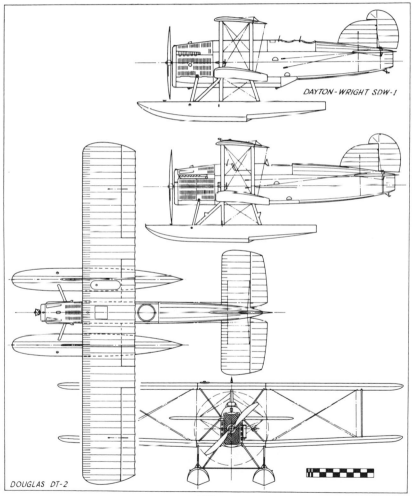

DAYTON-WRIGHT SDW-1

DOUGLAS DT-2

Dayton-Wright modified three of the latter as long-range scouts designated SDW-1s, with additional fuel in a deepened fuselage.

Four DT-4s were NAF conversions of the DT-2 with Wright T-2 engine, and the single DT-5 had a geared T-2. The DT-6 was a modified DT-2 with the Wright P-1 air-cooled radial engine, prototype of the famous 450 hp Cyclone, first flown at Murchio's Field, New Jersey, in 1925.

Although designed and purchased primarily as torpedo-bombers, the DT-2s served in several other roles, including scouting, observation, and aerial gunnery practice (Marine Squadron VJ-7) and experimental flying. Torpedo development was aided by test drops from DT-2s, and an aircraft of this type on floats was used for experimental launchings from the USS *Langley* using an early catapult in 1925. The type passed out of service in 1926.

TECHNICAL DATA (DT-2)

Manufacturer: Douglas Aircraft Company, Santa Monica, California; Naval Aircraft Factory, Philadelphia, Penn; L.W.F. Engineering Company, Inc., College Point, LI, NY.

Type: Torpedo-bomber.

Accommodation: Pilot and observer.

Power plant: One 400 hp Liberty.

Data for seaplane:

Dimensions: Span, 50 ft; length, 37 ft 7½ in; height, 15 ft 1 in; wing area, 707 sq ft.

Weights: Empty, 4,528 lb; gross, 7,293 lb.

Performance: Max speed, 99 mph at sea level; initial climb, 10 min to 3,850 ft; service ceiling, 7,400 ft; range, 274 st miles with torpedo.

Armament: One 1,835 lb torpedo.

Serial numbers:

DT-1: A6031.

DT-2 (Douglas): A6032–A6033; A6405–A6422; A6563–A6582 (A6581 to DT-6).

DT-2 (L.W.F.): A6583–A6602 (A6593, A6596–A6597 to SDW-1).

DT-2 (NAF): A6423–A6428 (A6423, A6424 to DT-4; A6427, A6428 to DT-4 then DT-5).

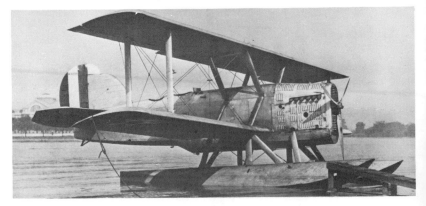

L.W.F.-built DT-2 modified to SDW-1 by Dayton-Wright, with deepened fuselage. (*US Navy*)

Douglas T2D-1 floatplane of VP-1. (*US Navy*)

Douglas T2D, P2D

Seeking better performance than could be provided by existing Navy torpedo-bombers, the Bureau of Aeronautics in 1925 designed a large twin-engine biplane for this role. It had two Wright R-1750 engines, one on each lower wing, and could operate on wheels or floats. A prototype was built at Philadelphia by the Naval Aircraft Factory as the XTN-1, and Douglas was given a contract for three identical aircraft designated T2D-1.

The XTN-1 and the first T2D-1 appeared early in 1927, and the Douglas biplanes were issued to Squadron VT-2 on May 25 the same year. This unit was attached to USS *Langley*, and its operation of T2D-1s from that experimental carrier in 1927 is believed to be the first occasion on which twin-engine aircraft were flown from an aircraft carrier. Location of the pilot well forward in the fuselage allowed him an uninterrupted view ahead, and this was of considerable advantage in deck landings.

Nine more T2D-1s were ordered from Douglas in 1927, but by the time these were delivered the mounting Army criticism of Navy operation of land-based bombers made it politic to use these aircraft as floatplanes, and the majority were assigned to VP-1D14 at Pearl Harbor for general duties. When a further 18 examples were ordered from Douglas in June 1930, the same reasoning caused them to be redesignated in the patrol category as P2D-1s, although they were almost identical with the earlier torpedo-bombers. Powered by 575 hp R-1820 engines, the P2D-1s flew with VP-3 from Coco Solo in the Panama Canal Zone until replaced by PBY-1s in 1937.

TECHNICAL DATA (T2D-1)

Manufacturer: Douglas Aircraft Company, Santa Monica, California.

Type: Ship or shore-based torpedo-bomber.

Accommodation: Pilot and two gunners.

Power plant: Two 525 hp Wright R-1750s.

Dimensions: Span, 57 ft; length, 44 ft 11½ in; height, 14 ft 7¾ in; wing area, 886 sq ft.

Weights: Empty, 6,011 lb; gross, 10,890 lb.

Performance: Max speed, 124 mph at sea level; initial climb, 5 min to 5,000 ft; service ceiling, 13,830 ft; range, 422 st miles.

Armament: One fixed forward and one flexible dorsal 0·30-in gun. One 1,618 lb torpedo or bomb.

Serial numbers:

XTN-1: A7027.   T2D-1: A7051–A7053; A7587–A7595.

P2D-1: A8644–A8661.

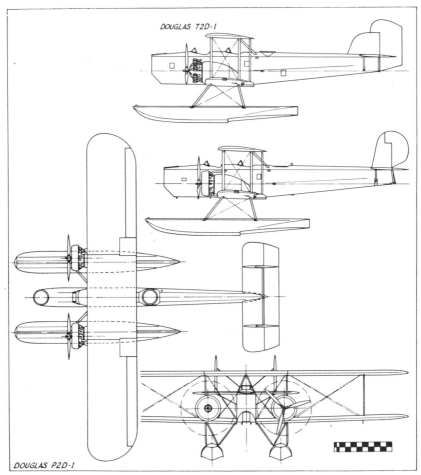

DOUGLAS T2D-1

DOUGLAS P2D-1

Douglas RD-4 of the US Coast Guard at San Francisco, in green and grey finish, June 1942.
(*Peter M. Bowers*)

Douglas RD

When the Douglas Aircraft Company introduced a twin-engine commercial amphibian in 1930, the armed services were quick to express their interest. The new model, named Dolphin, featured an all-metal hull with provision for eight people including pilot and co-pilot, a plywood-covered cantilever wing very similar to that standardized by Fokker, and two tractor engines mounted in separate nacelles above the wing. First procurement was by the US Coast Guard, which obtained three Dolphins in 1931 with the plain designation of RD for Multi-engine Transport, Douglas. The first of these, delivered on March 9, 1931, was not an amphibian, although it gave the appearance of being one. What appeared to be retractable landing gear was merely detachable beaching gear. The other two RDs were true amphibians, and the first was soon converted. These were originally given Coast Guard serial numbers 27, 28 and 29. For identification until 1935, the model designation was combined with the serial numbers, e.g. RD-29. These numbers were changed to 227, 128 and 129 in January 1935. The numbers were again changed in October 1936, to V106, V109 and V111.

The Navy, meanwhile, ordered a single amphibian as the XRD-1, which was essentially a duplicate of the Coast Guard model and the Army Y1C-21, and it was delivered in December 1931. This order was followed by an order for three RD-2s which differed mainly in being powered with 450 hp Pratt & Whitney R-1340-96 Wasp engines in place of the 400 hp Wright R-975E Whirlwinds of the RDs and the XRD-1. The first of these, delivered in February 1933, was fitted with extra-luxurious accommodation, ostensibly for the use of President Roosevelt, but there is no evidence that this machine was actually assigned as a Presidential aircraft. The last Navy order, for six RD-3s, was filled in March 1933. This was followed by a Coast Guard order for an additional ten, delivered in November 1934. These were officially designated RD-4 and carried Coast Guard serial numbers 130–139. On October 13, 1936, these were changed to V125–V134 with the complete redesignation of all Coast Guard aircraft.

163

While the Navy RDs were used primarily as personnel transports, the Coast Guard models were used extensively in search and rescue missions and as flying lifeboats, often flying far out to sea from the several Coast Guard Air Stations to rescue stricken mariners or rush seamen in need of urgent medical care to hospitals ashore. Upon US entry into World War II in December 1941, the Coast Guard was taken into the Navy and surviving RD-4s were assigned to security patrols along the American coastlines.

TECHNICAL DATA (RD-2)

Manufacturer: Douglas Aircraft Company, Santa Monica, California.
Type: Transport amphibian.
Accommodation: Up to eight seats.
Power plant: Two 450 hp Pratt & Whitney R-1340-96s.
Dimensions: Span, 60 ft; length, 45 ft 2 in; height, 15 ft 2 in; wing area, 592 sq ft.
Weights: Empty, 6,377 lb; gross, 9,387 lb.
Performance: Max speed, 153 mph at sea level; initial climb, 7·2 min to 5,000 ft; service ceiling, 15,900 ft; range, 770 st miles.
Serial numbers:

RD: 27–29 (227; 128; 129: then V106; V109; V111).
XRD-1: A8876.
RD-2: 9347–9349.
RD-3: 9528–9533.
RD-4: 130–139 (V125–V134).

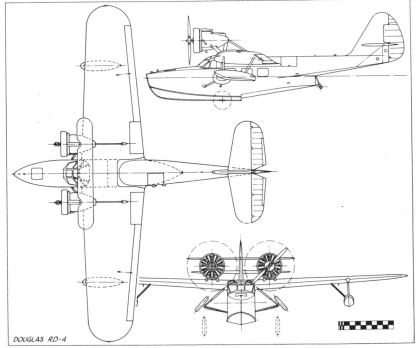

DOUGLAS RD-4

164

Douglas TBD-1 in service with VT-3, the first squadron to fly the type. (*Douglas photo*)

Douglas TBD Devastator

For operation from a new family of aircraft carriers, starting with the USS *Ranger* of 1934, the US Navy initiated development of a new torpedo-bomber during the same year. Prototypes were ordered on June 30 from both Douglas and the Great Lakes Company, the respective designs being for a monoplane and a biplane. In positive manner, therefore, this particular design competition marked the point of transition from biplane to monoplane in torpedo-bombers, for the Great Lakes XTBG-1 was the last of the biplanes in this category, and the Douglas XTBD-1 was the prototype of what was to become the Navy's first carrier-based monoplane in production.

The layout of the XTBD-1 was typical of its era. Associated with the low wing was a semi-retractable undercarriage, the main wheels protruding from the wings when retracted to give some protection in the event of a forced landing. Folding wings were no novelty on a carrier-based aeroplane, but the upward fold of the XTBD-1 was unusual, as was the provision for powered actuation of the wings. The crew of three was carried in tandem beneath a single long enclosure and the engine was an 800 hp Pratt & Whitney XR-1830-60. Armament comprised two 0·30-in guns and a 1,000 lb torpedo could be carried under the fuselage.

First flight of the XTBD-1 was made on April 15, 1935, and delivery to the Navy followed nine days later. Successful tests led to a contract on February 3, 1936, for 129 TBD-1s with 900 hp R-1830-64 engine, modified cowling and a raised canopy over the pilot's cockpit.

The first TBD-1 reached Navy Squadron VT-3 on October 5, 1937, and in the following year VT-2, VT-5 and VT-6 were all equipped with the new torpedo-bomber. These four squadrons remained fully operational with the Devastator until after the start of World War II and were in action early in 1942 against Japanese targets in the Marshall and Gilbert Islands, achieving a fair measure of success. Later the same year, however, during the Battle of Midway, one TBD squadron, VT-8, was destroyed in its entirety and another was decimated when Japanese Zero fighters intercepted them. Thereafter, the Devastator was withdrawn from operational use, but continued in service for a time as a trainer and station hack. One example was tested as a floatplane in 1939 with the designation TBD-1A.

TECHNICAL DATA (TBD-1)

Manufacturer: Douglas Aircraft Company, El Segundo, California.
Type: Carrier-based torpedo-bomber.
Accommodation: Pilot, torpedo officer/navigator, gunner.
Power plant: One 900 hp Pratt & Whitney R-1830-64.
Dimensions: Span, 50 ft; length, 35 ft; height, 15 ft 1 in; wing area, 422 sq ft.
Weights: Empty, 6,182 lb; gross, 10,194 lb.
Performance: Max speed, 206 mph at 8,000 ft; cruising speed, 128 mph; initial climb, 720 ft/min; service ceiling, 19,700 ft; range, 716 st miles with 1,000 lb bomb.
Armament: One fixed forward-firing 0·30-in gun and one flexible dorsal 0·30-in gun. One 1,000 lb torpedo externally under fuselage.
Serial numbers:
  XTBD-1: 9720.
  TBD-1: 0268–0381; 1505–1519.

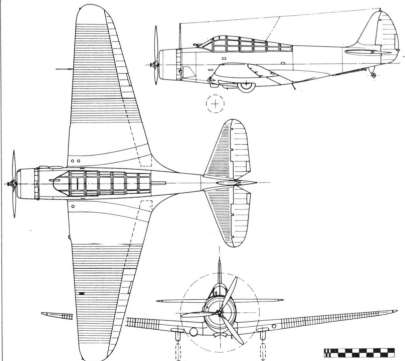

DOUGLAS TBD-1

166

Douglas SBD-1 of Marine Squadron VMB-1 showing the perforated wing flaps and unusual tapering of the rudder stripes. (*Douglas photo*)

Douglas SBD Dauntless

From December 1941, when Japan's attack on Pearl Harbor brought the US fully and finally into World War II, until the atomic bomb was unleashed against Japan to bring the war to an end, the US Navy was primarily engaged in operations against Japanese forces in the Pacific. For most of that time, the principal carrier-based bomber available to Navy squadrons was the Douglas SBD Dauntless, a refinement of the Northrop BT-1 that had been developed in 1935 (see page 357).

Prototype of the SBD was a production Northrop BT-1, redesignated XBT-2 for development purposes. By the time this machine had undergone engine and structural changes that made it virtually a new aeroplane, the Northrop company had become the El Segundo Division of Douglas; it was therefore logical to give the new scout- and dive-bomber a Douglas, rather than Northrop, identity in its production form. Production orders were placed on April 8, 1939, for 57 SBD-1s and 87 SBD-2s. All the SBD-1s were sent to Marine Corps units, starting with VMB-2 late in 1940 and VMB-1 in 1941; these units were renumbered VMSB-232 and VMSB-132 at the end of 1941. The SBD-2s, with added armament and fuel

Douglas SBD-3 at El Segundo, California, prior to delivery. (*Douglas photo*)

167

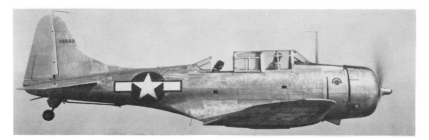

Douglas SBD-5 stripped of its camouflage finish at war's end. (*William T. Larkins*)

capacity, equipped Navy squadrons VB-6 and VS-6 aboard the USS *Enterprise* and VB-2 aboard the *Lexington* by the end of 1941.

Whereas the SBD-1 and SBD-2 had two 0·30-in guns in the cowling, plus one or two for rear defence, the SBD-3 which appeared in March 1941 had 0·50s in the cowling. Self-sealing tanks were introduced, as well as protective armour, and the engine was the 1,000 hp R-1820-52. Production, stepped up after Pearl Harbor, totalled 584 and by December 1941 the SBD-3 was already equipping VS-2 (*Lexington*), VS-6 (*Enterprise*), VB-5 and VS-5 (*Yorktown*) and VB-3 and VS-3 (*Saratoga*). In the course of the following year, many Marine squadrons, as well as additional Navy units, received the SBD-3, and its successor on the El Segundo production line, the SBD-4. The latter differed only in going over to a 24-volt system; 780 were built for the Navy. Camera-equipped versions of the Dauntless were designated SBD-1P, SBD-2P and SBD-3P.

Almost identical to the SBD-4 was the SBD-5 which was built at a new Douglas plant set up at Tulsa; this version had a 1,200 hp R-1820-60 engine and 2,409 were built for the Navy. Finally, Tulsa produced 451 SBD-6s with 1,350 hp R-1820-66 engines. In addition to these production totals were 168 SBD-3A, 170 SBD-4A, and 615 SBD-5A built for the USAAF, as the A-24, A-24A and A-24B respectively. When the last SBD-6

Douglas XSBD-6 in July 1943. (*Douglas photo*)

was delivered, Douglas had built 5,321 examples, excluding the original Northrop machines.

TECHNICAL DATA (SBD-5)

Manufacturer: Douglas Aircraft Company, El Segundo, California, and Tulsa, Oklahoma.

Type: Carrier-based scout/dive-bomber.

Accommodation: Pilot and observer/rear gunner.

Power plant: One 1,200 hp Wright R-1820-60.

Dimensions: Span, 41 ft 6¼ in; length, 33 ft 0⅛ in; height, 12 ft 11 in; wing area, 325 sq ft.

Weights: Empty, 6,675 lb; gross, 10,855 lb.

Performance: Max speed, 245 mph at 15,800 ft; cruising speed, 144 mph; initial climb, 1,190 ft/min; service ceiling, 24,300 ft; range, 1,100 st miles.

Armament: Two fixed forward-firing 0·50-in guns; two flexible dorsal 0·30-in guns. Up to 1,600 lb external under fuselage and two 325 lb under wings.

Serial numbers:

SBD-1: 1596–1631; 1735–1755.	SBD-5: 10807–10956; 10957–11066;
SBD-2: 2102–2188.	28059–28829; 28831–29213;
SBD-3: 4518 4691; 03185–03384;	35922–36421; 36433–36932;
06492–06701.	54050–54599.
SBD-4: 06702–06991; 10317–10806.	SBD-5A: 09693–09752.
	XSBD-6: 28830.
	SBD-6: 54600–55049.

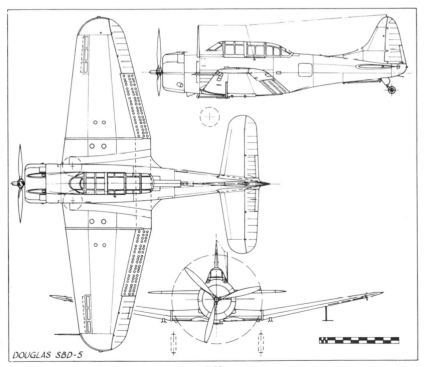

DOUGLAS SBD-5

Douglas R4D-1 in early 1942 markings. (*Douglas photo*)

Douglas R4D Skytrain, Skytrooper

Although the Douglas DC-3 variants served the Navy and Marine Corps in much smaller numbers than with the Air Force, they achieved an equal reputation for dependability and versatility. First procured in 1941 for Navy use, versions of the famous Douglas twins were still operating 30 years later, and it was indicative of the widespread use of this type that no fewer than 16 Navy variants qualified for redesignation in the unified system introduced in 1962.

Most of the 568 transports of this type procured for the Navy during World War II came from Army contracts, and the seven main variants in the R4D series each had a USAAF equivalent, and carried the same popular names. The R4D-1, powered by two Pratt & Whitney R-1830-82 engines, was a cargo transport counterpart of the C-47 Skytrain, 100 being acquired plus one odd specimen from the RAF by 'reverse lease–lend'. Two R4D-2s were ex-airline DC-3s (USAAF C-49) and were later designated R4D-2F and then R4D-2Z to indicate their VIP (flagship) interiors; they were the only R4Ds with Wright R-1820 engines. The 20 R4D-3s were diverted from USAAF contracts for C-53s (two actually being C-53Cs). The 10 R4D-4s were PAA DC-3As requisitioned in production and therefore equivalent to the C-53C, and seven R4D-4Rs were ex-commercial DC-3s. One other DC-3A acquired later from PAA apparently had no Navy designation. Major Navy cargo transport variants were the R4D-5 with R-1830-92 engines and 24-volt electric system, matching the C-47A Skytrain, and of which 238 were diverted to the Navy from USAAF contracts; the R4D-6 with R-1830-90B engines, 148 being diverted from USAAF contracts for C-47Bs (plus four more C-47Bs transferred from USAF in 1961); and the R4D-7, which matched the TC-47B navigation trainer, 43 being acquired.

The R4Ds provided basic equipment for the Naval Air Transport Service, created five days after Pearl Harbor on December 12, 1941. In the hands of NATS, the Douglas transports were soon at work, flown by transport Squadrons VR-1, VR-2 and VR-3 on airline-like scheduled

Douglas R4D-5 in service with US Marines, China, 1945. (*Peter M. Bowers*)

operations between points in the USA and nearby overseas territories. During 1942 another new unit, the South Pacific Combat Air Transport Service, began using R4Ds to ferry supplies right into the combat zones and casualties out. In a single month this unit carried 22,000 passengers, 3,300,000 lb of freight and 941,000 lb of mail. Other similar units later appeared in other parts of the Pacific, while the Marine Corps paratroopers began to operate in R4D-3s and R4D-4s.

In addition to serving as plain cargo or troop transports, the R4Ds were adapted for a variety of more specialized roles, with suitable equipment added. Such versions included, for radar countermeasures, the R4D-1Q, R4D-5Q and R4D-6Q; with special electronic equipment, the R4D-5E and R4D-6E; for air–sea warfare training, the R4D-5S and R4D-6S; for navigation training the R4D-5T and R4D-6T; and winterized for operation in the Antarctic (usually on skis), the R4D-5L and R4D-6L. Executive or VIP versions were the R4D-5Z and R4D-6Z, as well as the R4D-2Z already mentioned, while basic cargo aircraft equipped to carry passengers were the R4D-5R and R4D-6R. Redesignation of many of these variants in the basic C-47 series in 1962 produced the following new designations: C-47H (R4D-5), EC-47H (R4D-5Q), LC-47H (R4D-5L), SC-47H (R4D-5S), TC-47H (R4D-5R) and VC-47H (R4D-5Z); C-47J

Douglas R4D-6Q radar-operator trainer. (*E. M. Sommerich*)

(R4D-6), EC-47J (R4D-6Q), LC-47J (R4D-6L), SC-47J (R4D-6S), TC-47J (R4D-6R) and VC-47J (R4D-6Z); TC-47K (R4D-7). Subsequently, new modification programmes resulted in the appearance of the C-47L and the C-47M, equipped to support the activities of US missions overseas.

In 1951 the Navy successfully evaluated the prototype Super DC-3 which Douglas had developed as a private venture attempt to rejuvenate the Grand Old Lady of the transports. The Super DC-3 had new outer wing panels with modest sweepback, a longer and strengthened fuselage, enlarged tail unit, modified nacelles to enclose the undercarriage totally, up-rated R-1820-80 engines and many other changes. After USAF trials with a prototype as the YC-47F, it was evaluated by the Navy as the R4D-8, and a modernization programme was approved to convert 98 earlier R4D models to the same standard. These included winterized R4D-8Ls, some R4D-8T trainers, and R4D-8Z staff transports. Some R4D-8s operated in Korea during the war there, primarily with the Marine

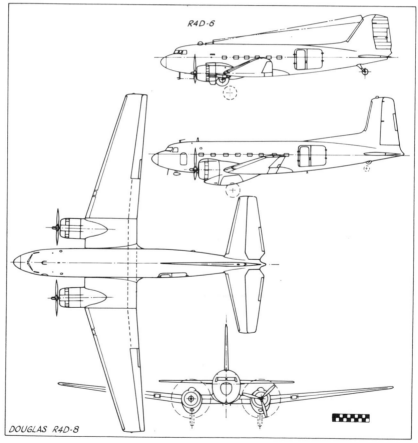

172

Corps including VMF(N)-513 which used them for close air support by night as flare carriers. Other R4D-8s operated in the Antarctic on skis. They became C-117D (R4D-8), LC-117D (R4D-8L), VC-117D (R4D-8Z) and TC-117D (R4D-8T) in 1962 and continued serving in small numbers into the mid-'seventies.

TECHNICAL DATA (R4D-5)

Manufacturer: Douglas Aircraft Company, Santa Monica, California.
Type: Troop and personnel transport.
Accommodation: Crew of three and up to 27 passengers or 10,000 lb of cargo.
Power plant: Two 1,200 hp Pratt & Whitney R-1830-92s.
Dimensions: Span, 95 ft; length, 63 ft 3 in; height, 17 ft; wing area, 987 sq ft.
Weights: Empty, 16,578 lb; gross, 29,000 lb.
Performance: Max speed, 227 mph at 7,500 ft; cruising speed, 135 mph; initial climb, 940 ft/min; service ceiling, 22,500 ft; range, 1,975 st miles.
Serial numbers (allocations):

R4D-1: 3131–3143; 4692–4706; 01648–01649; 01977–01990; 05051–05072; 12393–12404; 30147; 37660–37680; 91104.
R4D-2: 4707–4708.
R4D-3: 05073–05084; 06992–06999.
R4D-4: 07000–07003; 33815–33820.
R4D-4R: 33615–33621.

R4D-5: 12405–12446; 17092–17248; 39057–39095.
R4D-6: 17249–17291; 39096–39098; 39100; 39109; 50740–50839; 150187–150190.
R4D-7: 39099; 39101–39108; 39110–39136; 99824–99857.
DC-3A: 99099.

TECHNICAL DATA (R4D-8)

Manufacturer: Douglas Aircraft Company, Santa Monica, California.
Type: Personnel transport.
Accommodation: Crew of three and up to 35 troops, 30 passengers or 27 stretchers.
Power plant: Two 1,475 hp Wright R-1820-80s.
Dimensions: Span, 90 ft; length, 67 ft 9 in; height, 18 ft 3 in; wing area, 969 sq ft.
Weights: Empty, 19,537 lb; gross, 31,000 lb.
Performance: Max speed, 270 mph at 5,900 ft; cruising speed, 238 mph at 10,000 ft; initial climb, 1,300 ft/min; range, 2,500 st miles.
Serial numbers:
138820 (prototype, originally allocated 138659) plus 98 conversions with original numbers.

Douglas R4D-8 showing new tail, wing and undercarriage details. (*Peter M. Bowers*)

173

Douglas R5D-3 with white top to fuselage. (*Gordon S. Williams*)

Douglas R5D Skymaster

Over 200 examples of the Douglas DC-4 transport went into service with Navy and Marine units during World War II; many were still in service 20 years later. Each of five main versions matched USAAF variants designated in the C-54 series, and performed similar missions. As R5Ds, these aircraft were operated primarily by units of the Naval Air Transport Service between 1941 and 1948; subsequently they were divided between Navy squadrons assigned to MATS and to the Fleet Logistics Air Wing and Marine units.

The R5D-1, with R-2000-7 engines, was equivalent to the C-54A, 58 being acquired by the Navy; the 30 R5D-2s (C-54) had revised fuel systems which dispensed with two tanks in the cabin and had additional capacity in the wings. Some R5D-1Cs were modified to have this fuel system; other aircraft, used as flagships with plush interiors, were R5D-1F and R5D-2F (later R5D-1Z and R5D-2Z). The 98 R5D-3s had R-2000-11 engines but were otherwise similar to the R5D-2, matching the C-54D, while 20 similar R5D-4s were equivalent to the C-54E. Another small engine change, to R-2000-9, plus a revised interior, identified the R5D-5 (conversions of the R5D-2 and -3); the R5D-6 designation was reserved for a Naval equivalent of the cancelled C-54J with airline-type interior.

After the war, administrative versions with improved interiors were designated R5D-3Z and R5D-5Z, and basic cargo aircraft modified as personnel carriers were designated R5D-4R and R5D-5R. After the NATS had been disbanded in July 1948, three Navy squadrons flying R5Ds (VR-3, VR-6 and VR-8) were assigned to the inter-service Military Air Transport Service; three other squadrons were flying Skymasters for the Fleet Logistics Air Wing in 1950, these being VR-1, VR-5 and VR-21, while two Marine R5D units were VMR-152 and VMR-352.

174

Nine versions of the Skymaster were still serving the Navy in 1962, when they were redesignated as follows: VC-54N (R5D-1Z), C-54P (R5D-2), VC-54P (R5D-2Z), C-54Q (R5D-3), VC-54Q (R5D-3Z), C-54R (R5D-4R), C-54S (R5D-5), VC-54S (R5D-5Z) and C-54T (R5D-5R). A few R5D-3s were transferred to the Coast Guard as RC-54V. The last Navy Skymaster, a C-54Q used by the Navy Test Pilot School, was retired in 1974.

TECHNICAL DATA (R5D-1)

Manufacturer: Douglas Aircraft Company, Santa Monica, California.

Type: Personnel, supply and staff transport.

Accommodation: Crew of four and up to 30 passengers.

Power plant: Four 1,350 hp Pratt & Whitney R-2000-7s.

Dimensions: Span, 117 ft 6 in; length, 93 ft 11 in; height, 27 ft 6¼ in; wing area, 1,462 sq ft.

Weights: Empty, 37,040 lb; gross, 65,000 lb.

Performance: Max speed, 281 mph at 15,200 ft; cruising speed, 210 mph; initial climb, 1,010 ft/min; service ceiling, 22,700 ft; range, 2,290 st miles.

Serial numbers:
 R5D-1: 39137–39181; 50840–50849; 57988–57989; 91105.
 R5D-2: 50850–50868; 90385–90395.
 R5D-3: 50869 50878; 56484–56549; 91994–92003.
 R5D-4: 90396–90415.

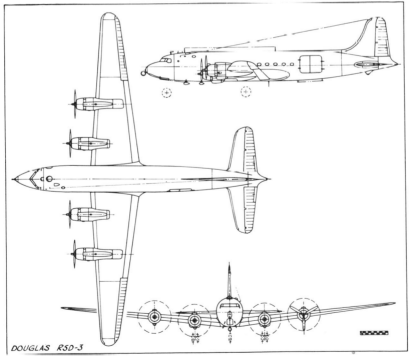

DOUGLAS R5D-3

Douglas AD-1 in overall blue finish. (*William T. Larkins*)

Douglas AD-1 Skyraider

Developed to a World War II specification for a carrier-based dive-bomber and torpedo-carrier—the first single-seater in this category—the Skyraider was too late for service in that conflict, but proved to be one of the most versatile air weapons available to the US forces in both the Korean War and the Vietnam War. Although production ended in 1957, 12 years after the line was first established, serious proposals were made as late as 1966 for the re-establishment of production in order to meet Vietnam needs. By that time the Skyraider had become widely known as the Spad, a name which reflected the affection in which it was held by its pilots who likened its reliability and versatility to the famed World War I fighter.

Ordered by the US Navy on July 6, 1944, as the XBT2D-1, the Douglas design showed the influence of war-time experience which indicated that the ability to carry a large and varied load was of overriding importance. Unlike the SBD Dauntless (see page 167) and other scout and torpedo-bombers, the XBT2D-1 was laid out as a single-seater, the weight saved being available for load carrying. Although the airframe was by no means small—it was designed around the big, 2,500 hp Wright R-3350 engine—no provision was made for internal weapon stowage; this again saved airframe weight and complexity and allowed ordnance loads of up to 8,000 lb to be carried on 15 external strong points under the wings and fuselage. In all, four designs in this new single-seat category were ordered during 1944, one other besides the Douglas project achieving production status as the Martin AM-1 Mauler (see page 321), but the Skyraider was clearly the best of the competing projects.

The initial Navy order was for 25 aircraft to production standards although bearing the X designation, and the name Destroyer II was used for a time, the Destroyer I having been the unsuccessful BTD-1. On March 18, 1945, two weeks before US forces invaded Okinawa as they pushed the Japanese back across the Pacific, the XBT2D-1 made its first flight, and a

Douglas AD-3Q with radar operator's cockpit in fuselage. (*Douglas photo*)

month later the Navy ordered 548 production examples to be designated AD-1 in the then new 'attack' category.

Acceptance of the XBT2D-1s by the Navy began during 1945, and within a year pilots from VA-3B and VA-4B had completed carrier qualification tests on the USS *Sicily*. The production contract was cut back to 277 after V-J Day and delivery of these began in December 1946 with VA-19A becoming the first unit equipped. While the AD-1s were being produced and delivered to Navy and Marine attack units, a number of design developments were under way, and six of the XBT2D-1s were converted to new configurations. These included two XBT2D-1Ns which carried radar and searchlights under the wings and two radar operators in the fuselage, to evaluate the aircraft's potential in night attack operations; an XBT2D-1P reconnaissance version with fuselage-mounted cameras and the XBT2D-1Q with ECM equipment and an operator in the fuselage behind the pilot. The success of this prototype resulted in production of 35 similarly equipped AD-1Qs, in addition to 242 AD-1s. Another XBT2D-1 became the XAD-1W with a large 'inverted mushroom' radome under the fuselage for airborne early-warning, and one was converted to the XAD-2 with strengthened inner wing, increased internal fuel capacity, higher operating weights and 2,700 hp R-3350-26W engine in place of the R-3350-24W in the original version. After this prototype had crashed, another AD-1 was converted to XAD-2 configuration, and Douglas then built 156 AD-2s plus 21 AD-2Qs with ECM equipment, deliveries starting in 1948. In 1949 a single AD-2QU was delivered with a target-towing conversion-kit, and two aircraft were converted to AD-2Ds for drone-control duties.

First delivered in 1948, the AD-3 had further structural strengthening, a longer stroke main undercarriage oleo and a redesigned cockpit canopy. Production covered four variants: 124 plain AD-3s, 15 night-operating AD-3Ns, 23 countermeasures AD-3Qs and 31 AD-3Ws with AEW radar as first fitted on the single XAD-1W. In addition, two of the N models became AD-3S with special submarine attack equipment to complement

177

Douglas AD-4W with ventral radome and extra fins. (*Douglas photo*)

Douglas AD-5W with larger fin and rudder, and revised cockpit details. (*US Navy*)

Douglas AD-5N with searchlight pod under port wing. (*Douglas photo*)

Douglas A-1H, redesignated from AD-6. (*US Navy*)

two submarine-search AD-3Es converted from AD-3Ws.

As service use of the AD was gradually extended to most attack units of the Navy and Marines, further development of the airframe brought still more versions into production. The AD-4, which replaced the AD-3 on the production line during 1949 and began to reach squadrons in 1950, had APS-19A radar in place of the earlier APS-4, a P-1 autopilot and other small modifications. Production of the basic AD-4 totalled 372, plus 165 AD-4Bs with provision for carrying tactical nuclear weapons and four wing-mounted cannon; 307 AD-4Ns for night operation; 39 AD-4Qs and 168 AD-4Ws. Also, 29 AD-4s were converted to AD-4B standard and 63 were winterized as AD-4Ls, 159 AD-4Ns were stripped for day-attack duties as AD-4NA in addition to 23 built as such, and another 37 were winterized as AD-4NL.

Within three days of the US undertaking to back the United Nations resolution in support of the Republic of Korea on June 30, 1950, Skyraiders were in action flying from the USS *Valley Forge* against targets north of Seoul. Thereafter, the AD was seldom out of Korean skies, and its ability to carry rockets, bombs, napalm, torpedoes, mines, depth-charges, and cannon allowed it to be used against almost all Korean targets. Its achievement was such that the Skyraider was described as 'the best and most effective close support airplane in the world' by the commander of Task Force 77, Rear Adm John W. Hoskins.

One result of the Korean War was to keep the Skyraider in production longer than might otherwise have been the case, the version most affected being the AD-5. First flown on August 17, 1951, this was a major re-design of the Skyraider which incorporated all the lessons learned in five years of operations with the wide range of variants described above. The forward fuselage was widened to seat two side by side and was lengthened by 2 ft, there being a compensatory increase in vertical fin area. The wing armament of four 20-mm cannon used on the AD-4B became standard, and a number of equipment changes were made. Most important innovation in the AD-5 was that it came complete with a conversion kit which

179

allowed any one of the three basic production versions to be quickly modified to serve as a transport (up to 12 seats), freighter, ambulance or target-tug. The production versions were the plain AD-5 (212 built); AD-5N (239 built) and AD-5W (218 built); 54 AD-5Qs were converted from AD-5Ns, and a single AD-5S was built. While these aircraft were being built, Douglas also produced 713 single-seat AD-6s with equipment for low-level bombing but other features of the AD-5. The rigours of low-level operation imposed considerable strain on the airframe, and Skyraider production ended with 72 AD-7s, similar to the AD-6 but with strengthened wings and landing gear, and R-3350-26WD engine. The last of 3,180 Skyraiders was completed in February 1957.

In September 1962, Skyraiders still in service were redesignated in the A-1 series. A few AD-4NAs then became A-1Ds; the AD-5 became A-1E and, in target-tug configuration, UA-1E; AD-5W and AD-5Q became

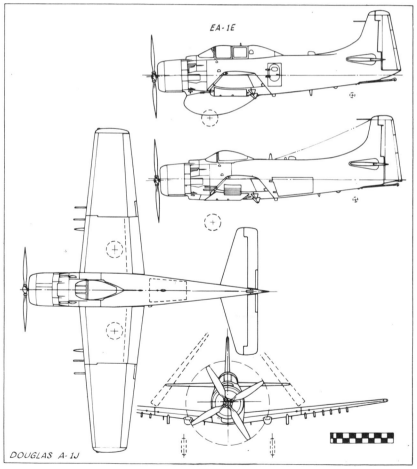

EA-1E

DOUGLAS A-1J

180

EA-1E and EA-1F respectively; the AD-5N became A-1G, and the AD-6 and AD-7 became A-1H and A-1J respectively. Under these designations, the type became operational in Vietnam, not only with Navy and Marine units but also with the USAF, which acquired a number of A-1E, G, H and J models, and by the Vietnamese Air Force. The final version in Navy service was the EA-1E.

TECHNICAL DATA (AD-2)

Manufacturer: Douglas Aircraft Company, El Segundo, California.
Type: Carrier-borne attack-bomber.
Accommodation: Pilot only.
Power plant: One 2,700 hp Wright R-3350-26W.
Dimensions: Span, 50 ft; length, 38 ft 2 in; height, 15 ft 5 in; wing area, 400 sq ft.
Weights: Empty, 10,546 lb; gross, 18,263 lb.
Performance: Max speed, 321 mph at 18,300 ft; cruising speed, 198 mph; initial climb, 2,800 ft/min; service ceiling, 32,700 ft; range, 915 st miles.
Armament: Two fixed forward-firing 20 mm guns. Up to 8,000 lb externally beneath the wings and fuselage.
Serial numbers:

XBT2D-1 (XAD-1): 09100–09109; 09085–09099.
AD-1: 09110–09386 (cancelled to 09392, and 21742–22006).
XAD-2: 09195; 09108.
AD-2: 122210–122365.
AD-2Q: 122366–122387.
AD-3: 122729–122852.
AD-3Q: 122854–122876.
AD-3W: 122877–122907.
AD-3N: 122908–122922.
AD-4: 122853; 123771–124005*; 127844–127853; 127861–127865; 127867; 127873–127879; 128917–128936; 128944–128970; 128979–129016.
AD-4B: 127854–127860; 127866; 127868–127872; 128937–128943; 128971–128978; 132227–132391.
   * Converted to AD-5N later.

AD-4Q: 124037–124075.
AD-4N: 124128–124156; 125707–125741; 126876–127018; 127880–127920.
AD-4NA: 125742–125764.
AD-4NL: 124725–124760.
AD-4W: 124076–124127; 124761–124777; 125765–125782; 126836–126875; 127921–127961.
XAD-5: 124006.
AD-5: 132392–132476; 132478; 132637–132686; 133854–133929.
AD-5N: 132477; 132480–132636; 134974–135054.
AD-5S: 132479.*
AD-5W: 132729–132792; 133757–133776; 135139–135222; 139556–139605.
AD-6: 134466–134637; 135223–135406; 137492–137632; 139606–139821.

TECHNICAL DATA (AD-7)

As AD-2 except:
Dimensions: Span, 50 ft 9 in; length, 38 ft 10 in; height, 15 ft 8¼ in; wing area, 400 sq ft.
Weights: Empty, 10,550 lb; gross, 25,000 lb.
Performance: Max speed, 318 mph at 18,500 ft; cruising speed, 188 mph at 6,000 ft; initial climb, 2,380 ft/min; service ceiling, 32,000 ft; range, 900 st miles.
Armament: Four fixed forward-firing 20 mm cannon. Up to 8,000 lb external.
Serial numbers:

XAD-7: 142010.
AD-7: 142011–142081.

Douglas F3D-2Q countermeasures aircraft of Marine Squadron VMCJ-3. (*Clay Jansson*)

Douglas F3D Skyknight

On April 3, 1946, the US Navy contracted with the Douglas company for construction of three prototypes of a new all-weather fighter: the first in its class to use jet engines. Designated XF3D-1, the new design featured a straight wing, side-by-side seating for the pilot and radar operator, and side-by-side engine location under the wing centre section. Design of the tail unit closely resembled that of the D-558-1 Skyrocket, the first Douglas jet aeroplane, and a novel feature was the provision of a tunnel from the cockpit to the bottom of the fuselage to facilitate bale-out at high speed. Armament comprised four 20 mm cannon.

Engines in the prototypes were 3,000 lb static thrust Westinghouse J34-WE-22s. The first flight was made on March 23, 1948, and work proceeded on production of an initial batch of 28 F3D-1 Skyknights with J34-WE-24 engines. First flown on February 13, 1950, the F3D-1s began to reach VC-3 a year later, at the same time that testing began on the F3D-2. This model had been designed to use the 4,600 lb s.t. Westinghouse J46-WE-3 engines, but after the production of the J46 had been abandoned, J34-WE-36s rated at 3,400 lb s.t. each were substituted.

Operational use of the F3D-2 was exclusively by Marine squadrons. Deployed to Korea soon after the war there began, Skyknights were responsible for the destruction of more enemy aircraft than any other type of aircraft flown by the Navy or Marines. The first Skyknight success was recorded on November 2, 1952, this also being the first occasion when one jet aircraft had destroyed another (a MiG-15) during a night interception.

Production of the F3D-2 totalled 237, and several of these were later converted to special-duty variants. These included 16 F3D-2M missile-carriers and 30 F3D-2Qs for electronic reconnaissance and counter-measure duties with VMCJ units. A few Skyknights with special armament were designated F3D-2B in 1958, and a number of others became

182

F3D-2T and F3D-2T2 for training duties, primarily to instruct radar controllers in the air and on the ground. The F3D-1s also served eventually as radar trainers, and a few were modified as missile carriers as F3D-1Ms. Production of a sweptwing development, the F3D-3, was planned, but a contract for 187 was cancelled before the first had been completed.

Skyknights still on hand in September 1962 were redesignated: F3D-1 to F-10A; F3D-2 to F-10B; F3D-2Q to EF-10B; F3D-2M to MF-10B and F3D-2T2 to TF-10B. The F3D-2Q had been the first tactical jet aircraft in the electronic warfare role, and as such it played an important operational role in the Cuban crisis and later in Vietnam.

TECHNICAL DATA (F3D-2)

Manufacturer: Douglas Aircraft Company, El Segundo, California.
Type: All-weather carrier-based fighter.
Accommodation: Pilot and radar observer.
Power plant: Two 3,400 lb s.t. Westinghouse J34-WE-36/36As.
Dimensions: Span, 50 ft; length, 45 ft 6 in; height, 16 ft; wing area, 400 sq ft.
Weights: Empty, 18,160 lb; gross, 26,850 lb.
Performance: Max speed, 600 mph at 20,000 ft; cruising speed, 350 mph; initial climb, 4,500 ft/min; range, 1,200 st miles
Armament: Four fixed forward-firing 20 mm guns.
Serial numbers:

XF3D-1: 121457–121459.	F3D-2: 124595–124664; 125783–
F3D-1: 123741–123768.*	125882; 127019–127085.

* Twelve converted to F3D-1M.

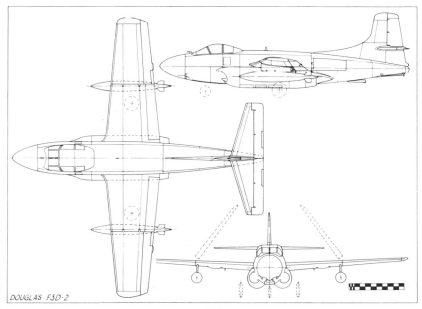

DOUGLAS F3D-2

Douglas F4D-1 of Navy Squadron VFAW-3 in 1956. (*Peter M. Bowers*)

Douglas F4D Skyray

German work on delta wing planforms, directed by Dr Alexander Lippisch, led to a US Navy proposal in 1947 for a short-range carrier-based interceptor fighter using a similar layout. Project studies were initiated by the Douglas design team led by Ed Heinemann with the object of producing a fighter optimized for a high rate of climb and capable of intercepting enemy aircraft before they reached their targets. These studies led to a design which, rather than being a pure delta, was a tailless aircraft with a sweptback wing of extremely low aspect ratio, and the Douglas designers thus followed Lippisch's own evolution of this layout for the Messerschmitt Me 163 target-defence interceptor.

Two prototypes of the Douglas design were ordered on December 16, 1948, as the XF4D-1, to be powered by the Westinghouse J40 turbojet. This engine suffered serious delays in development, and the XF4D-1s flew with 5,000 lb s.t. Allison J35-A-17 engines, the initial flight being made on January 23, 1951. Named Skyray, the F4D-1 was designed to carry an armament of four 20 mm cannon in the wings and, despite the interceptor role, was provided with six underwing strong points for bombs, rockets or drop tanks.

Both prototypes of the Skyray subsequently flew with the 7,000 lb s.t. XJ40-WE-6 and the 11,600 lb (with afterburner) XJ40-WE-8 engine, but the failure of the J40 engine programme led to a decision in March 1953 to switch to the 14,500 lb (with afterburner) Pratt & Whitney J57-P-2 in the production F4D-1s, orders for which had already been placed. This airframe/engine combination flew for the first time on June 5, 1954, when the first production F4D-1 was completed. Three more aircraft were built in 1954, and 32 came off the line in 1955, but the first delivery to a Navy unit (VC-3) was not made until April 16, 1956, several flight-test problems having delayed acceptance of the aircraft.

With production building up rapidly, to nearly two a week in 1956 and nearly three a week in 1957, introduction of the Skyray to Navy and Marine units continued rapidly, and the aircraft proved to be a highly effective fighter once in service. Production ended in December 1958 with a grand total of 420 built, all in the basic F4D-1 version.

The Skyray was redesignated F-6A in September 1962, by which time the type had been relegated to second-line duties and assigned to Reserve Wings. A development of the same basic airframe with a J57-P-14 engine and other changes was undertaken under the F4D-2 designation, but the four aircraft of this type which appeared in 1956 and 1957 were re-designated F5D-1 before the first flight.

TECHNICAL DATA (F4D-1)

Manufacturer: Douglas Aircraft Company, El Segundo, California.
Type: Carrier-based interceptor.
Accommodation: Pilot only.
Power plant: One 9,700 lb s.t. Pratt & Whitney J57-P-2 or 10,500 lb s.t. J57-P-8B turbojet with afterburner.
Dimensions: Span, 33 ft 6 in; length, 45 ft 8¼ in; height, 13 ft; wing area, 557 sq ft.
Weights: Empty, 16,024 lb; gross, 25,000 lb.
Performance: Max speed, 695 mph at 36,000 ft; initial climb, 18,000 ft/min; service ceiling, 55,000 ft; range, 1,200 st miles.
Armament: Four fixed forward-firing 20 mm cannon. Up to 4,000 lb of bombs, rockets, or other stores on six external points.
Serial numbers:
  XF4D-1: 124586–124587.
  F4D-1: 130740–130750; 130751 (static test); 134744–134973; 136163–136392 (cancelled); 139030–139207.

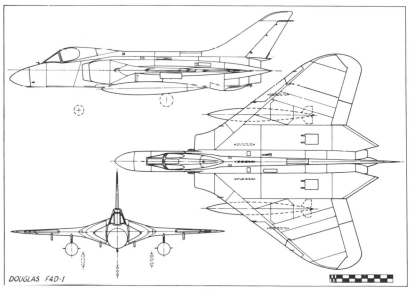

DOUGLAS F4D-1

185

Douglas A-3A with original radar turret in tail. (*Gordon S. Williams*)

Douglas A3D Skywarrior

Jet engines and nuclear weapons, taken together, opened up for the US Navy the possibility of adding true strategic strike capability to its other offensive roles, since it was no longer necessary, by 1946, to think in terms of giant superbombers for strategic operations. A bomber to operate from the very large carriers which were then in the planning stages was outlined by the Bureau of Aeronautics in 1947, and project design of a suitable aircraft was completed by Douglas two years later. The requirement was met with an aeroplane of some 60,000 lb gross weight, the largest and heaviest ever projected for carrier use. It had a high-mounted wing with 36 degrees of sweepback, a large internal weapons-bay with provision for 12,000 lb of conventional or nuclear weapons, and podded engines—originally intended to be Westinghouse J40s. The crew of three was grouped in a pressurized cockpit, tail defence being by means of a radar-controlled barbette with two 20 mm guns, these being the only guns carried.

Douglas was awarded a two-prototype contract on March 31, 1949, when the aircraft was designated XA3D-1. The first flight was made on October 28, 1952, this prototype being powered by two 7,000 lb s.t. XJ40-WE-3 engines; the J40 programme had to be abandoned, however, and the 9,700 lb s.t. Pratt & Whitney J57-P-6 engine was adopted instead.

Douglas RA-3B showing camera ports in fuselage and flight refuelling probe. (*Douglas photo*)

186

With these engines, the first example flew on September 16, 1953, and production deliveries began on March 31, 1956, to Heavy Attack Squadron One (VAH-1). The 50 aircraft of the A3D-1 (later A-3A) version built by Douglas were used primarily to evaluate the total concept of carrier-based strategic bombers and to explore problems of operating so large an aircraft at sea. A single example was converted to YA3D-1P (YRA-3A) as a proto-type for the photo-reconnaissance version produced subsequently, and a a few others became YA3D-1Q (YEA-3A) and A3D-1Q (EA-3A) with ECM equipment, four crew and bomb capabilities removed.

The definitive production version of the Skywarrior was the A3D-2 (initially A3D-1B, eventually A-3B), powered by J57-P-10 engines and with weapons-bay changes to permit the carriage of a wider range of stores, including mines, or a flight-refuelling pack with a 1,082 Imp gal fuel tank. The A3D-2 entered service in 1957 with VAH-2, and the 164 aircraft of this type built by Douglas eventually equipped eight Heavy Attack Squadrons for service aboard carriers of the *Essex* and *Midway* class. A few were subsequently modified as VA-3B high-speed personnel carriers.

Following trials with the YA3D-1P reconnaissance version of the Sky-

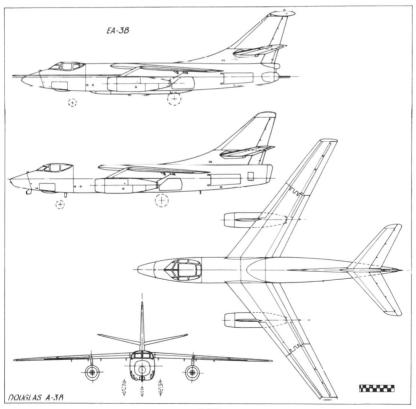

EA-3B

DOUGLAS A-3B

187

Douglas KA-3B in service with Tactical Electronic Warfare Squadron VAQ-130. (*US Navy*)

warrior, the Navy ordered 30 A3D-2Ps (RA-3B) to equip Heavy Photographic Squadrons VAP-61 and VAP-62, these two units providing detached flights aboard operational carriers as required. The entire fuselage was pressurized, and accommodation provided in the space of the weapons-bay for two reconnaissance specialists and as many as 12 oblique and vertical cameras. A related variant, the A3D-2Q (EA-3B), first flew on December 10, 1958, and also had a fully pressurized fuselage. As well as the flight crew, it carried four electronic specialists and a wide range of equipment in order to undertake radar countermeasures and electronic reconnaissance sorties. The 25 aircraft of this type equipped Fleet Reconnaissance Squadron VQ-1 for service as required.

Douglas also built 12 A3D-2Ts (TA-3B) with the pressurized fuselage and accommodation for six radar/navigation pupils, plus an instructor and the pilot. First flown on August 29, 1959, the A3D-2T served with training units VAH-3 and VAH-R123 alongside the A3D-1s.

Skywarriors served throughout the Vietnam War in the tanker role as KA-3B conversions of the A-3B, and about 30 became EKA-3Bs in TACOS (Tanker Aircraft/Countermeasures or Strike) configuration.

TECHNICAL DATA (A3D-2)

Manufacturer: Douglas Aircraft Company, El Segundo, California.
Type: Carrier-based attack-bomber.
Accommodation: Crew of three.
Power plant: Two 12,400 lb s.t. Pratt & Whitney J57-P-10 turbojets.
Dimensions: Span, 72 ft 6 in; length, 76 ft 4 in; height, 22 ft 9½ in; wing area, 812 sq ft.
Weights: Empty, 39,409 lb; gross, 82,000 lb.
Performance: Max speed, 610 mph at 10,000 ft; service ceiling, 41,000 ft; tactical radius, 1,050 st miles.
Armament: Two 20 mm guns in radar-controlled rear-turret. Internal stowage for up to 12,000 lb of bombs, depth-charges, etc.
Serial numbers:
XA3D-1: 125412–125413.   A3D-1: 130352–130363; 135407–135444.
A3D-2: 138902–138976; 142236–142255; 142400–142407; 142630–142665; 144626–144629; 147648–147668.   A3D-2T: 144856–144867.
A3D-2P: 142256; 142666–142669; 144825–144847; 146446–146447.
A3D-2Q: 142257; 142670–142673; 144848–144855; 146448–146459.

The Fairchild R4Q-1, identical with the C-119C, served with the Marine Corps. (*Fairchild photo*)

Fairchild R4Q Packet

To provide the Marine Corps with a tactical assault transport in place of the universally used R4Ds and other smaller types, the Fairchild C-119 was adopted in 1950. The C-119 with its distinctive twin-boom layout and capacious fuselage, which had earned it the nickname of Flying Boxcar, was conceived as early as 1941 and, in the original C-82 version, went into production for the USAAF. The considerably modified and improved C-119 appeared in 1947, with a wider fuselage, more power, higher operating weights and many other changes.

The 1950 production model of the Packet for the USAF was the C-119C, with 3,500 hp P. & W. R-4360-20WA engines and accommodation for 42 troops. This model was purchased virtually off-the-shelf by the Marine Corps and designated R4Q-1. Like the USAF C-119Cs, the R4Q-1s were fitted with new tail units after introduction into service, with dorsal and ventral fins added and the outer portions of the tailplane removed. Squadron VMR-252 received its first unmodified R4Q-1 early in 1950, and a total of 41 was delivered. They served primarily with Marine units and were used in Korea during the war there.

Fairchild R4Q-2s, with ventral fins and new engines, began to replace the R4Q-1s in 1952. (*Charles N. Trask*)

189

A second batch of 58 Packets began to reach the Navy and Marines in 1952 with the designation R4Q-2. These were similar to the USAF's C-119F and were redesignated as such in the unified designation shuffle of 1962. The R4Q-2 had 3,400 hp R-3350-36W engines, the switch from Pratt & Whitney to Wright engines being made because of the unsatisfactory service record of the former type. The R4Q-1s were discarded by Marine squadrons soon after the R4Q-2s became available; the latter were still in service in the early 'seventies.

TECHNICAL DATA (R4Q-2)

Manufacturer: Fairchild Engine and Airplane Corporation, Aircraft Division, Hagerstown, Md.
Type: Tactical troop and cargo transport.
Accommodation: Crew of five and 42 troops.
Power plant: Two 3,400 hp Wright R-3350-36WAs.
Dimensions: Span, 109 ft 3 in; length, 86 ft 6 in; height, 26 ft 6 in; wing area, 1,447 sq ft.
Weights: Empty, 40,000 lb; gross, 64,000 lb.
Performance: Max speed, 250 mph; cruising speed, 205 mph; initial climb, 820 ft/min; range, 2,000 st miles.
Serial numbers:
 R4Q-1: 124324–124333; 126574–126582; 128723–128744.
 R4Q-2: 131662–131719.

FAIRCHILD R4Q-2

The Ford RR-4 was a commercial Model 5-AT-C purchased by the Navy. (*Ned Moore*)

Ford JR, RR

Nine examples of the famed Ford Tri-Motor went into service with the US Navy and Marines between 1927 and 1931 and gave good service until 1935. Included in the total were examples of both the Model 4 and Model 5 versions of the Tri-Motor, which had its origins in the series of all-metal cantilever monoplanes designed and built by Bill Stout between 1923 and 1926.

After the Stout factory (and the 3-AT prototype) at Ford Airport had been destroyed by fire on January 17, 1926, the Ford Motor Company at once erected a new factory, and initiated work to develop a three-engine passenger transport along the lines of Stout's earlier types. The prototype of the new design took shape rapidly and, designated 4-AT, made its first flight on June 11, 1926. It was powered by three 200 hp Wright J-4 engines and had accommodation for eight passengers in the cabin. With a wing span of 68 ft 10 in, the 4-AT was the largest all-metal aircraft built in America up to that time, and it featured corrugated Alclad covering of the fuselage, wings and tail, and an internally braced cantilever wing.

The Ford Tri-Motor gained quick acceptance by airlines in the US and elsewhere, and deliveries from the production line began before the end of 1926. The first Navy contract was placed on March 9, 1927, for a single example of the Model 4-AT (actually the fourth aircraft built) designated XJR-1 as a personnel and cargo transport. It was built in 1927 and tested at Anacostia in 1928, then served until being written off in April 1930.

Two of the improved Model 4-AT-Es were purchased as JR-2s in 1929, and both went into service with the US Marines. They had 300 hp Wright J6-9 engines and increased performance. Three JR-3s were Ford Model 5-AT-Cs with an enlarged wing and Pratt & Whitney Wasp engines; the first went to the Navy and two others to the Marine Corps in 1930. Designations were changed to RR-2 and RR-3 shortly after, when the Navy adopted the R for transport aircraft, and a single RR-4 went to the Navy, being another Model 5-AT-C. Finally, two Model 5-AT-Ds were acquired as RR 5s, one each for the Marine Corps and Navy.

One other Tri-Motor was tested by the Navy but not taken on charge: this was the Model 5-AT-CS fitted with two 30-ft floats and operated as a seaplane at the Philadelphia Navy Yard, where its suitability as a torpedo-bomber was studied.

TECHNICAL DATA (JR-3)

Manufacturer: Ford Motor Company, Aircraft Division, Dearborn, Michigan.
Type: General purpose transport.
Accommodation: Two crew and up to 15 passenger seats.
Power plant: Three 450 hp Pratt & Whitney R-1340-88s.
Dimensions: Span, 77 ft 10 in; length, 50 ft 3 in; height, 13 ft 6 in; wing area, 835 sq ft.
Weights: Empty, 8,149 lb; gross, 13,499 lb.
Performance: Max speed, 135 mph at sea level; cruising speed, 122 mph; initial climb, 8,000 ft in 10 min; service ceiling, 18,000 ft; range, 505 st miles.
Serial numbers:

XJR-1: A7526. RR-4: A8840.
JR-2: A8273–A8274. RR-5: 9205–9206.
JR-3: A8457; A8598–8599.

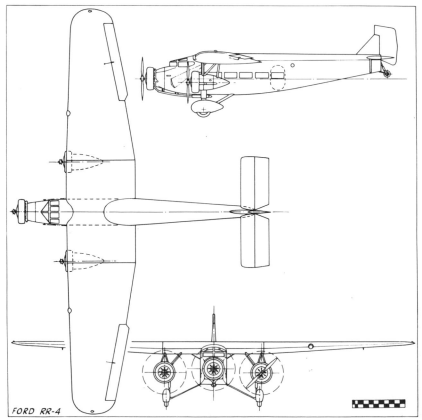

FORD RR-4

192

A Great Lakes BG-1 of VB-4 attached to the USS *Ranger* in 1937. (*Gordon S. Williams*)

Great Lakes BG-1

Although the Great Lakes Aircraft Corporation produced a number of interesting biplanes for the US Navy in the eight years of its existence from 1928 to 1936, only one of these, the BG-1, reached production. Before this type was developed, Great Lakes built a number of Martin T4M variants designated TG-1 and TG-2, design rights in this torpedo-bomber (see pages 310–313) having been acquired from Detroit Aircraft when the new company purchased the Martin factory at Cleveland after Martin had moved to Baltimore.

Experience with the TG-2 was used by Great Lakes to design a new dive-bomber for the Navy, under a contract issued on June 13, 1932. The aircraft, tested in competition with the Consolidated XB2Y-1, was required to carry a 1,000 lb bomb under the fuselage and to seat two. Powered by a Pratt & Whitney R-1535-64 Wasp radial engine, the XBG-1 was completed in mid-1933 and was tested by the Navy in its original open-cockpit form. The armament comprised two 0·30-in machine guns, one forward and one aft.

Proved superior to the XB2Y-1 in tests, the Great Lakes biplane was ordered into production in November 1933, and two subsequent contracts brought the total quantity to 60. Production models, which began to reach operational units in October 1934, had an enclosed canopy over the two cockpits but were in other respects similar to the prototype. First unit equipped was VB-3B aboard the USS *Ranger* (later, *Lexington*); in July 1937, this unit became VB-4 and transferred back to the *Ranger*, taking its

19 BG-1s with it. They were relinquished in 1938, having served operationally with only this one Navy unit, but remained in use for utility duties at shore bases for a few more years.

At least half of BG-1 production was assigned to the Marine Corps, equipping VB-4M in 1935 and VB-6M in 1936; these units were redesignated VMB-2 and VMB-1 respectively in July 1937 and retained their BG-1s until 1940, when VMS-1 at Quantico also flew the type for a short period.

TECHNICAL DATA (BG-1)

Manufacturer: Great Lakes Aircraft Corporation, Cleveland, Ohio.
Type: Carrier-based dive-bomber.
Accommodation: Pilot and observer/gunner.
Power plant: One 750 hp Pratt & Whitney R-1535-82.
Dimensions: Span, 36 ft; length, 28 ft 9 in; height, 11 ft; wing area, 384 sq ft.
Weights: Empty, 3,903 lb; gross, 6,347 lb.
Performance: Max speed, 188 mph at 8,900 ft; initial climb, 5·5 min to 5,000 ft; service ceiling, 20,100 ft; range, 549 st miles with 1,000 lb bomb.
Armament: One fixed forward-firing 0·30-in gun; one flexible 0·30-in rear gun.
Serial numbers:
  XBG-1: 9220.
  BG-1: 9494–9520; 9534–9550; 9840–9855.

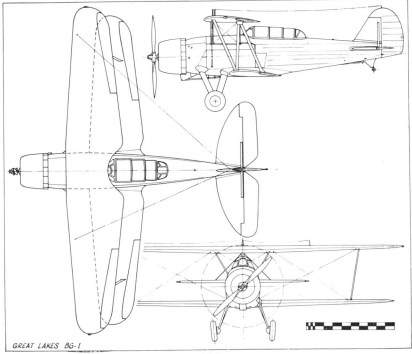

GREAT LAKES BG-1

194

A Naval Reserve Grumman SF-1, photographed at Vancouver, B.C., in 1936.
(Gordon S. Williams)

Grumman FF and SF

A contract dated April 2, 1931, signalled the start of a Grumman/US Navy association which was to continue without interruption for more than 40 years and was to produce some of the finest Naval aircraft of all time. The then young Grumman Aircraft Engineering Corporation had successfully designed some floats for landplane/seaplane conversions when it won the Navy order to build a prototype two-seat fighter. Designated XFF-1, the new design had several noteworthy features, especially in the method of undercarriage retraction, the two main wheels being raised almost vertically to lie flush in the fuselage sides ahead of the lower wing.

The XFF-1 was the first Navy fighter to have a retractable undercarriage. Also unusual for its era was the long canopy covering the two cockpits.

First flown towards the end of 1931, the all-metal XFF-1 was armed with a single 0·30-in fixed forward-firing gun and two Brownings of similar calibre in the rear cockpit. Initially powered by a 616 hp Wright R-1820E engine, it achieved 195 mph during service trials; later, a 750 hp R-1820F was installed, and the speed went up to 201 mph, better than any single-seat fighter then in service. A second prototype was completed with changed equipment and designated XSF-1 as a two-seat scout.

The Navy ordered 27 production examples of the fighter as FF-1s (Grumman G-5) and 33 SF-1 (G-6) scouts, powered respectively with the R-1820-78 and R-1820-84 engine. Deliveries of the FF-1, or 'Fifi', began on June 21, 1933, to Squadron VF-5B attached to USS *Lexington*. The SF-1 scouts were also assigned to *Lexington* serving with VS-3B; deliveries began on March 30, 1934. Both types had been withdrawn from Fleet service by the end of 1936 and were assigned to Naval Reserve units. Most of the fighters were modified to have dual control and were then designated FF-2.

TECHNICAL DATA (FF-2)

Manufacturer: Grumman Aircraft Engineering Corporation, Bethpage, LI, NY.
Type: Carrier-based fighter.
Accommodation: Pilot and observer/gunner.
Power plant: One 700 hp Wright R-1820-78.
Dimensions: Span, 34 ft 6 in; length, 24 ft 6 in; height, 11 ft 1 in; wing area, 310 sq ft.
Weights: Empty, 3,250 lb; gross, 4,828 lb.
Performance: Max speed, 207 mph at 4,000 ft; initial climb, 2·9 min to 5,000 ft; service ceiling, 21,000 ft; range, 921 st miles.
Armament: One fixed forward-firing and two flexible rear 0·30-in guns.
Serial numbers:

XFF-1: A8878.	XSF-1: A8940.
FF-1: 9350–9376.	SF-1: 9460–9492.

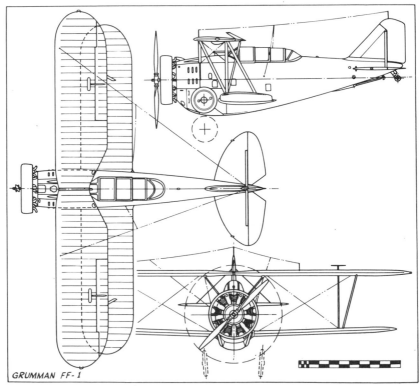

GRUMMAN FF-1

196

Grumman F2F-1 of VF-3, with fuselage painted light grey. (*William S. Swisher*)

Grumman F2F, F3F

Successful testing of the Grumman XFF-1 (see page 195) led to a new US Navy contract on November 2, 1932, for a prototype single-seat fighter designated XF2F-1. The configuration was the same as that for the FF-1, but as this was a single-seater, the overall dimensions were reduced. Location of the main wheel wells in the forward fuselage, however, gave the aircraft a very squat appearance which was to be a characteristic of Grumman types for many years to come.

First flown on October 18, 1933, the XF2F-1 had fabric-covered wings and a metal fuselage skin. It was powered by a two-row Pratt & Whitney R-1535-44 engine and carried an armament of two 0·30-in guns in the forward fuselage, firing through the propeller disc. Like the FF-1, the new prototype had an enclosed cockpit—the first Navy single-seater to enjoy this refinement. A production contract for 54 F2F-1s (Grumman G-8) soon followed the prototype's demonstration of a top speed of 229 mph at 8,000 ft and an initial rate of climb of more than 3,000 ft/min.

Deliveries of the F2F-1s, which had R-1535-72 engines, began early in 1935; by mid-year, VF-2B (the famous Fighting Two squadron) on USS *Lexington* and VF-3B on USS *Ranger* were flying the new biplane. Fighting Two, continuously on the *Lexington*, retained their Grumman Flying Barrels until 1940, VF-3B became VF-7B on USS *Yorktown* and then VF-5 on the *Wasp*, taking its F2Fs with it; they were replaced by F3Fs in 1939, and the older fighters passed into use as gunnery trainers attached to Naval Air Stations and patrol wings.

Grumman F2F-1 of VF-2 showing the distinctive tapered cowling. (*Grumman photos*).

Grumman F3F-1 of Marine Corps squadron VMF-2 showing the letter 'M' prefix to the squadron letter on the fuselage used by the Marines from July 1937 onwards. (*Art Green*)

The final aircraft on the F2F contract was completed as the XF3F-1 prototype, with a longer fuselage and increased wing span to improve manoeuvrability. It crashed during flight testing in May 1935, but was replaced by a new airframe with the same serial number. The higher weight of the XF3F-1 reduced the maximum speed, but this was improved by using the 650 hp R-1535-84 Twin Wasp in 54 production model F3F-1s ordered on August 24, 1935. Delivered in the first eight months of 1936, these fighters went into service with VF-5B on the USS *Ranger* and VF-6B on USS *Saratoga*; these two units became VF-4 and VF-3 respectively in July 1939, retaining their Grumman biplanes until 1940.

A prototype ordered from Grumman on July 25, 1936, was tested at Anacostia in January 1937 as the XF3F-2. The major change in this model was the use of the larger diameter, single-row 950 hp Wright R-1820-22 engine, changing the nose contours of the aircraft and increasing the performance. The Navy ordered 81 F3F-2s (Grumman G-37) on March 23, 1937, and used them to equip VF-6 on the USS *Enterprise* and Marine

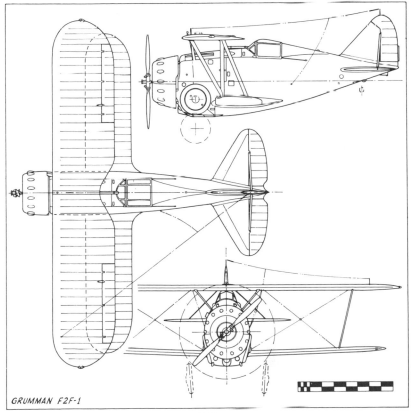

GRUMMAN F2F-1

199

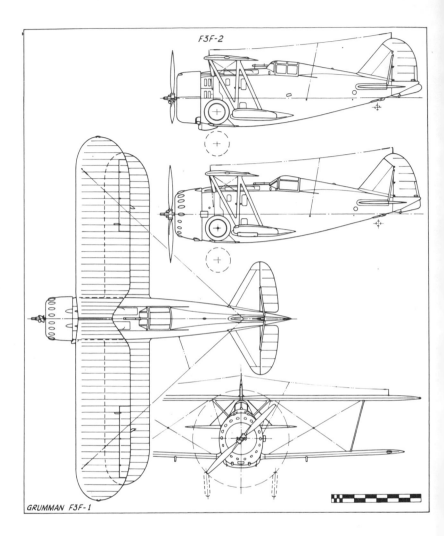

F3F-2

GRUMMAN F3F-1

Squadrons VMF-1 and VMF-2 during 1938. After the 65th production model F3F-2 had been fitted with an up-rated Cyclone as the XF3F-3, the Navy ordered 27 F3F-3s on June 21, 1938—the last of the Grumman fighter biplanes for the Navy and the last biplane fighters for any of the US Armed Services. Delivered from December 1938 onwards, they went to VF-5 on USS *Yorktown* but remained in front-line service for little more than a year. By the end of 1941, none of the Grumman biplanes were still in operational use, but 23 F2Fs and 117 F3Fs were distributed around a number of Naval Air Stations in the USA, where they remained in use as station hacks for some while longer.

TECHNICAL DATA (F2F-1)

Manufacturer: Grumman Aircraft Engineering Corporation, Bethpage, LI, NY.
Type: Carrier-based fighter.
Accommodation: Pilot only.
Power plant: One 700 hp Pratt & Whitney R-1535-72.
Dimensions: Span, 28 ft 6 in; length, 21 ft 5 in; height, 9 ft 1 in; wing area, 230 sq ft.
Weights: Empty, 2,691 lb; gross, 3,847 lb.
Performance: Max speed, 231 mph at 7,500 ft; initial climb, 2,050 ft/min; service ceiling, 27,100 ft; max range, 985 st miles.
Armament: Two fixed forward-firing 0·30-in guns.
Serial numbers:

XF2F-1: 9342.	F2F-1: 9623–9676; 9997.

TECHNICAL DATA (F3F-3)

Manufacturer: Grumman Aircraft Engineering Corporation, Bethpage, LI, NY.
Type: Carrier-based fighter.
Accommodation: Pilot only.
Power plant: One 950 hp Wright R-1820-22.
Dimensions: Span, 32 ft; length, 23 ft 2 in; height, 9 ft 4 in; wing area, 260 sq ft.
Weights: Empty, 3,285 lb; gross, 4,795 lb.
Performance: Max speed, 264 mph at 15,200 ft; initial climb, 2,750 ft/min; service ceiling, 33,200 ft; range, 980 st miles.
Armament: Two fixed forward-firing 0·30-in guns.
Serial numbers:

XF3F: 9727.	F3F-2: 0967–1047.
F3F-1: 0211–0264.	XF3F-3: 1031.
XF3F-2: 0452.	F3F-3: 1444–1470.

Grumman F3F-2 showing the large diameter, parallel chord cowling of the Wright Cyclone engine. (*David C. Cooke*)

201

Grumman J2F-1 attached to the aircraft carrier USS *Saratoga*, in the grey warpaint adopted in February 1941. (*Peter M. Bowers*)

Grumman JF, J2F Duck

At the same time that the newly founded Grumman company began work on its first aircraft, the FF-1 Navy fighter (see page 195), efforts were being made to improve the design of the amphibious floats used on various Naval types. This was a direct continuation of the experience gained by Leroy Grumman as an associate of Grover Loening in the latter's company. Until the early thirties, a series of Loening amphibians served the Navy in observation and utility roles, and it was to replace these that Grumman developed a new utility amphibian combining major features of the FF-1 with some characteristics of the Loening types. An important new feature was that the main wheels retracted into the sides of the single main float, an idea first used by Grumman in floats for other Navy types.

The first of the Grumman amphibians was designated XJF-1 and flew for the first time on May 4, 1933. Powered by a 700 hp Pratt & Whitney R-1830 engine, the XJF-1 was normally a two-seater, with an enclosure over the tandem cockpits. The type was intended for utility duties aboard aircraft carriers and to provide a ship–shore link, and an order for 27 production model JF-1s soon followed the prototype trials. Delivery began late in 1934, and, after replacing OL-9s, some JF-1s began to reach VJ squadrons in 1936, replacing patrol and torpedo types used previously. In addition, the US Coast Guard ordered 14 of these amphibians with special equipment and engine changed to the 750 hp Wright Cyclone. These were designated JF-2s (Grumman G-9); four were later transferred from the Coast Guard to the Navy, and five similarly engined JF-3s were ordered by the Navy.

The next version of the Grumman amphibian, appearing in 1937, was designated J2F-1 (G-15), but the differences from JF-1 were of a minor nature, the most obvious external clue being deletion of the inter-aileron strut between the wings. Equally small changes distinguished subsequent models procured up to 1941. After 20 J2F-1s came 21 J2F-2s, 20 J2F-3s and 32 J2F-4s. In addition, nine J2F-2As were produced, with machine guns and bomb racks, for operation by Marine Squadron VMS-3 at St Thomas, Virgin Isles.

Expansion of the Navy at the end of 1940 under the threat of imminent war brought Grumman an order for 144 J2F-5s and official confirmation of the name Duck, which was already in popular use. The new model had an 850 hp R-1820-50 engine and long-chord cowling, with an overall im-

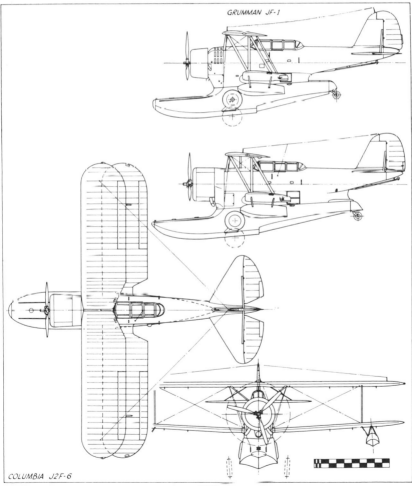

GRUMMAN JF-1

COLUMBIA J2F-6

203

'A silver-finished Grumman J2F-6 built by Columbia Aircraft Corporation. (*William T. Larkins*)

provement in performance. These were the last Ducks built by Grumman, but the Navy ordered another 330 after Pearl Harbor, and assigned production to the Columbia Aircraft Corporation at Valley Stream, LI. Standard designating practice, which would have required these amphibians to be called JL-1s, was ignored; instead, they were delivered with Grumman designation J2F-6. The engine was the 900 hp R-1820-54, and there was a further small gain in top speed.

In its different versions, the Duck served throughout World War II and provided Grumman with a solid basis for development of its later amphibians for the Navy and other users.

TECHNICAL DATA (J2F-5)

Manufacturer: Grumman Aircraft Engineering Corporation, Bethpage, LI, NY; and Columbia Aircraft Corporation, Valley Stream, LI, NY.
Type: Utility amphibian.
Accommodation: Pilot, observer and, optional, radio operator.
Power plant: One 850 hp Wright R-1820-50.
Dimensions: Span, 39 ft; length, 34 ft; height, 15 ft 1 in; wing area, 409 sq ft.
Weight: Empty, 4,300 lb; gross, 6,711 lb.
Performance: Max speed, 188 mph; cruising speed, 150 mph; initial climb, 1,500 ft/min; service ceiling, 27,000 ft; range, 780 st miles.
Serial numbers:

XJF-1: 9218.	J2F-2A: 1198–1206.
JF-1: 9434–9455; 9523–9527.	J2F-3: 1568–1587.
JF-2: V135–V148; (0266, 00371–2;	J2F-4: 1639–1670.
01647, to USN from USCG).	J2F-5: 00659–00802.
JF-3: 9835–9839.	J2F-6 (Columbia): 32637–32786;
J2F-1: 0162–0190.	33535–33614; 36935–37034.
J2F-2: 0780–0794; 1195–1197; 1207–	
1209.	

A Grumman F4F-4 in the early-1942 markings with rudder stripes and
large fuselage stars. (*US Navy*)

Grumman F4F Wildcat

Grumman's first monoplane and one of the outstanding Naval fighters
of World War II, the F4F design began in 1935 while the company's latest
biplane for the USN, the XF3F-1, was still undergoing its initial tests. In
November 1935 the USN initiated a formal design competition for a new
carrier-based fighter and during the early months selected Brewster and
Grumman designs for prototype testing. The Brewster F2A-1 (see page
71) was a monoplane—the first adopted by the Navy at the design stage—
while the Grumman XF4F-1 was a biplane, backed by the USN primarily
as an insurance against failure of the monoplane. Ordered on March 2,
1936, the XF4F-1 was an equal-span biplane of 4,500 lb gross weight with
a Pratt & Whitney R-1535-92 engine. However, data on the Brewster aeroplane soon showed that the biplane could not compete with a successful
monoplane, while steady development of the F3F improved its performance nearly to equal that of the projected XF4F-1. Further work on
the latter consequently seemed pointless, and an alternative monoplane
was designed by Grumman. This was ordered by the Navy on July 28,
1936, as the XF4F-2, and in this completely revised form the Grumman
design was destined for far greater success than ever achieved by its
Brewster-designed contemporary.

Powered by a 1,050 hp Pratt & Whitney R-1830-66 Twin Wasp engine,
the XF4F-2 (Grumman G-18) had a gross weight of 5,535 lb and a design
maximum speed of 290 mph. It was a mid-wing all-metal monoplane with
an armament of two 0·50-in guns in the fuselage and provision for two
more in the wings, or two 100 lb bombs beneath the wings. The main
wheels of the undercarriage retracted into the fuselage in typical Grumman
fashion. First flight was made from the Grumman Long Island factory at
Bethpage on September 2, 1937, and company and Navy testing occupied
the next nine months. Official trial results credited the XF4F-1 with a top
speed of 290 mph, 10 mph better than the F2A-1, but in several other

respects it was rated less satisfactory, and the Brewster aeroplane was declared winner of the design contest in June 1938.

The Grumman design had appeared too promising to neglect altogether, however, and in October 1938 the Navy contracted for a modified prototype to be powered by a version of the Twin Wasp with a two-stage two-speed supercharger, the XR-1830-76. In addition to the new engine installation, this modified XF4F-3 (Grumman G-36) had increased wing span with blunt tips, revised tail surfaces and higher gross weight. The first flight of the XF4F-3 was made on February 12, 1939, and in subsequent Navy trials a speed of 333·5 mph was recorded at 21,300 ft, ample evidence of the improvement bestowed by the two-stage supercharger. Although engine cooling proved to be a major problem, the potential of the Grumman fighter was now clearly established and in August 1939 the Navy ordered 54 F3F-3s. Work, in fact, had already been started by Grumman in anticipation of the order, and the first production F4F-3 was ready to fly in February 1940. The first two F4F-3s had two 0·30-in fuselage guns and two 0·50s in the wings; subsequently, four 0·50s in the wings became standard. Nos. 3 and 4 off the line were completed in June 1940 with Wright R-1820-40 engines and designated XF4F-5, while Nos. 5 and 8 had armour protection and strengthened landing gear. All production aircraft used a new high tailplane position. A few were subsequently equipped with cameras as F4F-3P reconnaissance fighters and one was fitted with twin floats as the F4F-3S.

Export orders for the Grumman G-36 had been placed in 1939, when France purchased 81 G-36As; this entire order was transferred to Britain in June 1940, and the first aircraft, with its British name of Martlet I, was delivered on July 27, 1940, ahead of F4F-3 deliveries to the USN. By December 1940, 22 F4F-3s had been accepted by the Navy, and initial deliveries were being made to VF-4 (USS *Ranger*) and VF-7 (USS *Wasp*) at Norfolk Naval Air Station in Virginia. During 1941, VF-42 and VF-71 were equipped with F4F-3s as well as Marine Squadrons VMF-121, 211 and 221. In addition to further contracts for the F4F-3, Grumman received a USN order for 95 F4F-3As, these being powered by R-1830-90 engines with single-stage superchargers. A prototype installation of this

Grumman F4F-3A in the overall light grey camouflage used in 1941. (*Grumman photo*)

Grumman F4F-4 in late-1942 markings after deletion of the red centre from the star. (*US Navy*)

Grumman F4F-4s with markings used from July to September 1943, including a red border round the national markings. (*US Navy*)

General Motors-built FM-2 in predominantly white colouring. (*US Navy*)

engine had been made in the single XF4F-6 late in 1940 when the two-stage blower was still giving trouble; production of the small batch of F4F-3A fighters was largely an insurance against failure of the newer engine. The F4F-3A was used by Navy Squadron VF-6 and Marine unit VMF-111.

By the end of 1941 the USN and USMC together had 183 F4F-3s and 65 F4F-3As. Most of these were in the States or aboard *Ranger* and *Wasp*, but two Marine squadrons were at Ewa, Hawaii, when the Japanese attack was launched and nine aircraft were lost on the ground. Later that day a detachment of the same squadron, VMF-211, lost seven more aircraft on the ground at Wake Island. In the ensuing battle for Wake, five remaining F4F-3s scored a number of victories over Japanese bombers and fighters before they fell to the vastly superior strength of the Japanese attack force. These were the first combat operations by Wildcats, the name adopted for the F4F by the USN, although RN Martlets had earlier been in action against German aircraft off the coasts of Britain.

Production of the F4F-3 totalled 285 (plus the 95 F4F-3As), all by Grumman. They were followed by 1,169 examples of the F4F-4 (G-36B), a progressive development which differed primarily in having folding wings. The prototype XF4F-4, final aircraft on the first F4F-3 contract, had hydraulic wing-folding but the production version had manual folding to avoid the complication of additional hydraulics in the aircraft. The XF4F-4 first flew on April 14, 1941, and five production aircraft were delivered the same year; the remainder followed in 1942. Grumman also produced a long-range unarmed reconnaissance version of the Wildcat, the F4F-7. With a gross weight of 10,328 lb, the F4F-7 had a range of well over 3,500 miles; the first of 100 on order flew on December 30, 1941, but production was cancelled after 20 more had been built during 1942.

Supplementing the F4F-3s in service during 1942, the F4F-4s were soon in action in the Pacific, participating notably in the battles of the Coral Sea and Midway. With the launching of the Marines' attack on Guadalcanal, the pace of the war in the air for the F4Fs hotted up still further, and Navy and Marine units, including VF-5 (USS *Saratoga*), VF-6 and VF-10 (*Enterprise*), VF-71 (*Wasp*), VF-72 (*Hornet*) and VMF-112, VMF-121, VMF-212, VMF-223 and VFM-224, were in constant action until the F4U and F6F began to appear during 1943. The US Navy also took the Wildcat into operation in North Africa in November 1942, with such units as VF-41 on *Ranger*, VGF-27, VGF-28 and VGF-30 on the escort carrier *Suwannee* and VGF-26 on *Sangamon*.

On April 18, 1942, Eastern Aircraft was made a second production source for the F4F with a contract for 1,800 examples of the F4F-4 designated FM-1, leaving Grumman free to concentrate upon the F6F Hellcat. Eastern, a group of five factories previously assembling General Motors cars, flew its first FM-1 on August 31, 1942, and delivered some 840 in the first 12 months, plus about 300 to Britain as Martlet Vs in 1942–3. The advent of the small CVE or escort carrier led to development of the final Wildcat version, the FM-2. This combined the more powerful

Wright R-1820-56 Cyclone engine with a lighter airframe to obtain improved take-off performance from the shorter carrier decks. Grumman built two prototypes as XF4F-8s, the first of these being flown on November 8, 1942, at Bethpage; a production contract, initially for 1,265 examples designated FM-2, was placcd with Eastern early in 1943. The FM-2 had a taller fin to counteract the increased engine power, and later aircraft in the

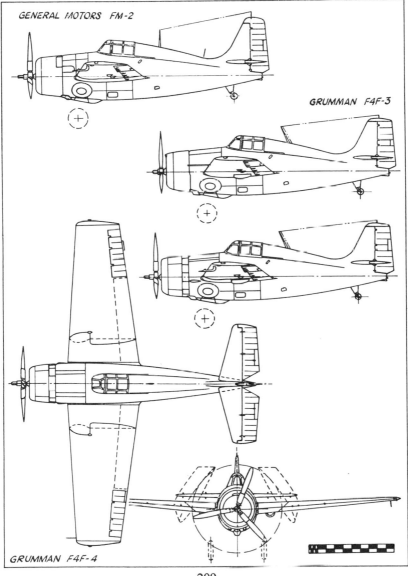

GENERAL MOTORS FM-2

GRUMMAN F4F-3

GRUMMAN F4F-4

batch had a water-injection system. Eventually, Eastern built 4,127 FM-2s for the USN and 340 for Britain as Wildcat VI (the original British name of Martlet was dropped in favour of Wildcat in March 1944). The FM-2s became standard equipment on the majority of the 114 escort carriers put into service by the end of the war.

A contemporary of the Japanese Zero, the F4F in fact had an inferior performance in several respects, yet proved capable of holding its own thanks to its superior armament, rugged construction and well-trained pilots. During 1942 the ratio of victories to losses for air combat for the F4F was 5·9 : 1, and for the whole war the official figure for the F4F/FM was 6·9 : 1. A large proportion of these victories was obtained against bombers and transports, but the figures show that the Wildcat was not unsuccessful, especially in the first half of the war in the Pacific, when the Grumman design was the sole carrier-based fighter operating with the USN.

TECHNICAL DATA (F4F-4)

Manufacturer: Grumman Aircraft Engineering Corporation, Bethpage, LI, NY.
Type: Carrier-based fighter.
Accommodation: Pilot only.
Power plant: One 1,200 hp Pratt & Whitney R-1830-86.
Dimensions: Span, 38 ft; length, 28 ft 9 in; height, 11 ft 10 in; wing area, 260 sq ft.
Weights: Empty, 5,785 lb; gross, 7,952 lb.
Performance: Max speed, 318 mph at 19,400 ft; cruising speed, 155 mph; initial climb 1,950 ft/min; service ceiling, 34,900 ft; range, 770 st miles.
Armament: Six fixed forward-firing 0·50-in guns.
Serial numbers:

XF4F-2, -3: 0383.
F4F-3: 1844–1845; 1848–1896; 2512–2538; 3856–3874; 3970–4057; 12230–12329.
F4F-3A: 3875–3969.
XF4F-4: 1897.

F4F-4: 4058–4098; 5030–5262; 01991–02152; 03385–03544; 11655–12227.
XF4F-5: 1846–1847.
XF4F-6: 7031.
F4F-7: 5263–5283.
XF4F-8: 12228–12229.

TECHNICAL DATA (FM-2)

As F4F-4 except as follows:
Manufacturer: General Motors Corporation, Eastern Aircraft Division, Trenton, NJ.
Power plant: One 1,350 hp Wright R-1820-56.
Dimensions: Length, 28 ft 11 in; height, 11 ft 5 in.
Weights: Empty, 5,448 lb; gross, 8,271 lb.
Performance: Max speed, 332 mph at 28,800 ft; cruising speed, 164 mph; initial climb, 3,650 ft/min; service ceiling, 34,700 ft; range, 900 st miles.
Armament: Four fixed forward-firing 0·50-in gun. Two 250 lb bombs or six 5-in rockets.
Serial numbers:

FM-1: 14992–15951; 46738–46837.

FM-2: 15952–16791; 46838–47437; 55050–55649; 56684–57083; 73499–75158; 86297–86973.

Grumman JRF-5 in end-of-war overall blue finish. (*Harold G. Martin*)

Grumman JRF Goose

The first of a long line of amphibian flying-boats produced by Grumman for military and commercial use appeared in 1937 as the G-21. Intended as a general purpose transport with six/seven seats, the G-21 was of typical flying-boat appearance with a high wing, two-step hull, fixed underwing floats and two 450 hp Pratt & Whitney engines. For land use, the aircraft had a tailwheel type undercarriage, all three wheels retracting into the hull. With increased weights and up-rated engines, the amphibian went into production for commercial use as the G-21A.

In 1938 the Navy ordered a G-21 for evaluation as the XJ3F-1, but when 20 more were ordered the designation was changed to JRF in the utility transport category. The first ten were JRF-1s, and delivery began late in 1939; five of these seven-seat transports were adapted for target-towing and photography as JRF-1As. Ten more intended to be similar became JRF-4s before delivery, and the Navy later acquired two more JRF-1s and two JRF-4s from miscellaneous sources. The JRF-4s had provision for two 250 lb bombs or depth-charges beneath the wings.

Several of the Grumman amphibians procured by the Navy were used by the Marine Corps, and ten more were purchased in 1939–40 for use by the Coast Guard. The latter comprised seven JRF-2s and three JRF-3s, the last being for use in Northern waters with anti-icing equipment and autopilot.

Starting in 1941, Grumman began to deliver JRF-5s, the principal production version of the Goose. This had later engines, cameras for air survey work and other refinements. Production totalled 185, including two

211

supplied to the Coast Guard and four to Britain as Goose I. For lend–lease delivery to Britain, the Navy also procured the JRF-6B, similar to the JRF-5 but equipped as a navigation trainer. At least 50 JRF-6Bs were assigned to Britain as Goose 1As, but Navy records indicate that only 37 of these were built; they served primarily at Piarco in the West Indies.

TECHNICAL DATA (JRF-5)

Manufacturer: Grumman Aircraft Engineering Corporation, Bethpage, LI, NY.
Type: Utility transport amphibian.
Accommodation: Crew of two–three and four–seven passengers.
Power plant: Two 450 hp Pratt & Whitney R-985-AN-6s.
Dimensions: Span, 49 ft; length, 38 ft 6 in; height, 16 ft 2 in; wing area, 375 sq ft.
Weights: Empty, 5,425 lb; gross, 8,000 lb.
Performance: Max speed, 201 mph at 5,000 ft; cruising speed, 191 mph at 5,000 ft; initial climb, 1,100 ft/min; service ceiling, 21,300 ft; range, 640 st miles.
Serial numbers:

XJ3F-1: 1384.
JRF-1: 1674–1677; 1680; 07004; 09782.
JRF-1A: 1671–1673; 1678–1679.
JFR-2: V174–176; V184–187.
JRF-3: V190–192.
JRF-4: 3846–3855; 09767; 35921.

JRF-5: 6440–6454;   04349–04358; 34060–34094;   37771–37831; 39747–39748;   48229;   84790–84818; 87720–87751.
JRF-6B: 66325–66361.

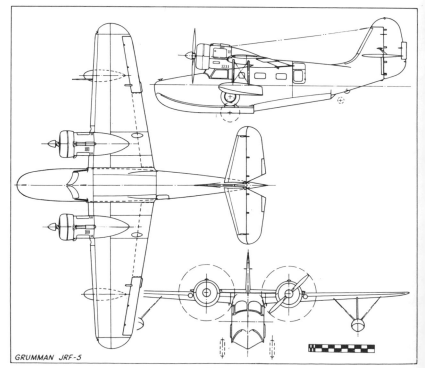

GRUMMAN JRF-5

212

Grumman TBF-1 in green and light grey finish. (*Grumman photo*)

Grumman TBF, TBM Avenger

Operational for the first time on June 4, 1942, in the midst of the Battle of Midway, the Grumman Avenger can scarcely be said to have opened its active career auspiciously. Of the six TBF-1s launched into battle from Midway early that morning, five failed to return; the sixth came back with only the trim tab for longitudinal control, with one wheel and the torpedo-bay doors hanging open and with one gunner dead and the other wounded. Despite this sad beginning, the Avenger was destined to become the Navy's standard torpedo-bomber throughout World War II and to remain in operational Fleet service in a variety of roles until 1954.

When Grumman received a Navy contract for two prototypes of a torpedo-bomber designated XTBF-1, on April 8, 1940, the company had no previous experience of an aircraft in this category, but had been specializing for several years in the design of Navy fighters. The TBF took shape as a chunky mid-wing monoplane with a weapons-bay large enough for a torpedo or 2,000 lb bomb, and a dorsal turret at the rear end of the long cockpit glasshouse. In addition to the 0·50-in gun in this turret, there was a forward-firing 0·50 in the engine cowling, and a 0·30-in tunnel gun.

First flight of the XTBF-1 was made on August 1, 1941, a production contract for 286 examples having been placed with Grumman the previous December. Following acceptance of the second prototype in December 1941, the first production TBF-1 appeared in January 1942, and in the first half of the year, Grumman delivered 145. Pilots from Torpedo Squadron Eight (VT-8) checked out on the TBFs at Norfolk Naval Air Station in May and flew the first six aircraft to Pearl Harbor. They were destined for the USS *Hornet*, on which the remainder of VT-8 was equipped with TBD-1s, but the ship was already at sea. On June 1, 1942, the six aircraft flew on across the Pacific to Midway and, as described above, all but one were destroyed four days later.

213

Grumman TBF-1C in 1944 colours. (*Grumman photo*)

To meet growing production requirements, General Motors Corporation was asked to establish a second source for Avengers at its Eastern Aircraft division, already building Grumman F4Fs. The first contract was placed with Eastern on March 23, 1942, and deliveries began in November the same year, this version being designated TBM-1. Grumman production continued until early 1944, with a total of 2,290. These were primarily of the TBF-1 or TBF-1C version, the latter having provision for two 0·50-in wing guns; also included in the total were 395 TBF-1Bs for lend–lease to Britain, and prototypes of the TBF-2 and TBF-3, respectively with XR-2600-10 and R-2600-20 engines. Special-purpose versions of the Grumman-built Avengers were introduced in 1944 and 1945 through modification programmes and included the TBF-1D with special radar, the TBF-1CP with a trimetrogen camera for reconnaissance, the radar-equipped TBF-1E, the TBF-1J equipped for bad-weather flying, and the TBF-1L with a searchlight in the bomb-bay.

Eastern produced 2,882 TBM-1s, and variants were designated similar to those of the TBF-1. Plans to produce the TBF-2 as a standardized model were abandoned, and Eastern went on to build 4,664 TBM-3s to bring the grand total of Avengers to 9,836. In all, Britain received 921 of these, and 63 went to the RNZAF. The TBM-3D, TBM-3E and TBM-3L

Post-war modification of a General Motors-built TBM-3W2 with ventral radome and rear turret removed.

214

were respectively comparable to the TBM-1D, TBM-1E and TBM-1L apart from the up-rated engine; the TBM-3P carried cameras, and the TBM-3H had special search radar. The final Avenger version was to have had a strengthened (5g) airframe but only a single prototype, the XTBM-4, appeared before the war ended, over 900 being cancelled.

Special anti-submarine detection radar in the TBM-3E made this

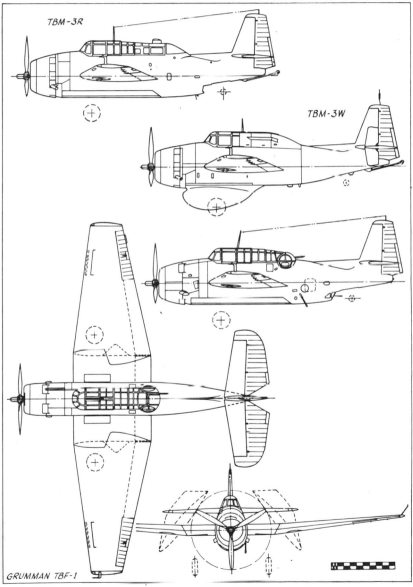

TBM-3R

TBM-3W

GRUMMAN TBF-1

variant of particular value to the Navy in the early post-war years, and it became the principal operational version of the Avenger after 1945. Improved search capability was bestowed by the APS-20 radar carried in a large ventral radome by the TBM-3W and TBM-3W-2, which had the rear turret removed and other changes. These versions were eventually paired with the TBM-3S and TBM-3S-2 submarine-strike Avengers, and these search-and-strike aircraft remained in operational service until June 1954.

Other post-war variants of the Avenger included the target-towing TBM-3U, the TBM-3N for special night operations, the TBM-3Q for radar countermeasures and the TBM-3R, a seven-seat transport for COD (carrier on-board delivery). Several of these variants flew on with the Navy for years after the front-line operational career of the Avenger had ended.

TECHNICAL DATA (TBF-1)

Manufacturer: Grumman Aircraft Engineering Corporation, Bethpage, LI, NY.
Type: Torpedo-bomber.
Accommodation: Pilot, gunner, radar operator.
Power plant: One 1,700 hp Wright R-2600-8.
Dimensions: Span, 54 ft 2 in; length, 40 ft; height, 16 ft 5 in; wing area, 490 sq ft.
Weights: Empty, 10,080 lb; gross, 15,905 lb.
Performance: Max speed, 271 mph at 12,000 ft; cruising speed, 145 mph; initial climb, 1,430 ft/min; service ceiling, 22,400 ft; range, 1,215 st miles.
Armament: One fixed forward-firing 0·30-in gun (two 0·50-in in TBF-1C); one dorsal 0·50-in; one ventral 0·30-in. Up to 1,600 lb in bomb-bay.
Serial numbers:

XTBF-1: 2539–2540.	TBF-1C: 24242–24340; 24342–24520;
TBF-1: 00373–00658; 01731–01770;	47638–48123.
05877–06491; 23857–24140;	XTBF-2: 00393.
24142–24241; 47438–47637.	XTBF-3: 24141; 24341.

TECHNICAL DATA (TBM-3E)

Manufacturer: General Motors Corporation, Eastern Aircraft Division, Trenton, NJ.
Type: Torpedo-bomber.
Accommodation: Pilot, gunner, radar operator.
Power plant: One 1,900 hp Wright R-2600-20.
Dimensions: Span, 54 ft 2 in; length, 40 ft 11½ in; height, 16 ft 5 in; wing area, 490 sq ft.
Weights: Empty, 10,545 lb; gross, 17,895 lb.
Performance: Max speed, 276 mph at 16,500 ft; cruising speed, 147 mph; initial climb, 2,060 ft/min; service ceiling, 30,100 ft; range, 1,010 st miles.
Armament: Two fixed forward-firing 0·50-in guns; one dorsal 0·50-in; one ventral 0·30-in. Up to 2,000 lb in bomb-bay.
Serial numbers:

TBM-1: 24521–25070.	TBM-2: 24580.
TBM-1C: 16792–17091; 25071–25174;	XTBM-3: 25175; 25521; 25700; 45645.
25176–25520; 25522–25699;	TBM-3: 22857–23656; 53050–53949;
25701–25720; 34102–34105;	68062–69538; 85459–86292.
45445–45644; 45646–46444;	91107–91752.
73117–73498.	XTBM-4: 97673–97675.

Grumman F6F-3 in mid-1943 with red-bordered markings. (*Grumman photo*)

Grumman F6F Hellcat

Destined to become the most significant fighter in the US Navy's armoury during World War II, the F6F Hellcat, on which work began during 1941, was a logical extrapolation of the same design team's F4F Wildcat (see page 205). At that time the first monoplane fighters were already in service with the Navy, and pilot experience was becoming available to Grumman designers. This, and some feed-back of information on air fighting in the European war theatre, helped to shape the new design, which the Navy ordered on June 30, 1941. Two prototypes were covered by the initial contract; respectively powered by a 2,000 hp Wright R-2800-10W and a supercharged R-2800-21, they were to be designated XF6F-1 and XF6F-2. Subsequent developments and design changes led to a change of designation to XF6F-3 for the first prototype before its first flight, and the second eventually appeared as the XF6F-4.

Construction of the XF6F-3 proceeded rapidly in the Bethpage, Long Island, factory and the first flight was made by Selden A. Converse on June 26, 1942. Large production orders had been placed in May, and the first production F6F-3 followed the prototype only five weeks later, making its first flight on July 30, 1942. Deliveries began early in 1943, with VF-9 receiving the first aircraft aboard the USS *Essex* only 18 months after the first experimental contract had been signed. Operational use of the Hellcat began on August 31, 1943, when VF-5, flying from USS *Yorktown*, took part in an attack on Marcus Island. The standard aircraft carried an armament of six 0·50-in guns in the wings with 400 rounds per gun, and a drop tank under the fuselage.

Re-equipment of Wildcat squadrons proceeded rapidly during 1943 and by mid-1944 the Hellcat and its Chance Vought contemporary the Corsair (see page 403) had become standard US Navy equipment throughout the Pacific. In June 1944, Hellcats achieved a major victory against Japanese forces in the Battle of the Philippine Sea, setting the seal upon their operational success. Production by that time, in a new Grumman plant built for the purpose, had totalled 4,423 F6F-3s; in this total there were 205 F6F-3N night fighters with APS-6 radar in a pod under the starboard wing, and 18 F6F-3Es similarly equipped with APS-4 radar.

During 1944 deliveries began of a new Hellcat version, the F6F-5, with a number of detail refinements and improvements. The R-2800-10W engine (with water injection) was retained, but the cowling was modified, and the windshield was also improved. Provision was made for 2,000 lb of bombs under the centre section and six rockets under the outer wings, and 20 mm cannon usually replaced the inner machine guns on this model. Production of this version totalled 6,681, plus 1,189 F6F-5Ns with APS-6 in the pod on the starboard wing. The Royal Navy received 252 F6F-3s and 930 F6F-5s which it operated as Hellcat I and Hellcat II respectively.

Official statistics show that US Navy carrier-based Hellcats were credited with the destruction of 4,947 enemy aircraft in air-to-air combat, with another 209 claimed by land-based Navy and Marine units, almost 75 per cent of all the Navy's air-to-air victories. Production ended in November 1945 with a grand total of 12,275. For an aircraft built in such large quantities, the Hellcat underwent remarkably little design development, and few variants reached the hardware stage. As already noted, the original second prototype eventually appeared in March 1943 as the XF6F-4 with an R-2800-27 engine. Two XF6F-6s were also built, with 2,100 hp R-2800-18W engines; the first of these flew on July 6, 1944.

The F6F-5s and F6F-5Ns remained in service for a number of years after the war's end with Navy and Reserve units, and some were modified with cameras as F6F-5Ps. A number also became F6F-5Ks as target drones for missile tests and some of these were used in little-publicized operations by Guided Missile Unit 90 during the Korean War. Based on the USS *Boxer*,

The night-fighter Grumman F6F-5N showing the starboard wing radome. (*Grumman photo*)

218

this unit launched six attacks with explosive-laden F6F-5Ks against North Korean targets, the first on August 28, 1952, with Douglas ADs flying as control aircraft. A few F6F-5D drone directors also appeared.

TECHNICAL DATA (F6F-5)

Manufacturer: Grumman Aircraft Engineering Corporation, Bethpage, LI, NY.
Type: Carrier-based fighter.
Accommodation: Pilot only.
Power plant: One 2,000 hp Pratt & Whitney R-2800-10W.
Dimensions: Span, 42 ft 10 in; length, 33 ft 7 in; height, 13 ft 1 in; wing area, 334 sq ft.
Weights: Empty, 9,238 lb; gross, 15,413 lb.
Performance: Max speed, 380 mph at 23,400 ft; cruising speed, 168 mph; initial climb, 2,980 ft/min; service ceiling, 37,300 ft; range, 945 st miles.
Armament: Six fixed forward-firing 0·50-in guns (or two 20 mm and four 0·50-in).
Serial numbers:

XF6F-1, -4: 02981.
XF6F-3: 02982.
F6F-3: 04775–04958; 08798–09047; 25721–26195; 65890–66244; 39999–43137 (including 125 F6F-3N and 18 F6F-3E).

XF6F-4: 02981.
F6F-5, -5N: 58000–58999; 69992–72991; 77259–80258; 93652–94521.
XF6F-6: 70188, 70913.

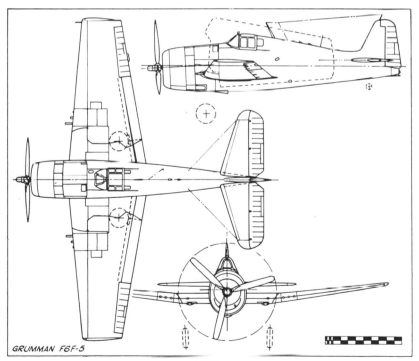

GRUMMAN F6F-5

219

Grumman F7F-3, the principal single-seat version of the Tigercat. (*Grumman photo*)

Grumman F7F Tigercat

Long experience in the production of fighters for the Navy gave Grumman a valuable lead in development of multi-engine fighters for carrier-deck operation. As early as June 30, 1938, the Navy had ordered a prototype twin-engine fighter from the company and, by the middle of 1941, preliminary flight-test data for this aircraft, the XF5F-1 (Grumman G-45) was available. Although the XF5F-1 itself suffered a number of shortcomings, it provided a useful basis for the development of a larger twin-engine fighter ordered from Grumman on June 30, 1941. Designated XF7F-1 (Grumman G-51) this new fighter was intended to be operated from the forthcoming 45,000 ton carriers of the USS *Midway* class. It was

An early production Grumman F7F-1, with large spinners. (*Grumman photo*)
220

The two-seat Grumman F7F-3N night-fighter with nose radome. (*Logan C. Coombs*)

destined to become the Navy's first twin built in production quantities, and the first carrier-based fighter to operate with a tricycle undercarriage.

Although classified as a fighter, the F7F-1 was designed to operate in a tactical ground-support role, for which it was heavily armed. Four 20 mm guns in the wing root had 200 rounds apiece and four 0·50-in guns in the nose had 300 rpg. Underwing strong points could carry 1,000 lb bombs, and provision was made for a standard Navy torpedo to be carried under the fuselage. All this was provided in a single-seater with a short broad-chord wing, powered by two 2,100 hp Pratt & Whitney R-2800-22W engines slung in large nacelles beneath the wings.

The first of the two XF7F-1s flew in December 1943, by which time the pressing needs of Marine Corps squadrons engaged in support of the land fighting in the Pacific had led to an order for 500 of the new fighters. Deliveries began in April 1944, but operational problems and changing requirements led to restrictions in the production programme and delays in deployment of the Tigercat, as the type was named by the Navy.

After 34 single-seat F7F-1s had been delivered, production switched temporarily to a two-seat night-fighting version, the F7F-2N; 65 of the latter were built including the XF7F-2N prototype. Grumman then built 189 F7F-3s, similar to the original single-seat model but with R-2800-34W engines. Further production on the original contract was cancelled, but later orders kept the Tigercat in production until late 1946 with 60 F7F-3Ns and 13 F7F-4Ns delivered. These were both two-seat night fighters, with longer radar-carrying nose, and larger fin. The nose guns were deleted, leaving only the four wing-root cannon, and the F7F-4N variant had arrester gear for carrier operations, the only Tigercat variant so fitted. Post-delivery modification also produced some specially equipped F7F-3Es and some camera-carrying F7F-3Ps. Too late for operational service in World War II, the Tigercat served with a few Marine squadrons after the war but was soon displaced by the advent of jet-powered fighters and fighter-bombers.

TECHNICAL DATA (F7F-3)

Manufacturer: Grumman Aircraft Engineering Corporation, Bethpage, LI, NY.
Type: Carrier-based fighter-bomber.
Accommodation: Pilot only.
Power plant: Two 2,100 hp Pratt & Whitney R-2800-34Ws.
Dimensions: Span, 51 ft 6 in; length, 45 ft 4 in; height, 16 ft 7 in; wing area, 455 sq ft.
Weights: Empty, 16,270 lb; gross, 25,720 lb.
Performance: Max speed, 435 mph at 22,200 ft; cruising speed, 222 mph; initial climb, 4,530 ft/min; service ceiling, 40,700 ft; range, 1,200 st miles.
Armament: Four fixed forward-firing 20 mm guns and four 0·50-in guns. Up to 1,000 lb under each wing; one torpedo under fuselage.
Serial numbers:

XF7F-1: 03549–03550. F7F-3: 80359–80547.
F7F-1: 80259–80260; 80262–80293. F7F-3N: 80549–80608.
XF7F-2: 80261. F7F-4N: 80548; 80609–80620.
F7F-2N: 80294–80358.

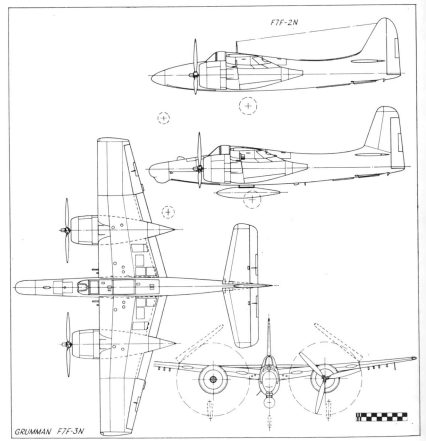

GRUMMAN F7F-3N

222

Grumman F8F-1 with its pilot's medal ribbons reproduced on the cowling.
(*William T. Larkins*)

Grumman F8F Bearcat

One of the best piston-engined fighters to see operational service with the US Navy, the Bearcat was the final development in the line of Grumman fighters started with the F4F Wildcat (see page 205). Although too late to serve in World War II, it played an important part in the war in Indo-China, surplus Navy Bearcats having passed into service with the French *Armée de l'Air* and the Royal Thai Air Force.

Work on the Bearcat, under the Grumman company designation G-58, began in 1943 with the object of providing a high performance derivative of the F6F Hellcat (see page 217) which could operate from the smallest aircraft carrier, primarily in the interceptor role. The overall configuration closely followed that of the Hellcat, but the new design was smaller, despite having the same Pratt & Whitney R-2800 engine. Lightweight design ensured that the G-58 would have an outstanding performance, especially in the climb, but range was necessarily sacrificed. An unusual design feature, which was eventually abandoned, was the provision of break points in the wings, plus explosive bolts, so that if the aircraft was handled too violently in flight, the tips would fail at selected points and a symmetric situation could be maintained.

Two prototypes of the new Grumman design were ordered by the Navy on November 27, 1943, and the designation XF8F-1 was assigned. The first aircraft flew on August 21, 1944, powered by an R-2800-22W engine and the hoped-for performance was confirmed in flight, with an initial rate of climb of 4,800 ft/min and a top speed of 424 mph. The armament comprised four 0·50-in guns in the wings, and wing racks could carry two

223

The Grumman F8F-1B introduced four wing-mounted 20-mm cannon. (*Harold G. Martin*)

1,000 lb bombs or two drop tanks. Contracts for the production of 2,023 F8F-1s were placed with Grumman on October 6, 1944, and General Motors was brought into the programme on February 5, 1945, with a contract for 1,876 Bearcats designated F3M-1. Production aircraft had the R-2800-34W engine and a small increase in fuel capacity.

Grumman began deliveries in February 1945 and the first operational squadron, VF-19, began to equip on May 21, 1945. The end of the war against Japan in August, however, brought cancellation of the GM programme and a cut in Grumman contracts to 770 aircraft. Re-equipment of Navy squadrons with the Bearcat continued through 1946 and 1947, a total of 24 units receiving the type by 1948. In addition to the original (reduced) order, Grumman received contracts for 126 F8F-1Bs, with 20-mm cannon replacing the wing guns; 15 of the original order were completed as F8F-1N night fighters with radome pod under one wing.

A general improvement programme produced the F8F-2 in 1948, with 20 mm cannon, revised engine cowling, taller fin and rudder and other changes. Grumman built 293 of the F8F-2 model, plus 12 night fighter F8F-2Ns and 60 photographic F8F-2Ps carrying only two cannon. Production of the Bearcat ended in May 1949 by which time the F8F-2 equipped a dozen Navy squadrons, with another 12 still flying the F8F-1.

A Grumman F8F-2 showing the heightened fin of this model. (*Harold G. Martin*)

Starting in mid-1949 the Navy began to withdraw its Bearcats from front-line units, the last F8F-2Ps going late in 1952. Some Bearcats were subsequently used as drone control aircraft designated F8F-1D and F8F-2D.

TECHNICAL DATA (F8F-1)

Manufacturer: Grumman Aircraft Engineering Corporation, Bethpage, LI, NY.
Type: Carrier-borne interceptor fighter.
Accommodation: Pilot only.
Power plant: One 2,100 hp Pratt & Whitney R-2800-34W.
Dimensions: Span, 35 ft 10 in; length, 28 ft 3 in; height, 13 ft 10 in; wing area, 244 sq ft.
Weights: Empty, 7,070 lb; gross, 12,947 lb.
Performance: Max speed, 421 mph at 19,700 ft; cruising speed, 163 mph; initial climb, 4,570 ft/min; service ceiling, 38,700 ft; range, 1,105 st miles.
Armament: Four fixed forward-firing 0·50-in machine guns.
Serial numbers:

XF8F-1: 90460–90461; 94753.
F8F-1: 90437–90459 (trials aircraft); 94752; 94754–95498.
XF8F-2: 95049, 95330.

F8F-1B: 122087–122152; 121463–121522.
F8F-2, -2N, -2P: 121523–121792; 122614–122708.

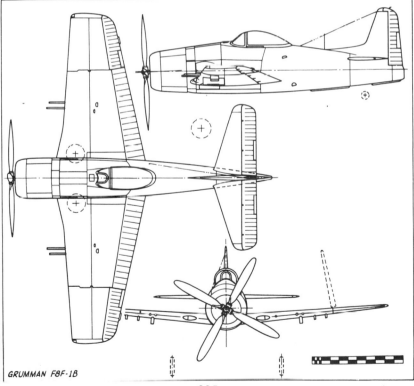

GRUMMAN F8F-1B

225

A Grumman AF-2W with radar for submarine detection.

Grumman AF Guardian

In attempts to evolve a replacement for the highly successful TBF Avenger, Grumman laid out two new projects during 1944. One was twin-engined and was related to the F7F Tigercat, with two R-2800-22 Wasp radials. Allocated the designation XTB2F-1 when the Navy officially recognized the project, it was cancelled a few months later in January 1945 in favour of the composite-powered Grumman G-70. Prototypes were ordered in February 1945 with the designation XTB3F-1.

Following the TBF formula, the XTB3F was a mid-wing design with a single 2,300 hp R-2800-46 Double Wasp engine in the nose. In place of defensive armament, however, the new torpedo-bomber had a Westinghouse 19XB turbojet in the tail to give it a high escape speed. Side-by-side seating was provided for the crew of two, and the weapons-bay could accommodate two torpedoes or a 4,000 lb bomb-load. Two 20 mm cannon in the wings increased the offensive power for attacks on shipping.

The first of two XTB3F-1s flew on December 19, 1945, but the effect of

A production model AF-2S operating in the strike role. (*US Navy*)

the jet boost did not justify continuation of this feature, although the proto-types proved of some value as test beds for jet engines such as the Allis-Chalmers J36 and Westinghouse J34. Removing the jet engine left some spare load-carrying capability which was utilized in a revision of the basic design as an anti-submarine search aircraft, the XTB3F-1S, with a large ventral radar set.

Production models were ordered in two configurations: as the radar-equipped AF-2W and the weapon-carrying AF-2S, operating in hunter-killer pairs. These versions had a 2,400 hp R-2800-48W engine, and the AF-2W carried two additional crew-men in the fuselage to operate the

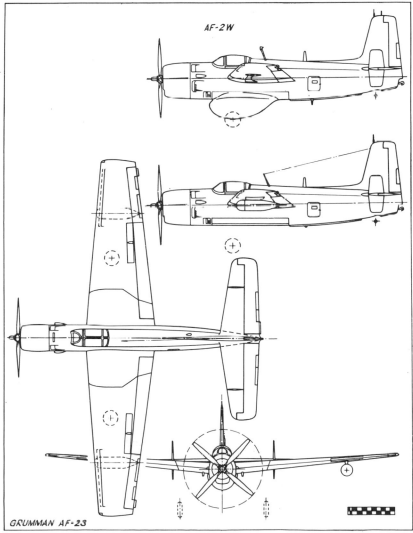

AF-2W

GRUMMAN AF-2S

An operational pair of Guardians with the AF-2W 'hunter' in the foreground and the AF-2S 'killer' behind. (*US Navy*)

radar detection gear. The AF-2S carried a smaller APS-30 radar under one wing, used to pinpoint the target once it had been detected by the accompanying AF-2W, and operated as a three-seater.

First flight of the AF-2 was made on November 17, 1949, and deliveries to VS-25 began in October 1950. Production of the AF-2W totalled 153 and of the AF-2S, 193. Deliveries ended in 1953, in which year the Navy also received 40 AF-3S models, with fuselage-mounted MAD gear. The Guardian teams remained in service from carriers of the US fleet for several more years before being replaced by S2Fs from the same manufacturer.

TECHNICAL DATA (AF-2S)

Manufacturer: Grumman Aircraft Engineering Corporation, Bethpage, LI, NY.
Type: Anti-submarine search or strike aircraft.
Accommodation: Crew of two (-2S) or four (-2W).
Power plant: One 2,400 h.p. Pratt & Whitney R-2800-48W.
Dimensions: Span, 60 ft 8 in; length, 43 ft 4 in; height, 16 ft 2 in; wing area, 560 sq ft.
Weights: Empty, 14,580 lb; gross, 25,500 lb.
Performance: Max speed, 317 mph at 16,000 ft; initial climb, 1,850 ft/min; service ceiling, 32,500 ft; range, 1,500 st miles.
Armament: (-2S only) One 2,000 lb torpedo or two 2,000 lb bombs or two 1,600 lb depth-charges in weapons-bay.
Serial numbers:
XTB3F-1: 90504–90506.
AF-2W: 123089–123117 (odd numbers only); 124187–124209 (odd numbers only); 124779–124849 (odd numbers only); 126738–126755; 126822–126835; 129258–129299; 130389–130404.
AF-2S: 123088–123116 (even numbers only); 124188–124210 (even numbers only); 124778–124848 (even numbers only); 126720–126737; 126720–126737; 126756–126821; 129196–129242.
AF-3S: 129243–129257; 130364–130388.

Grumman UF-1G of the US Coast Guard. (*Harold G. Martin*)

Grumman UF-1 Albatross

Work was started in 1944 on a new general purpose amphibian for the Navy, as a successor to the JRF Goose which served throughout the war years. The Grumman company had the benefit of more than ten years experience in the design and production of amphibians for the Navy, and the new type, as the Grumman Model G-64, was a continuation of the JRF design philosophy. It featured a conventional two-step hull into the sides of which the main wheels retracted, a high wing with fixed stabilizing floats, and a single tail unit.

When the Navy selected the G-64 for construction it assigned the designation XJR2F-1 to the prototype. The JR indicated utility transport, and JR2F-1 was the first of five successive designations applied to these Navy amphibians since the first order was placed. The prototype flew on October 24, 1947, and evaluation flying not only led the Navy to confirm its proposed order but also aroused USAF interest in the type for air–sea rescue duties. This interest eventually materialized in USAF orders under the designation SA-16. After the JR category had been abandoned by the Navy, the Albatross was tentatively redesignated PF-1 in the patrol category, and then became the UF-1 in the Navy's new utility series.

The Navy's first UF-1s were similar in most respects to the prototype, but they also included a small number of UF-1Ls which were winterized for Antarctic operation, and five UF-1Ts with full dual-control for use as trainers. These three types became HU-16C, LU-16C and TU-16C in 1962.

Following USAF practice with its SA-16As, the Navy initiated conversion of its UF-1s to UF-2 standard in 1957. Among the several modifications were an increase of $16\frac{1}{2}$ ft in the wing span, use of fixed leading-edge camber in place of slots, enlarged ailerons and a taller fin and rudder. A general clean-up programme was undertaken, to minimize the drag of

229

aerials. Fifty-one UF-1s were rebuilt to the new standard and were later designated HU-16D. In addition the US Coast Guard had 34 UF-1G converted to UF-2G standard and received a further 37 from the USAF; these aircraft were redesignated HU-16E.

TECHNICAL DATA (UF-2)

Manufacturer: Grumman Aircraft Engineering Corporation, Bethpage, LI, NY.
Type: General purpose amphibian.
Accommodation: Crew of four–six; up to ten stretchers.
Power plant: Two 1,425 hp Wright R-1820-76As or Bs.
Dimensions: Span, 96 ft 8 in; length, 61 ft 3 in; height, 25 ft 10 in; wing area, 1,035 sq ft.
Weights: Empty, 22,883 lb; gross, 35,700 lb.
Performance: Max speed, 236 mph; cruising speed, 150 mph; initial climb, 1,450 ft/min; service ceiling, 21,500 ft; range, 2,850 st miles.
Serial numbers:

XJR2F-1: 82853–82854.
UF-1, -1T: 124374–124379; 131889–131918; 141261–141288; 149822–149824 (MAP); 149836–149837.
UF-1L: 142428.
UF-1G: 137899–137933 (some UF-1); 142358–142362; 142429.
UF-2: 146426–146430 (MAP); 148240–148245; 148324–148329.

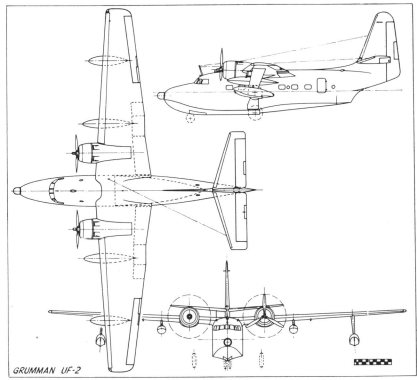

GRUMMAN UF-2

A Grumman F9F-5 with nose refuelling probe. (*Harold G. Martin*)

Grumman F9F-2/5 Panther

Grumman's first jet fighter design began somewhat unconventionally, largely because early jet engine development under Navy sponsorship made a false start. One of the engines under development and available to Grumman designers in the closing days of the war was the 1,500 lb s.t. Westinghouse J30, and in order to obtain the required performance four of these engines were considered necessary in the original Grumman design, which was for a night fighter. A contract for this design, designated XF9F-1, was issued on April 22, 1946, but as the design progressed, doubts arose about the method of installing the four engines in the wings.

Consequently, Grumman decided to use a single engine located more conventionally in the fuselage, and turned to the imported Rolls-Royce Nene which was then rated at 5,000 lb s.t. The new design, for a day fighter, was designated XF9F-2, and the original contract was amended to cover two prototypes.

The first of two XF9F-2s flew on November 24, 1947, and a third prototype was flown on August 16, 1948, as the XF9F-3 with a 4,600 lb s.t. Allison J33-A-8; the latter engine was regarded as an alternative to the Nene, which Pratt & Whitney built under licence as the J42. Production contracts were placed initially for 47 F9F-2s with J42-P-6 engines and 54 F9F-3s with J33-A-8s, the Grumman 'cat' family being perpetuated with selection of the name Panther. Production progressed simultaneously and both variants flew in November 1948, but the -2 proved superior, and J42 production proceeded on schedule, so the F9F-3s were converted to -2s and further contracts were placed to bring the total quantity to 437. First deliveries to an operational unit were made in May 1949, the receiving unit being VF-51. All production aircraft had permanent wingtip tanks, a feature not present on the prototypes.

Use of the Allison J33-A-C6 was planned in 73 F9F-4s, but these aircraft were absorbed in contracts for a total of 655 F9F-5s with the 6,250 lb s.t. J48-P-2, J48-P-4 or J48-P-6A engine. The F9F-5 (and the projected F9F-4) had a 2-ft fuselage extension and a taller fin. The first flew on December 21, 1949, and production included camera-equipped F9F-5Ps.

Flying off the USS *Valley Forge*, F9F-2s became the first Navy jet

231

fighter ever used in combat when they went into action over Korea on July 3, 1950, and on November 9 the same year an F9F pilot became the Navy's first to shoot down a jet aircraft when he destroyed a MiG-15.

After being succeeded by sweptwing developments of the same design (see page 233), the straight-wing Panthers were adapted for special duties including the F9F-5KD which was equipped for use as a target drone or drone controller. The remaining examples in service in 1962 became DF-9Es.

TECHNICAL DATA (F9F-5)

Manufacturer: Grumman Aircraft Engineering Corporation, Bethpage, LI, NY.
Type: Carrier-based fighter.
Accommodation: Pilot only.
Power plant: One 6,250 lb s.t. Pratt & Whitney J48-P-6A turbojet.
Dimensions: Span, 38 ft; length, 38 ft 10 in; height, 12 ft 3 in; wing area, 250 sq ft.
Weights: Empty, 10,147 lb; gross, 18,721 lb.
Performance: Max speed, 579 mph at 5,000 ft; cruising speed, 481 mph; initial climb, 5,090 ft/min; service ceiling, 42,800 ft; range, 1,300 st miles.
Armament: Four fixed forward-firing 20 mm guns.
Serial numbers (allocations):

XF9F-2: 122475, 122477.
F9F-2: 122560–122589; 127086–127215; 123397–123713.
XF9F-3: 122476.
F9F-3: 123016–123083; 123086.
XF9F-4: 123084.
XF9F-5: 123085.

F9F-5: 125414–125443; 125489–125499; 125533–125648; 125893–125912; 125949–126264; 126627–126669; 125080–125321 (some-5P); 125913–125948 (assigned as F9F-4).
F9F-5P: 126265–126290; 127471–127472.

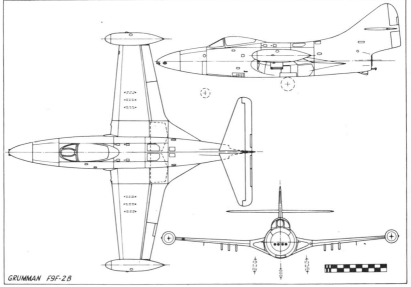

GRUMMAN F9F-2B

232

A Grumman F9F-7 from Navy Squadron VF-21 in 1953. (*US Navy*)

Grumman F9F-6/8 Cougar

Testimony to the excellence of Grumman's first jet fighter, the straight-wing F9F Panther (see page 231), was its successful development into a sweptwing fighter which prolonged production of the basic family for seven years.

Proposals to develop a sweptwing derivative of the Panther were first made a little over a year after the latter entered service, and were endorsed by the Navy with a contract dated March 2, 1951. The new variant, designated XF9F-6 (Grumman G-93) comprised the fuselage and tail unit of the Panther, an up-rated J48-P-8 of 7,250 lb s.t., and a completely new wing with 35 degrees sweepback. To provide the necessary control qualities at high and low speeds, the wing incorporated enlarged trailing-edge flaps and leading-edge slots, wing fences, and spoilers in place of ailerons. Armament, including underwing stores, was unchanged, but the tip tanks were not used.

The first flight of the XF9F-6 was made on September 20, 1951, and a new popular name—Cougar—was selected, despite the use of the same designation series for both straight and sweptwing versions of the design. Navy evaluation was completed during 1952, and the first operational unit, VF-32, began to receive F9F-6s in November that year. Production totalled 706 F9F-6s, plus 168 virtually identical F9F-7s which had Allison J33-A-16A engines. Some Cougars carried cameras for reconnaissance duties and were designated F9F-6Ps; later, a few surplus aircraft became target drones as F9F-6K and F9F-6K2 with different equipment standards, and others were drone directors, designated F9F-6Ds.

233

A prototype Grumman F9F-8P showing the camera installation in the nose. (*US Navy*)

Final single-seat version of the Cougar was the F9F-8, first flown on December 18, 1953. This had a lengthened fuselage (by 8 in) to accommodate additional fuel tanks, changes to the cockpit hood contours and modifications to the wing giving an effective increase of chord of 15 per cent. The 712 built included camera-equipped F9F-8Ps and an attack version, F9F-8B, with AAM or ASM armament. Finally came a two-seat trainer based on the F9F-8; designated F9F-8T, it had a 34-in longer fuselage and tandem seating for pupil and instructor. The first flight was made on April 4, 1956, and the 399 built brought the total quantity of Cougars for the Navy and Marine Corps to 1,985.

Cougars served with many Navy and Marine operational squadrons in succession to the earlier Panthers. They also were the first sweptwing

A Grumman F9F-8T trainer in standard white and orange colours. (*Grumman photo*)

aircraft used by the US Navy's famed Blue Angels aerobatic team, which flew them from 1955 to 1958. All Cougar variants were redesignated in the F-9 series in 1962 as follows: F9F-6 to F-9F, F9F-6D to DF-9F, F9F-6K to QF-9F, F9F-6K2 to QF-9G, F9F-7 to F-9H, F9F-8 to F-9J, F9F-8B to AF-9J, F9F-8P to RF-9J and F9F-8T to TF-9J. Subsequently a new drone conversion appeared as the QF-9J. Some of the -9J versions remained in second-line use by Reserve units or for target operations into the early 'seventies, by which time all earlier variants had been discarded or declared obsolete. The TF-9J training variant was retired by Squadron VT-4 in February 1974.

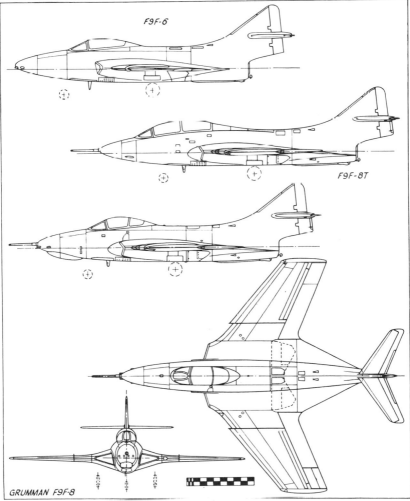

F9F-6

F9F-8T

GRUMMAN F9F-8

Wingtip radomes and new equipment in the nose identify this as an
F9F-6K target drone. (*Dustin W. Carter*)

TECHNICAL DATA (F9F-6)

Manufacturer: Grumman Aircraft Engineering Corporation, Bethpage, LI, NY.
Type: Carrier-based fighter.
Accommodation: Pilot only.
Power plant: One 7,250 lb s.t. Pratt & Whitney J48-P-8 turbojet.
Dimensions: Span, 36 ft 5 in; length, 41 ft 7 in; height, 15 ft.
Weights: Gross, 20,000 lb.
Performance: Max speed, 690 mph at sea level; initial climb, 7 min to 40,000 ft; service
ceiling, 50,000 ft; range, 1,000 st miles.
Armament: Four fixed forward-firing 20 mm guns; two 1,000 lb bombs under wings.
Serial numbers (allocations):

XF9F-6: 126670–126672.

F9F-6: 127216–127470 (some -6P);
128055–128294; 130920–131062.

F9F-6P: 127473–127492; 128295–
128310; 131252–131255;
134446–134465.

F9F-7: 130752–130919.

F9F-8: 131063–131251; 134234–
134244; 138823–138898;
141030–141229; 141648–
141666; 144271–144376.

F9F-8P: 141668–141727; 144377–
144426.

TECHNICAL DATA (F9F-8T)

Manufacturer: Grumman Aircraft Engineering Corporation, Bethpage, LI, NY.
Type: Operational trainer.
Accommodation: Pupil and instructor in tandem.
Power plant: One 7,200 lb s.t. Pratt & Whitney J48-P-8A turbojet.
Dimensions: Span, 34 ft 6 in; length, 44 ft 5 in; height, 12 ft 3 in.
Weights: Gross, 20,600 lb.
Performance: Max speed, 705 mph at sea level; initial climb, 8·5 min to 40,000 ft;
service ceiling, 50,000 ft; range, 600 st miles.
Armament: Two fixed forward-firing 20 mm guns.
Serial numbers:
141667; 142437–142532; 142954–143013; 146342–146425; 147270–147429.

A Grumman F11F-1 Tiger with drop tanks and Sidewinder missiles under the wings. (*Grumman photo*)

Grumman F11F Tiger

Last of the Grumman 'cat' family of fighters for the US Navy, the Tiger was the final development of the F9F design (see page 231) and concluded a quarter-century of Grumman fighter production which had started with the XFF-1 biplane ordered in 1931. Only 20 years separated the original biplane and the preliminary design of the Tiger, which originated as the Model G-98 and was an attempt to obtain the maximum possible performance from an aircraft based on the straight-wing F9F-2. A sweptwing development of the latter had already been built as the F9F-6, and the G-98, while continuing the same design configuration, was an almost wholly new design. Compared with the F9F-6, it had a thinner wing, fuselage-side intakes, fuselage-located landing gear, area-rule fuselage, low-mounted tailplane and afterburning Wright J65 engine.

In this form, the Navy ordered the G-98 on April 27, 1953, initially giving it the designation F9F-8 and changing this to F9F-9 when -8 was used for a later Cougar variant. Production contracts were placed for the -9 fighter and -9P reconnaissance versions. First flight of the YF9F-9 prototype was made on July 30, 1954, with a non-afterburning J65-W-7 engine fitted; the second prototype followed in October, and in January 1955 the latter made the first flights with an afterburner fitted, and a number of other changes. In April 1955 the designation was changed to F11F-1, the first three production aircraft having flown by this time— all without afterburners in the first instance. Recurrent problems with the engine led to production of two F11F-1Fs with General Electric J79-GE-3A engines which bestowed a Mach 2 performance.

After completion of 42 F11F-1s on the first contract (the -1P version having been cancelled) Grumman built 157 on a second order, these having a longer nose with provision for radar which in practice was never installed. Production was completed in December 1958.

Initial deliveries were made in March 1957 to Navy Squadron VA-156,

237

a day-fighter unit despite its Attack designation. Another five squadrons, two in AIRLANT (Naval Air Force, Atlantic Fleet) and the remainder, like VA-156, in AIRPAC (Naval Air Force, Pacific Fleet), were equipped with the Tiger, as was the Navy's Blue Angels aerobatic team. The F11F-1s began to phase out of front-line use during 1959, being reallocated to Advanced Training Command. In 1962, the Tiger was redesignated F-11A.

TECHNICAL DATA (F11F-1)

Manufacturer: Grumman Aircraft Engineering Corporation, Bethpage, LI, NY.
Type: Day fighter.
Accommodation: Pilot only.
Power plant: One 7,450 lb s.t. Wright J65-W-18 turbojet.
Dimensions: Span, 31 ft 7½ in; length, 46 ft 11¼ in; height, 13 ft 2¾ in; wing area, 250 sq ft.
Weights: Empty, 13,428 lb; gross, 22,160 lb.
Performance: Max speed, 750 mph at sea level; cruising speed, 577 mph at 38,000 ft; initial climb, 5,130 ft/min; service ceiling, 41,900 ft; range, 1,270 st miles.
Armament: Four fixed forward-firing 20 mm guns; four underwing Sidewinder 1A or 1C air-to-air missiles.
Serial numbers:
 F11F-1: 138604–138645; 141728–141884.   F11F-1F: 138646–138647.

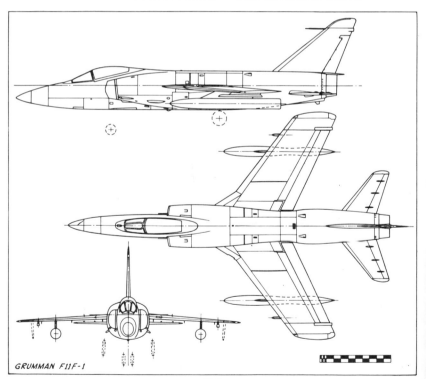

GRUMMAN F11F-1

An early production Grumman S2F-1 with radome and MAD gear extended. (*Grumman photo*)

Grumman S-2 Tracker, C-1 Trader, E-1 Tracer

The advent of missile-armed nuclear submarines had significant and far-reaching effects upon the nature and number of aircraft employed for anti-submarine duties. The key to defence against submarines is the ability to detect their presence, and the development of a range of sophisticated new detection equipment placed new demands upon the aircraft which carried it. At the same time, new weapons appeared, including air-to-underwater guided missiles, which had to be accommodated together with associated electronic systems.

These processes soon rendered out of date the anti-submarine hunter-killer teams of the immediate post-war years, such as the TBM-3S/TBM-3W and AF-2S/AF-2W. To succeed them, the Navy initiated on June 30, 1950, a programme to produce a new type combining both roles in one airframe. Key requirements were the ability to carry all the necessary weapons and equipment plus fuel for long search missions at low

Grumman S-2G Tracker of VS-31, with the MAD boom extended. (*US Navy*)

239

altitude, in an aeroplane suitable for operation from carrier decks. The Navy's choice to meet its specification was the Grumman G-89, a twin-engine high-wing monoplane powered by two 1,525 hp Wright R-1820-82WA piston engines. Weapons were carried in a fuselage bay, and bays in the rear of the engine nacelles carried sono-buoys. Detection equipment included APS-38 search radar in a retractable radome in the rear fuselage, a 70-million candlepower searchlight on the starboard wing and an ASQ-10 magnetic anomaly detector in a retractable fairing in the rear fuselage. Additional weapons or stores could be carried on three strong points under each wing.

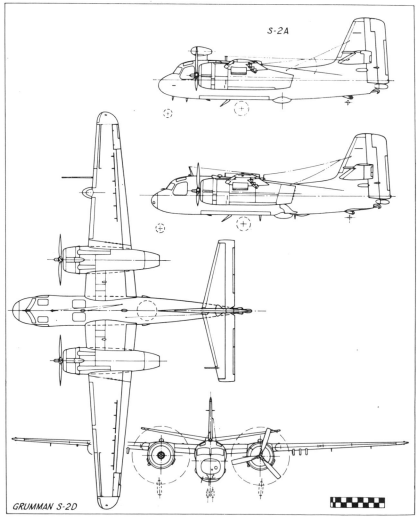

When first ordered, the prototype was designated XS2F-1, but all aircraft in the S2F Tracker series were redesignated in the S-2 series in 1962. The first flight was made on December 4, 1952, and the first production S-2A (S2F-1) entered service with Anti-Submarine Squadron VS-26 in February 1954. Production of this version totalled 755, including over 100 supplied to foreign nations under MAP arrangements. Post-delivery modifications produced a number of TS-2As (S2F-1T) for use by training squadrons, over 50 US-2A target tugs to replace the Navy's UB-26Js, and a quantity of S-2Bs (S2F-1S), which differed from the S-2A in having AQA-3 Jezebel passive long-range acoustic search equipment and its associated Julie explosive echo-sounding equipment.

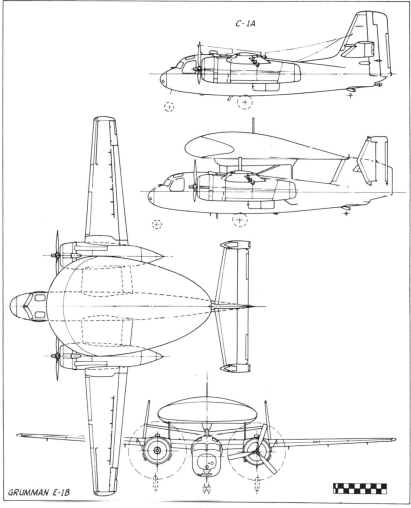

C-1A

GRUMMAN E-1B

241

This Grumman C-1A Trader bears the markings of VR-24; the US flag on the tail indicates that it is operating from a foreign base. (*Stephen P. Peltz*)

Sixty S-2Cs (S2F-2) delivered in 1954–5 had enlarged bomb-bays in order to carry a new type of homing torpedo, and the tail surfaces were enlarged to match the higher operating weights. A few of this batch, with cameras, were designated RS-2C (S2F-2P), and others used on utility duties were US-2C (S2F-2U). The second major production version was the Model G-121 S-2D (S2F-3), first flown on May 20, 1959. This had an enlarged front fuselage to improve working space and comfort for the two pilots and two radar operators; the span was increased, tail surfaces enlarged, fuel capacity increased and engine nacelles modified to carry 16 instead of eight sono-buoys each. Operational use of the S-2D began in May 1961, and the S-2E followed a little later, this being the S-2D with Julie-Jezebel equipment and a tactical navigation system. Production of the S-2D totalled 119, followed by 219 S-2Es.

Further up-dating of the early S-2Bs with the Julie/Jezebel installation resulted in the S-2F (S2F-1S1) designation being applied in 1962. Later utility conversions of the S-2A with some features of the S-2B/S-2F became US-2Bs. Finally, 50 S-2Fs were S-2Es converted to have AN/AQA-7 DIFAR sonobuoy processing equipment.

A derivative of the Tracker appeared in 1955 as the TF-1 (G-96) later becoming the C-1A Trader. This had a new fuselage with accommodation for nine passengers and was designed as a carrier on-board delivery transport. Production totalled 87 and included four G-125 EC-1As (TF-1Q) modified for electronic countermeasures missions.

A Grumman E-1B Tracer landing aboard the USS *Essex*. (*Stephen P. Peltz*)

242

Grumman US-2C, a version of the Tracker modified for utility duties.

To provide the Navy with an airborne early-warning aircraft capable of operating from aircraft carriers, Grumman began development of a version of the Tracker in 1954. Based on the S-2A, this was designated the WF-1 but was superseded by the WF-2, based on the C-1A and subsequently redesignated E-1B, more often being known as 'Willy Fudd' to its pilots. The first flight, on March 1, 1957, was by the G-117, a C-1A modified as an aerodynamic prototype carrying the massive dish-type radome above the fuselage. The other major external change consisted of a new tail unit with twin fins and rudders and a central fin. Delivery of 88 production model E-1Bs began in February 1958.

TECHNICAL DATA (S-2E)

Manufacturer: Grumman Aircraft Engineering Corporation, Bethpage, LI, NY.
Type: Anti-submarine search and strike.
Accommodation: Two pilots, two radar operators.
Power plant: Two 1,525 hp Wright R-1820-82WAs.
Dimensions: Span, 72 ft 7 in; length, 43 ft 6 in; height, 16 ft 7½ in; wing area, 499 sq ft.
Weights: Empty, 19,033 lb; gross, 26,867 lb.
Performance: Max speed, 253 mph at 5,000 ft; cruising speed, 149 mph at 1,500 ft; initial climb, 1,800 ft/min; service ceiling, 22,000 ft; range, 1,150 st miles.
Armament: Max weapon load, 4,810 lb. Fuselage weapons-bay for one depth-bomb or two torpedoes. Six underwing pylons for depth-bombs, torpedoes or rockets. Up to 32 sono-buoys in nacelles.
Serial numbers (allocations):

XS2F-1: 129137–129138.
S2F-1: 129139–129153; 133045–133328; 136393–136747;* 144696–144731;* 147549–147561; 147577; 147636–147645;* 148278–148303;* 149037–149049;* 149843–149844.*
S2F-2: 133329–133388.

* Total of 108 from these batches supplied to other nations through MAP.

S2F-3: 147531–147537; 147868–147895; 148717–148752; 149228–149256; 149257–149275.
S2F-3S: 149845–149892; 150601–150603; 151638–151685; 152332–152379; 152798–152845; 153559–153582.
TF-1: 136748–136792; 146016–146057.
WF-2: 145957–145961; 146303; 147208–147241; 148123–148146; 148900–148923.

243

A Grumman E-2B Hawkeye of VAW-116 from USS *Constellation*. (*US Navy*)

Grumman E-2 Hawkeye, C-2 Greyhound

Continuous development by the US Navy of airborne radar systems to detect targets beyond the line of sight of surface ships led by 1956 to a new concept of a Naval Tactical Data System. This system was designed to provide a task force commander with all the information on the disposition of ships and aircraft (both friendly and enemy) needed to control his forces. A key part in this concept was to be played by an airborne early-warning picket aircraft carrying, as well as long-range search radar, digital computers which would automatically detect targets and select the best available interceptor (in terms of location, fuel and armament state) to be despatched to meet each target.

Such an aircraft had to be designed for the task from the outset and the requirement was made the subject of an industry-wide design competition. This competition was won by Grumman on March 5, 1957, with what became the Hawkeye design. Like the WF-2 (see page 242) it was a high-wing monoplane with the stacked antennae elements in a 24-ft diameter Rotodome above the fuselage. Power was provided by two Allison T56-A-8 turboprops. The peculiar airflow over and around the enormous radome led to a multiple-surface tail unit.

Bearing its initial designation of W2F-1, the first Hawkeye flew on October 21, 1960; the designation was subsequently changed to E-2. The prototype was aerodynamically complete but did not have fully operative electronic systems, which were first airborne in a fully equipped Hawkeye on April 29, 1961. Deliveries to Navy units began on January 19, 1964, the first operational unit being VAW-11. This unit provided detachments as required aboard carriers in the Pacific, and was followed in 1965 by VAW-12 whose Hawkeyes served with units of the US fleet in the Atlantic area. Deliveries of the last of 62 production E-2As were made in 1967.

The E-2B was an improved version of the E-2A, the principal difference being that a Litton Industries L-304 micro-electric general-purpose computer was fitted. A prototype conversion of an E-2A was first flown on

244

20 February, 1969, and a retrofit programme was then undertaken to convert all operational E-2As to E-2Bs. By 1974, the later Hawkeye model equipped VAW-113, VAW-116, VAW-125 and VAW-126.

A major improvement in Hawkeye's operational capability was achieved with the introduction of the E-2C, the first prototype of which flew on January 20, 1971, followed by a second a year later. The avionics system of the E-2C was completely revised, with AN/APA-171 antenna system in the rotodome and AN/APS-120 search radar in the nose. In addition to the L-304 computer as used in the E-2B, the E-2C had an air data computer and a carrier aircraft inertial navigation system (CAINS), and uprated engines. The first production E-2C flew on September 23, 1972 and the type entered service with VAW-123 in November 1973. Over 40 had been included in annual defence budgets up to FY 1977.

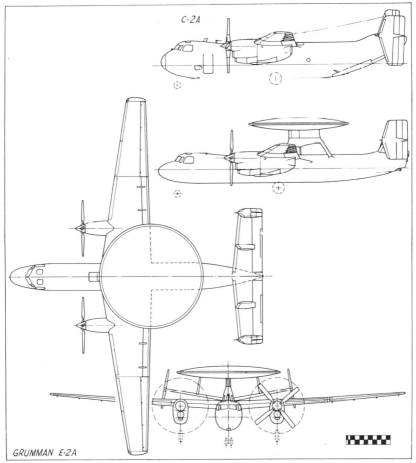

GRUMMAN E-2A

The prototype Grumman C-2A Greyhound. (*Grumman photo*)

A Grumman proposal for a transport derivative of the E-2 was made soon after the Hawkeye programme got under way, and the Navy ordered three prototypes (including one for static testing). This aircraft, designated the C-2A, used the same wings, power plant and tail unit as the E-2A, but had a new fuselage with a rear loading ramp and provision for up to 39 passengers, 20 stretchers or various items of cargo. Primarily intended for COD (carrier on-board delivery), the C-2A Greyhound was compatible with the largest US Navy carrier (CVS-10 and CVA-19). The first of two prototypes flew on November 18, 1964, and production of an initial batch began during 1965. These aircraft were delivered to Fleet Tactical Support Squadron· Fifty (VRC-50) which first operated the Greyhound in December 1966. A second order for 17 was placed subsequently but cancelled in part before completion.

TECHNICAL DATA (E-2C)
Manufacturer: Grumman Aerospace Corporation, Bethpage, LI, NY.
Type: Airborne early-warning picket.
Accommodation: Crew of five including two pilots, combat information centre officer, air control officer and radar operator.
Power plant: Two 4,910 shp Allison T56-A-422 turboprops.
Dimensions: Span, 80 ft 7 in; length, 57 ft 7 in; height, 18 ft 4 in; wing area, 700 sq ft.
Weights: Empty, 37,678 lb; gross, 51,569 lb.
Performance: Max speed, 374 mph; cruising speed, 310 mph; service ceiling, 30,800 ft; ferry range, 1,605 st miles.
Serial numbers:
  E-2A: 147263–147265; 148147–148149; 148711–148713; 149817–149819; 150530–150541; 151702–151725; 152476–152489.

TECHNICAL DATA (C-2A)
As for E-2A except as follows:
Type: Carrier on-board delivery transport.
Accommodation: Crew of three–four; 39 passengers.
Dimensions: Length, 56 ft 7½ in; height, 15 ft 11 in.
Weights: Empty, 31,250 lb; gross, 54,382 lb.
Performance: Max speed, 352 mph at 30,000 ft; cruising speed, 296 mph at 30,000 ft; initial climb, 2,330 ft/min; service ceiling, 28,800 ft; range, 1,655 st miles.
Serial numbers:
  C-2A: 152786–152797; 155120–155136 (some cancelled).

Grumman A-6A Intruder of VA-165 lands aboard the USS *Constellation* during China Sea operations. (*US Navy*)

Grumman A-6 Intruder

A 1956 requirement for a low-level, long-range strike aircraft for service with the US Navy produced 11 design proposals from which, in December 1957, that by Grumman, the G-128, was adopted. Born of the Navy's experience in the Korean War, the new aircraft was to have a high subsonic performance at tree-top height to permit under-the-radar penetration of enemy defences and be capable of finding and hitting small targets in any weather.

Grumman designed a mid-wing monoplane around two 8,500 lb s.t. Pratt & Whitney J52-P-6 engines, the jet-pipes of which were to be capable of being tilted down 23 degrees to shorten the take-off run. For maximum crew efficiency and best utilization of the information provided by the DIANE (Digital Integrated Attack Navigation Equipment), side-by-side seating was provided for the crew of two. Provision was made for an external load of 15,000 lb, including up to four AGM-12B Bullpup ASMs.

Initial orders for the A-6A (initially designated A2F-1) were placed in March 1959 and were for eight development aircraft. The first of these flew on April 9, 1960, and the remainder of the batch followed quickly. The first two production orders, in 1962 and 1963, totalled 69 A-6As and by late 1962 the first ten of these were flying. Deliveries totalled 83 by the end of 1964, and the production rate reached over 70 a year in 1966. Production aircraft did not have the swivelling engine jet-pipes, which had a permanent downward angle of 7 degrees instead. J52-P-8A engines were used in later aircraft.

Deliveries to the first Navy squadron to operate the Intruder, VA-42, began on February 1, 1963, and VA-75, VA-85 and VA-65 were among the other Navy units soon equipped to fly the A-6A. The first Marine Corps unit to become operational on the A-6A was VMA(AW)-242, at NAS Cherry Point in October 1964. During 1965, Intruders began operating in support of US forces in Vietnam, flying initially from USS *Independence*,

247

Grumman EA-6A, showing fin radome and underwing electronic pods. (*Grumman photo*)

and both Navy and Marine Corps squadrons were heavily engaged throughout the remainder of that campaign.

Production of the A-6A totalled 482 and was completed in December 1969. Of the total built, 19 were converted to A-6B, with capability of carrying Standard ARM missiles, to equip one USN squadron, and 12 more modified to A-6C with FLIR (forward-looking infra-red) and LLTV (low-light level television) equipment in a ventral fairing. On May 23, 1966, a prototype conversion of an A-6A to an in-flight refuelling tanker, the KA-6D, made its first flight. With a hose-and-reel unit in the rear fuselage, the KA-6D operated as a buddy refuelling tanker for operational Intruders; although production contracts were cancelled, 54 A-6As were converted to KA-6Ds, and three of the early Intruder test models were similarly equipped, as NA-6As, for use in the F-14 flight test programme.

Succeeding the A-6A in production, the A-6E first flew on February 27, 1970, and differed in having a completely new avionics fit with greatly increased capability. Production was expected to total 84 with a final quantity of 12 in the 1976 Fiscal Year defense budget. In addition, a total of 192 A-6As were programmed for conversion to A-6E standard. All the new and modified A-6Es were to be fitted in due course with TRAM (target recognition multi-sensor), a prototype of which first flew on March 22, 1974.

A version of the Intruder was developed to provide ECM facilities for

A Grumman KA-6D tanker serving with VA-165 aboard the USS *Constellation*. (*US Navy*)

248

A-6A squadrons. Although retaining some strike capability, this EA-6A (originally A2F-1Q) had some of the bombing/navigation equipment deleted to make room for over 30 different antennae used to detect, locate, classify, record and jam enemy transmissions. In all, 27 EA-6As were built, and could be recognized externally by a radome atop the fin, the aerials and, on most operations, ECM pods under the wings. The EA-6B, similarly equipped for ECM duties, is separately described on page 250.

TECHNICAL DATA (A-6E)

Manufacturer: Grumman Aerospace Corporation, Bethpage, LI, NY.
Type: Carrier-based attack-bomber.
Accommodation: Pilot and bombardier/navigator.
Power plant: Two 9,300 lb s.t. Pratt & Whitney J52-P-8A or -8B turbojets.
Dimensions: Span, 53 ft; length, 54 ft 7 in; height, 16 ft 2 in; wing area, 529 sq ft.
Weights: Empty, 25,630 lb; gross, 60,400 lb.
Performance: Max speed, 648 mph at sea level; cruising speed, 482 mph at 35,000 ft; initial climb, 8,600 ft/min; service ceiling, 44,660 ft; range with max load, 1,080 miles at sea level.
Armament: Five external stores positions with 3,600 lb capacity each; max combined load, 17,280 lb.
Serial numbers:
 A-6A*: 147864–147867; 148615–148618; 149475–149486; 149935–149958; 151558–151600; 151780–151827; 152583–152646; 152891–152964; 154124–154171; 155531–155721; 156994–157029; 158041–158052.
 EA-6A: 156979–156993.
 A-6E: 158528–158539; 158787–158798 (plus later contracts).
  * Including some conversions to EA-6A, A-6B, A-6C, KA-6D and A-6E.

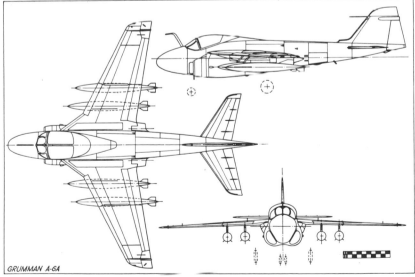

GRUMMAN A-6A

249

Grumman EA-6B Prowler, in service with tactical electronics squadron VAQ-129. (*US Navy*)

Grumman EA-6B Prowler

Following development of an electronic countermeasures version of the Grumman A-6 Intruder for use by US Marine Corps squadrons as the EA-6A (see page 249), the US Navy initiated, in the autumn of 1966, a new ECM version with considerably enhanced operational performance. Whereas the EA-6A, like the attack versions of the Intruder, was a two-seater, this new variant had a redesigned front fuselage that was 3 ft 4 in longer and carried a crew of four, the two additional crewmen, side-by-side in the new rear cockpit, operating the specialized avionics carried by the aircraft internally and externally.

In this new guise, the ECM aircraft was designated EA-6B, and named Prowler. Its primary mission was to support strike aircraft by degrading enemy defense systems, and its secondary mission was to defend ships of the fleet at sea, both actively by tactical support jamming and passively by the surveillance of enemy electromagnetic emissions. The Prowler's jamming systems were carried in pods—one under the fuselage and two beneath each wing.

Each of the tracking jammer pods contained two transmitters with steerable antenna and a track receiver, or exciter, to control the transmitter. Electric power was generated in each pod by a ram air turbine, with a small windmilling propeller on the nose. When the EA-6B first entered service, the pods provided coverage of the four wavebands relevant to the radars then in use in SE Asia; subsequently, the so-called EXCAP version of the EA-6B was introduced, with expanded capability to cover four other wavebands.

The EA-6B prototype first flew on May 25, 1968, and it was followed by four pre-production aircraft. The Navy planned to procure 88 production aircraft (plus two prototypes) with successive purchases up to FY 1979, and all aircraft except the first 23 were to EXCAP configuration. The basic (early) aircraft were subsequently modified to have an improved on-board system, following testing of a prototype that began in 1975. Three squadrons saw action during the Vietnam conflict, playing a key role in support of Navy strikes against land targets. Following the end of that conflict, the Prowler force was established at 36 aircraft in Navy Service—one unit to serve aboard each of the US Navy's 12 aircraft carriers—plus a training unit of five aircraft and 15 EA-6Bs to support three Marine Air Wings.

TECHNICAL DATA (EA-6B)

Manufacturer: Grumman Aerospace Corporation, Calverton, Long Island, NY.
Type: Carrier-borne electronic warfare aircraft.
Accommodation: Pilot and three electronic countermeasures operators.
Power plant: Two 9,300 lb s.t. Pratt & Whitney J52-P-8A or -8B or -408 turbojets.
Dimensions: Span, 53 ft 0 in; length, 59 ft 5 in; height, 16 ft 3 in; wing area, 529 sq ft.
Weights: Empty, 34,580 lb; typical mission weight, 57,000 lb; normal take-off weight, 58,500 lb; max overload, 63,177 lb.
Performance: Max speed, 599 mph at sea level; average cruising speed, 466 mph; service ceiling, 38,000 ft; ferry range (external tanks), 2,475 miles.
Armament: None.
Serial numbers:
    EA-6B: 156478–156482; 158029–158040;   158799–158817
          158540–158547; 158649–158651;   (plus later contracts)

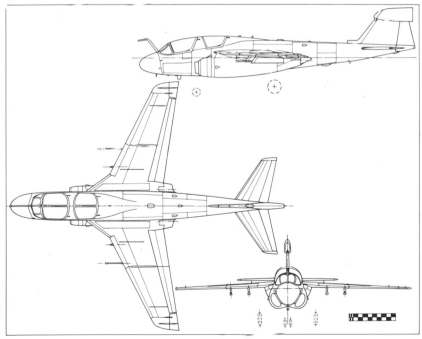

Grumman F-14A of VF-1, the first US Navy squadron to convert to the Tomcat. (*US Navy*)

Grumman F-14 Tomcat

Failure of the General Dynamics F-111B to meet US Navy requirements for an advanced carrier-based air superiority fighter left a significant gap in the Navy's inventory. The F-111B had been based on the concept of commonality with the USAF's F-111A and was consequently compromised in some aspects of the Naval role; following its cancellation in April 1968, the US Navy launched a new design contest, VFX, in which the finalists were McDonnell Douglas and Grumman. The latter company proposed a variable geometry, two-seat, twin-engined aircraft, the Model 303, and this was named the winning contender on 15 January, 1969.

Designated F-14 and eventually named Tomcat, the Grumman aircraft was the first wholly new Navy fighter to be designated in the 'new' post-1962 designation series. Procurement began in Fiscal Year 1969 with an order for six development aircraft, total US Navy plans at that time embracing 469 aircraft; subsequently the programme increased to a projected 722 but was then cut back to 313 before being established at 334, only to be increased in 1975 to a planned 403 by the early 'eighties.

The F-14 design benefited from Grumman's experience in designing and building the first US variable-sweep combat aeroplane, the XF10F-1 Jaguar (ordered by the USN but built only as a prototype) and with the F-111B experience that placed Grumman among the world leaders in vg design. Since rapid development was essential to meet the US Navy timescale, Pratt & Whitney TF30 engines—similar to those used in the F-111—were chosen for the first F-14 production standard.

First flight of the first R & D F-14A was made from Grumman's Bethpage airfield, Long Island, on December 21, 1970, but this aircraft was lost on its second flight on December 30 after a total hydraulic failure. Flight testing resumed with the second aircraft on May 24, 1971, and seven more of the trials batch flew in 1971. Deliveries to the US Navy began in late 1972, when VF-124 assumed responsibility for crew training and familiarization. First Navy units equipped to fly the Tomcat were VF-1 and VF-2, both based at NAS Miramar for their working-up period and followed by VF-14 and VF-32. Although production plans for the F-14B were dropped when the F-401-PW-400 engine difficulties arose, two F-14A airframes (Nos. 7 and 31) were converted to serve as F-14B prototypes, and the first of these flew on September 12, 1973.

TECHNICAL DATA (F-14A)

Manufacturer: Grumman Aerospace Corporation, Calverton, LI, NY.

Type: Carrier-borne air superiority fighter.

Accommodation: Pilot and Naval Flight Officer (NFO).

Power plant: Two 20,900 lb s.t. (with afterburning) Pratt & Whitney TF30-P-412A turbofans.

Dimensions: Span (max spread), 64 ft 1½ in; span (fully swept), 38 ft 2 in; span (overswept for ship stowage), 33 ft 3½ in; length, 61 ft 11¾ in; height, 16 ft 0 in; wing area, 565 sq ft.

Weights: Empty, 37,500 lb; normal take-off (internal fuel, four Sparrow AAMs), 55,000 lb; max permitted take-off, 72,000 lb.

Performance: Max design speed, Mach 2·34.

Armament: One M-61A1 20 mm cannon in forward fuselage, port side. Recesses beneath fuselage for four AIM-7E Sparrow AAMs plus four AIM-9G Sidewinders; or six AIM-54A Phoenix AAMs plus two Sidewinders. Optional provision for MK-82, MK-83 or MK-84 bombs up to total of 14,500 lb.

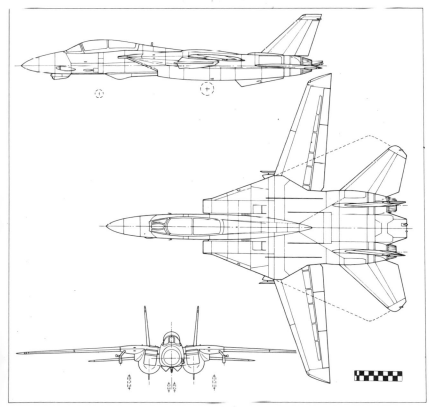

Hall PH-3 flying-boat in early war-time green and grey camouflage on beaching gear at San Francisco Coast Guard station in June 1942. (*Peter M. Bowers*)

Hall PH

Elaborating on the long series of flying-boat designs originated at the Naval Aircraft Factory with the PN-7 in 1924, the Hall Aluminum company produced an XPH-1 prototype in December 1929 on a Navy contract placed with the company on December 29, 1927. It was closely related to the Navy PN-11 with a similar hull and the same wings, but had a large fin and rudder and 537 hp Wright GR-1750 engines with close cowlings. Like the Navy boats, the XPH-1 had open cockpits for a gunner in the bows, two pilots side by side and another gunner behind the wings.

On June 10, 1930, Hall received a contract for nine PH-1 flying-boats and the first of these appeared in October 1931 with a number of changes from the prototype. In particular, the PH-1s had 620 hp Wright R-1820-86 radials with short-chord cowlings, giving a 10 mph improvement in top speed despite a considerable increase in gross weight. A rudimentary enclosure was provided over the pilots' cockpit. Testing of the PH-1 began in October 1931, and the nine boats of this type equipped Navy Patrol Squadron Eight from 1932 to 1937.

The Hall boat went back into production in June 1936 when the company received a contract for seven PH-2s to be used by the US Coast Guard for air–sea rescue duties. These had 750 hp Wright R-1820F-51 engines and special equipment for the Coast Guard role, but were otherwise similar to the PH-1s. They remained in service until 1941. They were survived by the seven PH-3s ordered for the US Coast Guard in 1939, with the same engines but long-chord NACA cowlings similar to those of the prototype XPH-1 and a more refined enclosure for the cockpit. After Pearl Harbor, the PH-3s were given standard Navy finish in place of the natural metal, and some were adapted for anti-submarine patrols for a time.

When the PH-3s eventually went out of service, the era of the biplane flying-boat ended too. These boats could trace their ancestry back to the Curtiss biplanes of 1914 and had given the Navy good service until they were rendered obsolete by the larger monoplane boats which were in service in time for operations in World War II.

TECHNICAL DATA (PH-3)

Manufacturer: Hall Aluminum Aircraft Corporation, Bristol, Pa.
Type: Patrol and air–sea rescue flying-boat.
Accommodation: Crew of six.
Power plant: Two 750 hp Wright R-1820F-51s.
Dimensions: Span, 72 ft 10 in; length, 51 ft; height, 19 ft 10 in; wing area, 1,170 sq ft.
Weights: Empty, 9,614 lb; gross, 16,152 lb.
Performance: Max speed, 159 mph at 3,200 ft; cruising speed, 136 mph; service ceiling, 21;350 ft; range, 1,937 st miles.
Armament: Four flexible 0·30-in Lewis guns.
Serial numbers:

XPH-1: A8004.
PH-1: A8687–A8695.

PH-2: V164–V170.
PH-3: V177–V183.

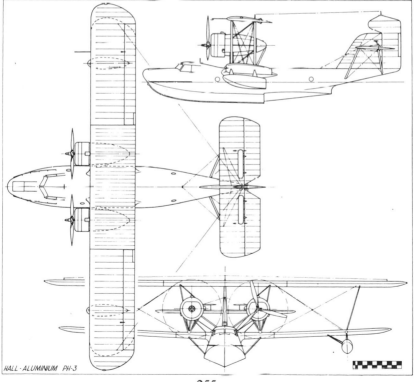

HALL · ALUMINIUM PH-3

255

Kaman UH-2A utility helicopter serving with HC-2 aboard the USS *Saratoga*.
(*Stephen P. Peltz*)

Kaman HU2K/H-2 Seasprite

To meet its requirements for a high-performance, all-weather helicopter operating a wide range of utility missions, the Navy held a design competition during 1956. Regarded primarily as a search and rescue helicopter, the new type was to be capable also of a variety of other missions, including such things as round-the-clock all-weather plane guard duties aboard aircraft carriers, gun-fire observation, reconnaissance, courier services, personnel transfer from ship to ship or ship to shore, casualty evacuation, and wire laying and tactical air controller operations.

Kaman Aircraft Corporation was announced winner of the design competition in 1956 and on November 29, 1957, received a formal contract for four prototypes and an initial production batch of 12. The original designation of HU2K-1 was later changed to UH-2A, and the name Seasprite was adopted. The first flight was made on July 2, 1959.

Deliveries of the Seasprite began on December 18, 1962, to Helicopter Utility Squadron Two at Lakehurst Naval Air Station. HU Squadron One received its first HU2K-1s a few weeks later at Ream Field Naval Auxiliary Air Station, near San Diego. These two units provided detachments aboard ships in the US Atlantic and Pacific Fleets respectively, for plane guard duties, and they relinquished H-25s (see page 461) when the Seasprites were delivered.

A total of 190 Seasprites were delivered, including 102 HU2K-1Us (UH-2B) with simplified equipment for VFR operations only, but these were all later converted to UH-2A standard with full IFR capability. In March 1965, one UH-2 was modified to a twin-engine configuration, two T58-GE-8B engines being located externally on the sides of the rotor pylon. A November 1965 contract called for refinement of the twin-engine version, designated UH-2C, and construction of two prototypes. These, with small changes from the original demonstrator, made their first flights on 14 March and 20 May, 1966, respectively, and the US Navy ordered conversion of 40 earlier Seasprites to UH-2C standard.

256

After the Navy began operating Seasprites on rescue missions in the Vietnam combat zone, in which role they proved extremely valuable, six early models were converted to armed HH-2C configuration, with one 7·62-mm Minigun in a nose turret and two in waist positions, plus armour protection and other special features. They also had dual main-wheels, four-bladed tail rotor, uprated transmission, 1,350 shp T58-GE-8F engines and 12,500-lb gross weight. All these features except the armament were then introduced on a new batch of conversions from single to twin-engined configuration, designated HH-2D, and about 70 such conversions were made.

In October 1970, the US Navy adopted a further modified Seasprite variant to meet its interim requirement for a Light Airborne Multi-Purpose System (or LAMPS) to provide an over-the-horizon search and strike capability for anti-submarine destroyers. In this new guise, the helicopter was identified as the SH-2D, carrying a high power search radar in a radome 'tub' under the nose, plus suitable equipment for either anti-submarine warfare (ASW—including two Mk 46 torpedoes) or anti-

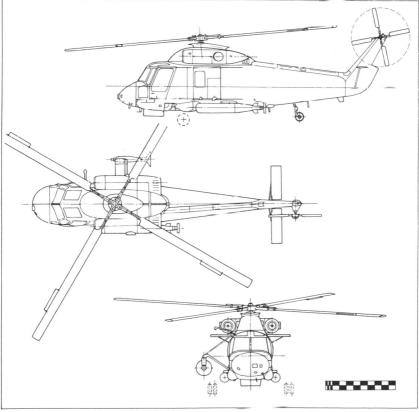

Kaman SH-2F Seasprite prototype, showing forward location of the tail wheel.
(*Kaman photo*)

ship missile defence (ASMD—including electronic jamming gear). After an initial batch of 10 SH-2D conversions had been ordered, the USN announced in March 1971 that all 115 Seasprites in the inventory would in due course be committed to the LAMPS programme and a second batch of 10 was ordered. The first flight was made on March 16, 1971, and first deliveries, to Helicopter Combat Squadron Four (HC-4) at Lakehurst, began in September, with first deployment at sea aboard the USS *Belknap* on December 7, 1971.

After delivery of 20 SH-2Ds, the standard changed to SH-2F, with an improved rotor, tailwheel located farther forward, improved avionics and a 13,300 lb gross weight. Two YSH-2Es, flown in 1972, were used to test a different nose radar and other new avionics. The programme of converting earlier Seasprites to SH-2F configuration was continuing in 1975 and had included original single-engined variants as well as UH-2C, HH-2C and HH-2D models.

TECHNICAL DATA (SH-2F)

Manufacturer: Kaman Aircraft Corporation, Bloomfield, Connecticut.
Type: Anti-submarine and anti-ship-missile defense helicopter.
Accommodation: Two pilots and one senior operator.
Power plant: Two 1,350 shp General Electric T58-GE-8F engines.
Dimensions: Rotor diameter, 44 ft 0 in; overall length, 52 ft 7 in; height, 15 ft 6 in; rotor disc area, 1,520 ft.
Weights: Max gross, 12,800 lb; overload, 13,300 lb.
Performance: Max speed, 168 mph; cruising speed, 150 mph; initial rate of climb 2,440 ft/min; service ceiling, 22,500 ft; max range, 445 miles at 5,000 ft; max endurance, 3·7 hours at sea level.
Serial numbers:
UH-2A: 147202–147205;   147972–   UH-2B: 150139–150186;   151300–
    147983;   149013–149036;   151335; 152189–152206.
    149739–149786.

Lockheed PV-1 with 'Donald Duck' insignia behind the fuselage star. (*Lockheed photo*)

Lockheed PV Ventura, Harpoon

Reversing earlier policy to use only flying-boats for overwater patrols, the Navy began to seek land-based patrol bombers early in 1942 as the vulnerability of flying-boats to Japanese fighters became apparent. The need for higher performance and better armament was too pressing to await development of a new type for the Navy, and a formal agreement was concluded on July 7, 1942, for the transfer of certain USAAF types then in production to Navy ownership. Primarily, this agreement covered the North American PBJ Mitchell, Convair PB4Y Liberator and Lockheed Vega PV Ventura. Use of the 'P' designation alone for the last-named was

Lockheed PV-1 with non-standard star marking on forward fuselage in April 1943. (*US Navy*)

an interesting differentiation indicating that it was wanted for patrol rather than bombing duties.

The Ventura, as Lockheed Model 37, had been developed initially from the commercial Model 18 to a British specification for an improved successor to the Hudson. Powered by two 2,000 hp Pratt & Whitney GR-2800-S1A4G engines, it was larger and heavier than the Hudson, and the RAF ordered 875 in 1940. First flight was made on July 31, 1941, and under lend–lease arrangements further contracts were placed by the USAAF. The latter retained a quantity of the Venturas originally ordered by Britain as well as some of the lend–lease aircraft, which had R-2800-31 engines and a larger bomb-bay. After the Navy had arranged to acquire a batch of the Venturas in production for the USAAF as B-34s and had allocated the designation PV-1 to these aircraft, 27 Lockheed Model 37s were requisitioned from a British lend–lease batch in September 1942 for training and familiarization, and these were designated PV-3. The PV-2 designation had meanwhile been allocated to a long-range version being developed.

Delivery of PV-1s to the Navy began in December 1942, with the first aircraft going to Squadron VP-82 to replace the PBO Hudsons (see page 447). All subsequent production was for the Navy, which procured a total

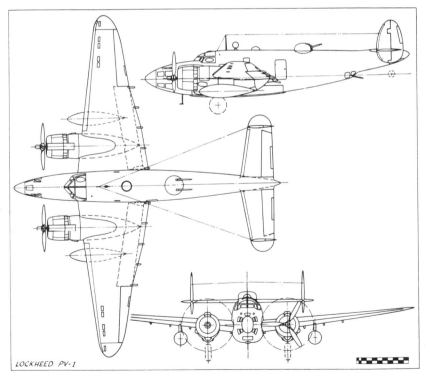

LOCKHEED PV-1

of 1,600 PV-1s, including 388 for the RAF on lend–lease. During 1945, the designation PV-1P was allocated to some Venturas with cameras.

The PV-2, ordered by the Navy on June 30, 1943, differed in several respects from the Ventura, and was renamed Harpoon although still designated in the PV series. The power plant and general configuration remained unchanged, but wing span and fuel capacity were increased, larger fins and rudders were fitted, and the armament was improved. Delivery of a batch of 500 began in March 1944, and these were followed by 33 PV-2Ds in which the nose armament was increased from five to eight 0·50-in guns. Because of difficulties in sealing the integral fuel tanks in the wings, the first 30 Harpoons were assigned to training duties with the outer wing tanks sealed off, and were designated PV-2Cs. Subsequent aircraft had leak-proof cells in the integral tanks.

After serving primarily in the Pacific area for the last year of the war, the Harpoons were withdrawn from front-line service but continued in use with eleven USN Reserve wings for several more years.

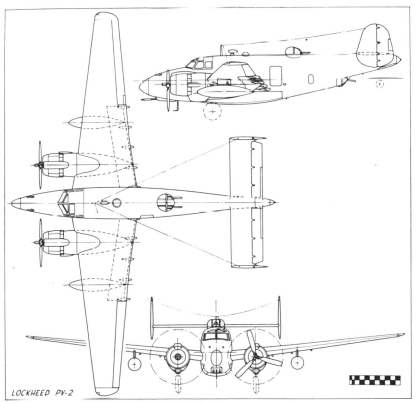

LOCKHEED PV-2

Lockheed PV-2 Harpoon in USN Reserve markings and with upper nose guns installed. (*E. M. Sommerich*).

TECHNICAL DATA (PV-1)

Manufacturer: Lockheed (Vega) Aircraft Division, Burbank, California.
Type: Patrol-bomber.
Accommodation: Crew of four or five.
Power plant: Two 2,000 hp Pratt & Whitney R-2800-31s.
Dimensions: Span, 65 ft 6 in; length, 51 ft 9 in; height, 11 ft 11 in; wing area, 551 sq ft.
Weights: Empty, 20,197 lb; gross, 31,077 lb.
Performance: Max speed, 312 mph at 13,800 ft; cruising speed, 164 mph; initial climb, 2,230 ft/min; service ceiling, 26,300 ft; range, 1,660 st miles.
Armament: Two 0·50-in machine guns each in nose and dorsal turret, two 0·30-in ventral guns. Six 500 lb bombs or one torpedo internal; up to two 1,000 lb bombs under wings.
Serial numbers:
PV-1: 29723–29922; 33067–33466; 34586–34997; 48652–48939; 49360–49659.
PV-3: 33925–33951.

TECHNICAL DATA (PV-2)

Manufacturer: Lockheed (Vega) Aircraft Division, Burbank, California.
Type: Patrol-bomber.
Accommodation: Crew of four or five.
Power plant: Two 2,000 hp Pratt & Whitney R-2800-31s.
Dimensions: Span, 74 ft 11 in; length, 52 ft 0½ in; height, 11 ft 11 in; wing area, 686 sq ft.
Weights: Empty, 21,028 lb; gross, 36,000 lb.
Performance: Max speed, 282 mph at 13,700 ft; cruising speed, 171 mph; initial climb, 1,630 ft/min; service ceiling, 23,900 ft; range, 1,790 st miles.
Armament: Five fixed forward-firing 0·50-in guns in nose; two flexible 0·50-in each in dorsal turret and ventral mount. Up to four 1,000 lb bombs internal and two 1,000 lb external.
Serial numbers:
PV-2: 37065–37534.
PV-2C: 37035–37064.
PV-2D: 37535–37550; 37624–37634; 84057–84064.

Lockheed P2V-2 serving with Navy Photographic Squadron One on aerial survey of Alaska in November 1948. (*US Navy*)

Lockheed P2V Neptune

Mainstay of US Navy land-based patrol squadrons for 15 years, from 1947 to 1962, the Neptune had a protracted early design life because of the need to concentrate upon more immediately available types during World War II. Initial design studies for the Neptune were made in 1941 by the Lockheed Vega subsidiary under the leadership of Mac V. F. Short, vice-president—engineering. Close contact with Navy patrol squadrons had led Mac Short to the conclusion that a new aircraft was needed with greater range and load-carrying ability, plus better field performance— and this almost before the Navy had accepted that landplanes could better fulfil the patrol mission than flying-boats. As a private venture, the Lockheed company formally initiated work on the new project, Model 26, on December 6, 1941—the day before the attack on Pearl Harbor—and appointed Jack Wassall as project engineer. Early drawings show an aircraft of similar layout to the Neptune, with·gun turrets above and below the fuselage just behind the cockpit, a high wing and the tailplane mounted above the fuselage on the base of the fin.

Preoccupation with war-time production reduced work on the Model 26 to a low rate until 1944, by which time the need for a new land-based patrol bomber had grown urgent. On April 4, 1944, the US Navy placed a contract with Lockheed for two prototypes and 15 production examples. The Vega division had been absorbed into the parent company at the end of 1943, but the Vega origin of the Model 26 design was reflected in the Navy designation P2V. Unlike the PV-1 and PV-2, which had been adapted from commercial designs (see page 259), the P2V was able to offer a fuselage optimized for operational efficiency, carrying a crew of seven, a range of electronic equipment and a weapons-bay large enough for two torpedoes or 12 depth charges.

The first XP2V-1 flew on May 17, 1945, powered by two 2,300 hp

Lockheed P2V-4 with 'solid' nose and wing tip tank and searchlight. (*Lockheed photo*)

Wright R-3350-8 engines and carrying pairs of 0·50-in machine guns in nose, dorsal and tail turrets. The 15 P2V-1s were similar, with underwing provision for 16 rocket projectiles. During September 1946, a specially modified P2V-1, named *Truculent Turtle*, set a world distance record of 11,236 miles. The first operational unit to receive P2V-1s was VP-ML-2, equipped with the new type in March 1947.

Eight months after the first Navy contract for Neptunes, a second order was placed, and further contracts followed regularly to keep the aircraft in production until April 1962. Navy contracts totalled 843 aircraft (apart from Neptunes ordered for MAP and export), and these appeared in seven main varieties and many minor variations. The P2V-1 was followed by the P2V-2, the prototype of which first flew on January 7, 1947, with R-3350-24W engines. This version had some features of the *Truculent Turtle* including a lengthened nose, and six forward-firing guns; the ninth and subsequent aircraft had 20 mm tail guns in place of 0·50-in calibre.

Lockheed P2V-6 with enlarged tip tanks and nose turret. (*Lockheed photo*)

Production of the P2V-2 (Lockheed 26 and 126) totalled 81, including two ski-equipped P2V-2Ns (first flight October 18, 1949) and a single P2V-2S (Model 226) with a prototype search radar in a ventral installation.

Forty P2V-3s which followed the P2V-2s were Lockheed Model 326 and differed only in having 3,200 hp R-3350-26 engines; the first was flown on August 6, 1948, and deliveries were completed by January 1950. With these same engines, two P2V-3Zs had special interiors and armour protection for use as staff transports in battle areas. Eleven P2V-3Cs were equipped for operation off carrier decks, following trials with a modified P2V-2, and on March 7, 1949, one of these aircraft took-off from the USS *Coral Sea* at a then record weight of 74,000 lb. Two P2V-3Cs were later converted to P2V-3B configuration. The final 30 aircraft in the -3 series were P2V-3Ws, with APS-20 search radar and a large radome under the forward fuselage; first flight was made on August 12, 1949.

The fourth major Neptune version, the P2V-4, first flew on November 14, 1949, and was similar to the P2V-3W with the addition of long-range tanks under the wingtips. Later aircraft in the bath of 57 of this model (Lockheed Model 426) had 3,250 hp R-3350-30W Turbo-Compound engines. This was the first Neptune variant to carry a new designation in the 1962 reallocation, when it became P-2D, although it was already passing out of service.

Major design changes were introduced in the P2V-5 which first flew on

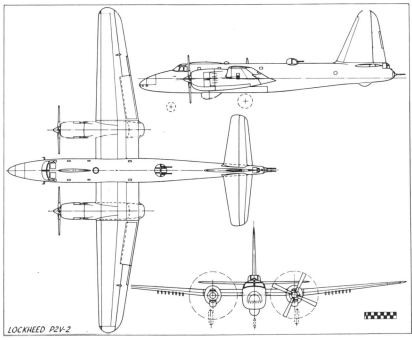

LOCKHEED P2V-2

Lockheed SP-2H, the final production version of the Neptune, in overall blue finish.
(*Gordon S. Williams*)

December 29, 1950. An Emerson ball turret with two 20 mm cannon was installed in the nose, and the wingtip tanks were enlarged and raised to the wing centre line. A searchlight, set to illuminate the target for the nose guns, was located in the nose of the starboard tank. To meet the requirements of the Korean War, production of the P2V-5 expanded rapidly, the Navy eventually receiving 348. Most of these became P2V-5F (P-2E) when modified after delivery to have a 3,400 lb s.t. Westinghouse J34-WE-34 turbojet under each wing, to improve take-off performance and the speed over the target. Other post-production modifications to the P2V-5s deleted the ventral armament and the Emerson nose turret and introduced magnetic anomaly detection gear in a lengthened rear fuselage. Another programme added Julie/Jezebel active and passive detection systems in a version designated P2V-5FS (SP-2E) and, in a more limited form, in the P2V-5FE (EP-2E). Other Neptunes of this series were equipped for target-drone launch and control as P2V-5FD (DP-5E).

Lockheed SP-2H in light grey and white colours current in 1968. (*Stephen P. Peltz*)
266

A further change in nose configuration plus more flexibility in load carrying marked the P2V-6 (Lockheed 626), first flown on October 16, 1952, and later designated P-2F. The 53 examples of this variant for the US Navy had R-3350-36W engines and could carry mines as well as torpedoes and depth-bombs; they also had smaller tip tanks and a smaller ventral radome. Some served as trainers as P2V-6T (TP-2F), and others carried two Fairchild AUM-N-2 Petrel missiles under the wings, when they were designated P2V-6M (MP-2F). The aircraft serving as patrol bombers later had Westinghouse jet pods added beneath the wings, when they became P2V-6F (P-2G).

Final production version of the Neptune was the P2V-7 (P-2H), first flown on April 26, 1954. This was the only production Neptune to have the underwing jet pods, adopted after trials with a modified P2V-5. A number of other refinements were also introduced, including features of the P2V-5 and P2V-6 series. Introduction of Julie/Jezebel detection gear produced the P2V-7S (SP-2H), and a single example was winterized and modified for use by VX-6 in the Antarctic in the reconnaissance role as P2V-7LP (LP-2J). Other P2V-7s operating on skis in the Antarctic were not redesignated. Production for the Navy totalled 212.

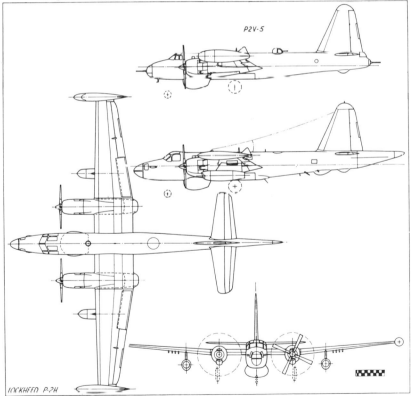

P2V-5

LOCKHEED P-2H

By the end of 1966, P-3s had replaced P-2s in many of the operational patrol squadrons, 19 of which were flying P-2Es and P-2Hs two years earlier. Neptunes were continuing in service in a number of second-line roles and with US Navy Air Reserve Training Units. Post-1962 modification programmes resulted in versions designated OP-2E and AP-2H, the latter for special duties in Vietnam.

TECHNICAL DATA (P2V)

Manufacturer: Lockheed Aircraft Corporation, Burbank, California.
Type: Patrol and anti-submarine search.
Accommodation: Crew of nine–ten.

	P2V-1	P2V-3	P2V-5	P2V-7
Power plant	2 × 2,300 hp R-3350-8	2 × 3,200 hp R-3350-26W	2 × 3,250 hp R-3350-30W	2 × 3,500 hp R-3350-32W 2 × 3,400 lb J34-WE-34
Dimensions:				
Span, ft in	100　0	100　0	102　0	103　10
Length, ft in	75　4	77　10	81　7	91　4
Height, ft in	28　6	28　1	28　1	29　4
Wing area, sq ft	1,000	1,000	1,000	1,000
Weights:				
Empty, lb	33,720	34,875	41,754	47,456
Gross, lb	61,153	64,100	76,152	75,500
Performance:				
Max speed, mph/ft	303/15,300	337/13,000	341	345/10,000
Cruising speed, mph	176/1,500	180/1,500	—	207/8,500
Climb, ft/min	1,050	1,060	—	—
Service ceiling, ft	27,000	28,000	29,000	22,000
Range, miles	4,110	3,930	4,750	2,200
Armament	6 × 0·50-in guns	8 × 20 mm cannon	6 × 20 mm cannon	No guns

Serial numbers (allocations):
XP2V-1: 48237–48238.
P2V-1: 89082–89096.
P2V-2: 39318–39368; 122438–122467.
P2V-3: 122923–122951; 122964–122987.
P2V-3W: 124268–124291;　124354–124361.
P2V-4: 124211–124467.
P2V-5*: 124865–124909;　127720–127782;　128327–128442; 131400–131543;　133640–133651;　134671–134676; 134718–134723.

P2V-6*: 126514–126547;　131544–131566; 134638–134663.
P2V-7*: 135544–135621;　140151–140160;　140430–140443; 140962–140986;　141231–141251;　142542–142545; 143172–143183;　144675–144692;　145900–145923; 146431–146438;　147562–147569;　147946–147971; 148330– 148362;　149070–149081;　149089–149130; 150279–150283;　153611–153616.

* Total of 193 from these batches supplied to other nations through MAP.

Lockheed TV-2 with partial white and orange trainer colours. (*Peter M. Bowers*)

Lockheed TO, TV, T2V Seastar

The first Navy operation of Shooting Stars, Lockheed's famous jet fighter, came when three P-80As were obtained from the USAF; one, with an arrester hook, was used for deck-landing trials aboard the USS *Franklin D. Roosevelt*. Although these P-80s carried USN serial numbers (29667–29668, 29689) they had no Navy designation.

During 1948 the Navy acquired a further batch of 50 single-seat F-80Cs from the Air Force for use as jet advanced trainers. Designated TO-1 initially and TV-1 after the Navy changed Lockheed's symbol from O to V, they had no arrester gear. Despite their training role, 16 of the TO-1s were used to equip a front-line Marine fighter squadron, VMF-311, late in 1948. During 1949 the Navy began procurement of the two-seat trainer developed for the USAF as TF-80C and produced as the T-33A. An

One of three Lockheed P-80As used by the US Navy, photographed in April 1946 during catapult trials. (*US Navy*)

Lockheed T-33B in standard trainer colours in 1967. (*Stephen P. Peltz*)

initial order for 26 of these aircraft took the Navy designation TO-2; this subsequently changed to TV-2 and procurement continued for several years. Total Navy and Marine procurement of the TV-2 series amounted to 699, for use as advanced and instrument trainers and for other miscellaneous duties. In 1951 a few TV-2s were modified for guidance and control duties in connection with missile and target trials, designated TV-2D, while in 1956 a few more to a new standard became TV-2KDs as drones or control aircraft. New designations assigned in 1962 to these aircraft were T-33B (TV-2), DT-33B (TV-2D) and DT-33C (TV-2KD). In 1968 the Navy received 252 QT-33A drones from USAF.

The TV-2 series proved generally unsuitable for carrier operations, and to overcome this Lockheed developed a private venture prototype during 1952 as the L-245. First flown on December 15, 1953, the L-245 differed from the TV-2 in having a humped cockpit to give a better forward view from the rear seat, and a number of general changes to improve overall performance and low-speed characteristics. The lower landing and take-off speeds needed for carrier operation were achieved by a combination

Lockheed T2V-1, final development of the Shooting Star design. (*Peter M. Bowers*)

270

of leading- and trailing-edge flaps and a blown-flaps system of BLC (Boundary Layer Control). In place of the 5,200 lb s.t. J33-A-20 engine of the TV-2, the L-245 used a 6,100 lb s.t. J33-A-24.

A US Navy contract for the new trainer quickly followed its first flight, and the name T2V-1 Seastar was allocated. Production deliveries began in 1956, and a total of 150 was supplied. The designation was changed to T-1A in 1962.

TECHNICAL DATA (T2V-1)

Manufacturer: Lockheed Aircraft Corporation, Burbank, California.
Type: Deck-landing trainer.
Accommodation: Pilot and instructor in tandem.
Power plant: One 6,100 lb s.t. Allison J33-A-24 or -24A turbojet.
Dimensions: Span, 42 ft 10 in; length, 38 ft 6½ in; height, 13 ft 4 in; wing area, 240 sq ft.
Weights: Empty, 11,965 lb; gross, 15,800 lb.
Performance: Max speed, 580 mph at 35,000 ft; initial climb, 6,330 ft/min; service ceiling, 40,000 ft; range, 970 st miles.
Serial numbers:

P-80A: 29667–29668, 29689.	L-245: N125D.
TV-1: 33821–33870.	T2V-1: 142261–142268; 142397–142399; 142533–142541; 144117–144216; 144735–144764.
TV-2: 124570–124585; 124930–124939; 126583–126626; 128661–128722; 131725–131888; 136793–136886; 137934–138097; 138977–139016; 141490–141558; 143014–143049.	QT-33A: 155918–156169.

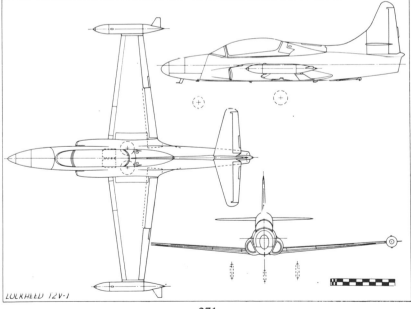

LOCKHEED T2V-1

271

Lockheed EC-121K operating with VW-13 at Argentia, Newfoundland, in 1964.
(*US Navy*)

Lockheed WV, R7V, C-121 Constellation, Warning Star

US Navy used variants of the Lockheed Constellation in two major roles: as transports and for airborne early-warning duties. Transport versions were first acquired in 1945 under the designation R7O-1; equivalent to USAAF C-69s, they served with VPB-101 in the Atlantic Fleet. Matching the larger commercial model L-1049 and the USAF C-121C, the R7V-1 was obtained primarily for the Navy contribution to the Military Air Transport Service, and of 51 R7V-1s acquired, 32 were transferred to USAF control while serving in MATS, being redesignated C-121G. The remaining Navy Super Constellation transports became C-121Js in 1962.

One R7V-1 was equipped with cameras in 1962 to serve in the reconnaissance role with VX-6 in the Antarctic, designated R7V-1P. Two R7V-2s acquired in 1955 (first flight on September 1, 1953) had Pratt & Whitney YT34-P-12A turboprops and were used, with two others transferred to USAF as YC-121Fs, to gain experience with this type of engine.

Use of the Constellation for patrol and airborne early-warning duties was first investigated in 1949, when the Navy acquired two Model 749s, under the designation PO-1W (later WV-1). These carried large, long-range radar requiring massive radomes both above and below the fuselage,

Lockheed C-121J based at Keflavik, Iceland.

272

and the size of the fins was increased because of the additional side area presented by these radomes. The PO-1Ws having proved the feasibility of the project to put large radar sets into aircraft, the Navy proceeded to place production orders for a similar version based on the commercial L-1049. Lockheed flew an aerodynamic prototype (actually the L-1049 demonstrator modified) before starting delivery of these new PO-2Ws or WV-2s in 1954. Powered by 3,400 hp Wright R-3350-34 or -42 engines, they carried a crew of 26 compared with 22 in the PO-1Ws.

A total of 142 WV-2 Warning Stars were delivered. Of these, one was modified in 1957 to WV-2E with a large dish radome above the fuselage, but this particular requirement was met by the Grumman WF-2 (see page 241). With changed electronics for countermeasures duties, some of the Warning Stars became WV-2Qs, and eight were converted to WV-3s for weather reconnaissance, with newer equipment, no tip tanks and a crew of eight. The WV-2s served, together with Air Force RC-121s, to provide

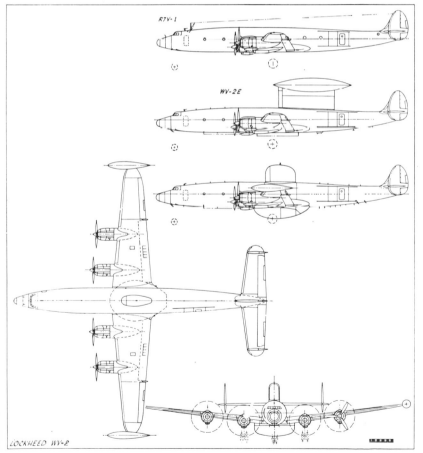

One of two Lockheed R7V-2s used by the Navy to gather experience of turboprop operation. (*Lockheed photo*)

Early Warning Barriers over the Atlantic and Pacific. In 1962, they were redesignated EC-121K (WV-2), EC-121L (WV-2E), EC-121M (WV-2Q) and WC-121N (WV-3). New search equipment introduced in 1963 changed the designation of some EC-121Ks to EC-121P.

TECHNICAL DATA (R7V-1)

Manufacturer: Lockheed Aircraft Corporation, Burbank, California.
Type: Personnel and cargo transport.
Accommodation: 72 troops.
Power plant: Four 3,250 hp Wright R-3350-91s.
Dimensions: Span, 123 ft; length, 116 ft 2 in; height, 24 ft 8 in; wing area, 1,650 sq ft.
Weights: Empty, 72,815 lb; gross, 145,000 lb.
Performance: Max speed, 368 mph at 20,000 ft; cruising speed, 259 mph at 10,000 ft; initial climb, 1,100 ft/min; service ceiling, 22,300 ft; range, 2,100 st miles.
Serial numbers:
  R7V-1: 128434–128444;   131621–   R7V-2: 131630–131631;   131660–
     131629;   131632–131659;   131661.
     140311–140313.

TECHNICAL DATA (WV-2)

Manufacturer: Lockheed Aircraft Corporation, Burbank, California.
Type: Long-range airborne early-warning patrol.
Accommodation: Crew of 26.
Power plant: Four 3,400 hp Wright R-3350-34s or -42s.
Dimensions: Span, 126 ft 2 in; length, 116 ft 2 in; height, 27 ft; wing area, 1,654 sq ft.
Weights: Empty, 80,611 lb; gross, 143,600 lb.
Performance: Max speed, 321 mph at 20,000 ft; initial climb, 845 ft/min; service ceiling, 20,600 ft; range, 4,600 st miles.
Serial numbers:
  WV-1(PO-1W): 124437–124438.   WV-3: 137891–137898.
  WV-2(PO-2W):   126512–126513;
     128323–128326;   131387–131392;
     135746–135761;   137887–137890;
     141289–141333;   143184–143230.

274

A Lockheed KC-130F tanker refuels two Marine Corps A4Ds. (*Lockheed photo*)

Lockheed C-130 Hercules

US Navy interest in the Lockheed Hercules, originally produced to USAF requirements and first flown on August 23, 1954, crystallized during 1960 when the USAF demonstrated the aircraft's potential in Arctic and Antarctic conditions, operating as a ski-plane. One USAF squadron was flying ski-equipped C-130Ds, primarily in support of DEW-Line construction sites in the Arctic, and five of these aircraft (C-130A derivatives) were despatched to the Antarctic to assist Navy operations in *Deep Freeze 60*. Their contribution was so outstanding that the US Navy purchased four ski-equipped C-130Bs for use by VX-6 Squadron in *Deep Freeze 61* and subsequently. Initially operated as C-130BLs they were redesignated LC-130Fs in 1962, and six later LC-130Rs had higher gross weight.

Independent of the Navy interest in the Hercules, the US Marine Corps became aware of its potential as a flight refuelling tanker during 1957, and two USAF C-130As were modified for evaluation. This led to a USMC order for 46 tankers, carrying 3,600-US gal fuel tanks in the fuselage and hose/drogue pods under each wing tip. The first flight was made on January 22, 1960, at which time the type was designated GV-1, this later being changed to KC-130F. During 1963, one KC-130F made a series of landings and unassisted take-offs from the USS *Forrestal*, at a gross weight of 120,000 lb. Max flight weight of the KC-130F was 135,000 lb, but four improved KC-130Rs ordered in 1974 (plus four more later) were based on the C-130E airframe at a weight of 155,000 lb. The US Navy bought seven Hercules transports similar to the Marine's tankers but with refuelling equipment removed; originally GV-1Us, these became C-130Fs in 1962.

Another specialized role for which the Navy picked the Hercules was that of communications relay aircraft to support the world-wide operations of the Fleet Ballistics Missile submarines. Four aircraft acquired

Ski-equipped Lockheed LC-130F Hercules of US Navy. (*E. M. Sommerich*)

in 1965, based on the C-130E airframe, were designated C-130G and later EC-130G and were equipped with VLF radio communications equipment known as Tacomo III. During 1973, the first of 10 improved EC-130Qs, also with Tacomo III permanently installed, began to reach Navy VQ-4 Squadron. Two modified USAF C-130As were also acquired by USN as DC-130A drone carriers, with four underwing pick-up points.

Versions of the Hercules have also been procured through US Navy channels by the Coast Guard, starting in 1959, in the air-sea rescue role. Under the original designation R8V-1G and then SC-130B, the Hercules

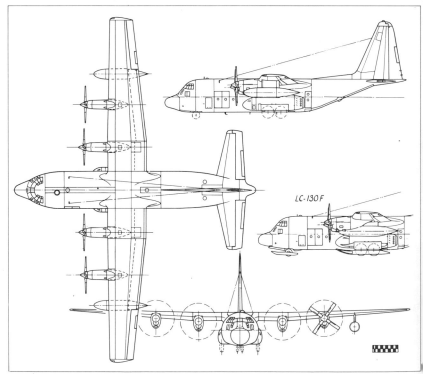

LC-130F

was modified to have additional crew positions, observation 'blister' windows and an observation door which could be rolled down in place of the usual paratroop door when the aircraft was not pressurised. Twelve of these aircraft, based on the C-130B airframe, were purchased by the Coast Guard, with first delivery on December 1, 1959; subsequently, they were redesignated HC-130B, and three HC-130Hs were acquired, similar to the USAF model but with the nose crew recovery gear removed. Finally, a single EC-130E used by the Coast Guard was specially equipped to calibrate LORAN A and C at locations throughout the world.

All USN and USMC versions of the Hercules except the C-130F were powered by 4,910 eshp T56-A-16 engines, Navy equivalent of the variant used in the USAF's C-130H model. The C-130F retained the 4,050 eshp T56-A-7s used in the C-130B and C-130E.

TECHNICAL DATA (KC-130F)

Manufacturer: Lockheed-Georgia Company, Marietta, Georgia.
Type: Flight refuelling tanker and transport.
Accommodation: Crew of seven (tanker) or five (transport). Up to 92 troops or 74 stretchers.
Power plant: Four 4,910 eshp Allison T56-A-16 turboprops.
Dimensions: Span, 132 ft 7 in; length, 97 ft 8 in; height, 38 ft 3½ in; wing area, 1,745 sq ft.
Weights: Operating weight empty, tanker, 74,454 lb; operating weight empty, transport, 70,491 lb; max take-off weight, 135,000 lb; max payload, 35,000 lb.
Performance: Max cruising speed, 357 mph; long-range cruising speed, 343 mph at 25,000 ft; refuelling speed, 250 mph at 120,000 lb weight at 20,000 ft; range on standard fuel capacity, up to 3,000 miles with full payload.
Serial numbers:

KC-130F and C-130F:
    147572–147573;   148246–148249;
    148890–148899;   149787–149816;
    150684–150690.
LC-130F: 148318–148321.
HC-130B: 1339–1342; 1344–1351.
C-130G: 151888–151891.

LC-130R: 155917; 159129–159131.
EC-130Q: 156170–156177; 159348;
    159469.
DC-130A: 158228–158229.
EC-130E: 1414.
HC-130E: 1452–1454.
KC-130R: 160013–160021.

A Lockheed HC-130B attached to the Coast Guard at San Francisco.
(William T. Larkins)
277

Lockheed P-3A Orion in standard grey and white colours. (*Lockheed photo*)

Lockheed P-3 Orion

To provide a replacement for the widely used P-2 Neptune (see page 263), the US Navy called for design proposals in August 1957 for a new high-performance anti-submarine patrol aircraft. To save both time and cost the Navy suggested that manufacturers should seek to meet the requirement with a variant of an aircraft already in production. Following this directive, Lockheed adapted the commercial Electra turboprop, winning the design contest with this proposal on April 24, 1958. The Lockheed Model 186 retained the wings, tail unit, power plant and other components of the Electra, as well as much of the fuselage structure. The fuselage was shortened by about 7 ft and a weapons-bay was incorporated, together with extensive new electronics and other systems.

Following receipt of a research and development contract, Lockheed modified an Electra as an aerodynamic test vehicle with a mock-up of the MAD radome in the rear and a simulated bomb-bay. This first flew on August 19, 1958, but made only eight flights. Then followed a prototype of the anti-submarine aircraft, flown on November 25, 1959, and designated YP3V-1. The name Orion was adopted late in 1960, and the P3V designation was changed to P-3 in 1962.

The Orion was very fully ·equipped for its ASW role, with extensive electronics in the fuselage plus stowage for search stores, and a 13-ft long unpressurized bomb-bay equipped to carry torpedoes,. depth-bombs, mines or nuclear weapons. Ten external pylons under the wings could carry mines or rockets. A searchlight was located under the starboard wing.

The first pre-production P-3A flew on April 15, 1961, and after flight-testing, evaluation and acceptance trials with six aircraft, deliveries to operational units began on August 13, 1962. Patrol Squadrons VP-8 and VP-44 were the first to receive the Orion, other units converting from the P-2 steadily thereafter; the 100th P-3A was delivered in November 1964.

The sole Lockheed RP-3D Orion, used in US Navy 'Project Magnet' investigation of earth's magnetic field. (*Lockheed photo*)

Of the total of 157 P-3As built for the USN, the final batch of 48 had the so-called Deltic system installed, with more sensitive ASW detection devices and improved tactical display equipment. Post-delivery modifications resulted in the delivery of four WP-3A weather-reconnaissance aircraft in 1970 to replace WC-121Ns, and 10 P-3As were later converted to EP-3E configuration to serve with VQ-1 and VQ-2 in place of EC-121s in the electronic reconnaissance role; they carried special radar, with radomes in long fairings above and below the fuselage and an additional ventral radome forward of the wing.

Following the P-3A in production, the P-3B introduced 4,910 ehp Allison T56-A-14 turboprops in place of 4,500 shp T56-A-10Ws, but was otherwise similar to the earlier model. Production totalled 125 for the USN and 20 for other nations, with one USN P-3B also being diverted to the RAAF as a replacement. Two became EP-3Bs as prototype electronic reconnaissance aircraft, being later modified further to EP-3Es together with 10 P-3As. One P-3B was converted to the YP-3C prototype, flown on September 18, 1968, and fitted with the A-NEW data processing system with a digital computer and generally improved operational capability.

The P-3C entered service in 1969, and the US Navy began an acquisition programme designed to replace all P-3As and P-3Bs in front-line service with the P-3C. By 1973, orders totalled about 160 and production was continuing at the rate of about 12 a year towards a planned total of 233.

A Lockheed EP-3E Orion, showing the revised dorsal and ventral radomes. (*Lockheed photo*)

279

One P-3C airframe was converted during manufacture to RP-3D, for use by VXN-8 on a five-year survey of the Earth's magnetic field. The designation P-3F applied to an export model for Iran, procured through US Navy channels.

TECHNICAL DATA (P-3C)

Manufacturer: Lockheed-California Company, Burbank, California.
Type: Anti-submarine patrol bomber.
Accommodation: Crew of 10, including five in the tactical compartment.
Power plant: Four 4,910 ehp Allison T56-A-14 turboprops.
Dimensions: Span, 99 ft 8 in; length, 116 ft 10 in; height, 33 ft 8½ in; wing area 1,300 sq ft.
Weights: Empty, 61,491 lb; normal take-off, 135,000 lb; max overload, 142,000 lb.
Performance: Max level speed, 473 mph; economical cruising speed, 378 mph, at 25,000 ft; patrol speed, 237 mph at 1,500 ft; initial rate of climb, 1,950 ft/min; service ceiling, 28,300 ft; mission radius, 3 hrs on station at 1,500 ft, 1,550 miles.
Armament: Internal stowage for bombs, mines, nuclear depth bomb, torpedoes, etc. Ten underwing pylons for bombs, mines, guided missiles, etc. Modified aircraft have provision for Harpoon missiles.
Serial numbers:

YP-3A: 148276.
P-3A: 148883–148889; 149667–149678; 150494–150529; 150604–150609; 151349–151396; 152140–152187.
P-3B: 152718–152765; 152886–152890;* 153414–153458; 154574–154605; 155291–155300;* 156599–156003.*

P-3C: 156507–156530; 157310–157332; 158204–158227; 158563–158574; 158912–158935; 159318–159329; 159503–159514; (plus later contracts).
WP-3D: 159773; 159875

* Exported through MAP/FSP.

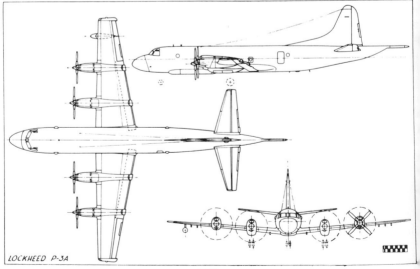

LOCKHEED P-3A

Lockheed S-3A Viking in service with VS-21 aboard the USS *John F. Kennedy*. (*US Navy*)

Lockheed S-3A Viking

The US Navy began its search for a new carrier-based, anti-submarine search and strike aircraft to replace the Grumman S-2 family during 1967, issuing a Request for Proposals to industry under the acronymn VSX. The final proposals were submitted at the end of 1968, and the Lockheed/ LTV project was selected as the winning VSX design in August 1969, the designation S-3A then being assigned.

The S-3A was designed as a compact, high-wing monoplane powered by two General Electric TF34 turbofan engines in underwing pods. The engine was developed from scratch primarily for its application in the VSX, and its flight development began in a B-47 test-bed in January 1971, prior to first flight of the S-3. With a crew of four, the S-3A was designed to carry a comprehensive array of avionics including passive and active sensors, computers, navigation and communications equipment. Sensors included a variety of sonobuoys up to a combined total of 60 and used in connection with a sonobuoy reference system; a search and navigation radar (AN/APS-116) in the nose; forward looking infra-red (FLIR) in a retractable ventral cupola and magnetic anomaly detector (MAD) in a retractable tail 'sting'. An electronic support measures (ESM) system (AN/ALR-47) had its antennas located in small wing-tip fairings.

The US Navy programme for the S-3A embraced a total of 179 pro-duction aircraft, following a batch of eight research and development air-frames ordered in two lots of four each in FY 70 and FY 71. The pro-duction batches were ordered in quantities of 13, 35, 45 and 45 in FY 72 to FY 75 respectively, with purchase of a final batch of 41 planned for FY 76. The first of the R & D S-3As flew on November 8, 1971, at Burbank, at which time the name Viking was formally bestowed upon it; second and third Vikings flew on May 19 and July 17, 1972, respectively, the latter being the first to have the full avionics system and APU fitted.

First carrier landings were made by an S-3A in November 1973, aboard the USS *Forrestal,* and deliveries to the Air Antisubmarine (Training) Squadron 41 (VS-41) began in February 1974 for crew training to begin at San Diego. One 10-aircraft squadron was to serve aboard each of 12 multi-purpose aircraft carriers. A transport version was also developed as the US-3A for COD use, and the Navy planned to procure 30.

TECHNICAL DATA (S-3A)

Manufacturer: Lockheed-California Company, Burbank, California (in association with Vought Systems Division of LTV).

Accommodation: Flight crew of four (pilot, co-pilot, tactical co-ordinator and sensor operator).

Power plant: Two 9,275 lb s.t. General Electric TF34-GE-2 turbofans.

Dimensions: Span, 68 ft 8 in; folded span, 29 ft 6 in; overall length, 53 ft 4 in; height, 22 ft 9 in; wing area, 598 sq ft.

Weights: Empty weight, 26,554 lb; internal fuel weight, 13,142 lb; max design gross weight, 52,539 lb; max carrier landing weight, 37,695 lb.

Performance: Max level speed, 514 mph; max cruise, 426 mph; sea level loiter speed, 184 mph; max combat range (internal fuel), 2,900 miles; radius for high altitude search of 4·5 hours at 426 mph, 530 miles; sea level loiter endurance, 7·5 hours; service ceiling, 40,000 ft.

Armament: Internal provision for four MK- 46 torpedoes, or bombs, or mines or depth bombs. One strong point under each wing for rockets or similar stores.

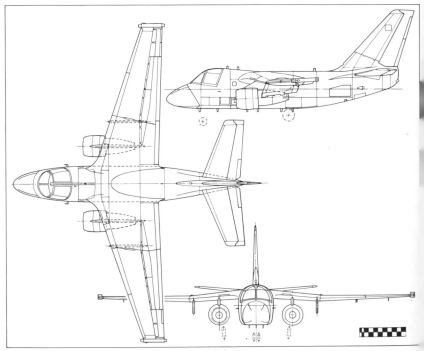

Loening LS-1 seaplane with the unusual Richardson floats which had flat inner faces. (*US Navy*)

Loening M Series

The Loening monoplanes were a daring innovation in their day, since the field of military aviation in 1918 was dominated almost entirely by biplane-minded pilots, engineers and procurement personnel. Grover C. Loening, who had acquired an aeronautical engineering degree from Columbia University in 1911 by virtually inventing the course, was able to see the inherent speed and structural simplicity advantages of the monoplane and had the tenacity to fight for his beliefs against adamant opposition. Loening got his chance in 1918, by which time he was a thoroughly experienced engineer, having worked for the Wright Brothers, been chief engineer of the Army Flying School in San Diego, and then chief engineer of the Sturtevant Aeroplane Co. After forming his own company in 1918, Loening was asked to design a two-seat fighter that would out-perform the famous British Bristol Fighter. The result was the M-8, a strut-braced high-wing monoplane built around the new 300 hp Hispano-Suiza engine just then going into production in the United States as the Wright H-3.

In addition to being a monoplane in a biplane's world, the M-8 had a number of advanced features. One was the radiator installation, which was mounted in a tunnel beneath the engine instead of surrounding the propeller shaft or being a separate bolted-on item outside the fuselage lines, as was customary. This feature was widely adopted by other high-performance military aircraft in the post-war years.

The parallel lift-struts connecting the wing to the lower longeron were fitted with wide fairings of aerofoil section that were expected to contribute lift. This feature was to become a trademark of the famous Bellanca monoplanes of 1925–40. With the upper longerons at the level of the wing, the rear gunner had an excellent field of fire for his twin 0·30-in Lewis guns. The pilot, of course, had the same unobstructed view above for fighting and could also see under the wing through windows below the longerons. The performance of the M-8 was such that a contract for 5,000 was placed, but the wholesale cancellations that followed the Armistice kept production models from being built for the Army.

The Navy, having taken a low-cost sample of Loening's basic monoplane design with three ultra-light Kittens (A442–A444) that were designed as ships' planes and were considered for quick disassembly and stowage aboard submarines, took an interest in the full-size fighter design after the war and ordered a single naval version as the M-8-0 (sometimes written as M-80). This was followed by orders for 46 production models designated M-8-0 and M-8-1 (M-81) which, although designed as fighters, were used for observation purposes. An additional six were ordered as M-8-1S twin-float seaplanes, and at least one of these was tested as an amphibian with wheels built into the bottoms of the floats.

Three practically identical models, designated LS for Loening seaplane, were also ordered, but the last two were cancelled. The single LS (A5606) was used to test the unique Richardson Pontoon which was in effect a standard float of somewhat greater than standard width split along the centreline. These halves were then moved apart to standard twin-float positions beneath the seaplane. The vertical inside face of each separate float was supposed to improve the water-handling characteristics of the combination, but nothing seems to have come of it and the experiment was dropped.

The Navy attempted to capitalize on the loudly proclaimed speed advantages of the monoplane by fitting one M-81 (A5791) with a special set of small wings for its entry in the 1920 Pulitzer race. Span was reduced by nearly 5 ft, the chord was reduced by 2 ft, and the wide lifting-struts were replaced by thin streamlined steel tubes. The racer, flown by Lt B. G. Bradley of the US Marine Corps, developed a water leak early in the race which forced it out in the last lap after achieving a speed of 160 mph around the closed course. It was ironic for Loening that after this misfortune the Navy pinned its racing hopes on biplanes and even kept a pair of triplanes in service for such purposes through the 1923

Loening M-81 landplane, illustrating the unusual ailerons at the wingtips.

season (see page 118). He was able to sell a handful of monoplane fighters and racers to the Army in the years 1920–2, but had to fall back on the traditional biplane (see page 286) in order to win significant orders and did not introduce another monoplane for over a decade.

TECHNICAL DATA (M-8-0)

Manufacturer: Loening Aeronautical Engineering Company, New York.
Type: Observation monoplane.
Accommodation: Pilot and observer.
Power plant: One 300 hp Hispano-Suiza.
Dimensions: Span, 32 ft 9 in; length, 24 ft; height, 6 ft 7 in; wing area, 229 sq ft.
Weights: Empty, 1,623 lb; gross, 2,068 lb.
Performance: Max speed, 145 mph; climb, 10 min to 13,900 ft; service ceiling, 22,000 ft; endurance 5·5 hrs.
Armament: Two flexible 0·30-in Lewis guns.
Serial numbers:

M-8: A5631.	M-8-0: A5637–A5646.
M-8-1: A5701–A5710; A5761–A5786.	M-8-1S: A5788-A5793.

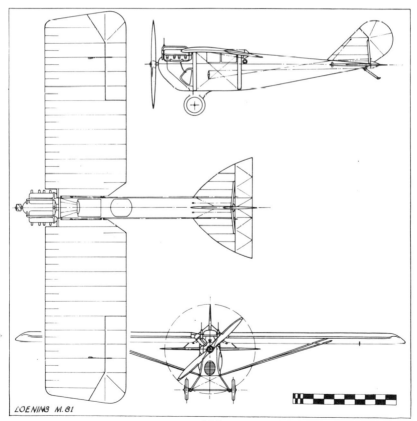

LOENING M.81

285

Loening OL-2 *Polar Bear* serving with the 1925 Arctic Expedition.
(*Kadel and Herbert*).

Loening OL

Of the numerous designs produced by the Loening Aeronautical Engineering Company between 1917 and 1928, the series of Army and Navy amphibians which first appeared in 1923 became the best known. Grover C. Loening, founder of the firm which bore his name, made a number of contributions to aircraft design (see page 283), and the distinctive amphibian was characteristic of his original approach. It resulted from an attempt to match contemporary landplane performance with an amphibian of similar horsepower.

To achieve this aim Loening adopted the novel approach of using a large single float under the fuselage, rather than a flying-boat hull as adopted in other amphibians. The float was faired into the fuselage, and wheels were located on each side, arranged so that they could be swung up out of the way when operating on water. The wings of this design were deliberately made as conventional as possible, and in fact were so similar to those of the DH-4 that they were interchangeable, although of better efficiency.

The first of the Loening amphibians were powered by inverted Liberty engines and were adopted in the observation role by the Army Air Corps as COA-1s. Five identical aircraft were purchased by the Navy in 1925 for use in connection with the Arctic Expedition that year, and designated OL-2. The OL-1 designation had already been applied by the Navy to two Loening amphibians under construction with 440 hp Packard 1A-1500 engines and a third cockpit in tandem. The second of these two aircraft incorporated a number of improvements compared with the first, and the Navy ordered four more to this same standard as OL-3s. Another six with the Liberty engine instead of Packard were designated OL-4s.

No examples of the OL-5 were built. The OL-6 reverted to the Packard engine and had a new, more angular tail shape; 28 were purchased and widely used by Navy and Marine units. One of these became the XOL-7 with experimental thick-section wings, and another became the XOL-8 with the first radial engine used in the Loening series, a Pratt & Whitney Wasp. This same engine was used in the 20 production model OL-8s, which were two-seaters and in 20 OL-8As which had arrester gear for carrier deck operation. Finally, the Navy acquired 26 OL-9s, outwardly similar to the OL-8; by this time, the company had merged with the Keystone Aircraft Corporation, soon to be acquired by the Curtiss-Wright Corporation.

The Navy had two other examples of the same basic aircraft for ambulance duties; designated XHL-1, they had a single open cockpit for the pilot and carried six passengers in the fuselage.

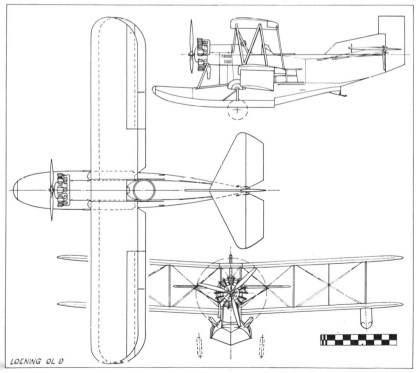

LOENING OL 8

Loening OL-9 amphibian from the Naval Academy, as indicated by the 'NA7' marking. (*Warren D. Shipp*)

TECHNICAL DATA (OL-3)

Manufacturer: Loening Aeronautical Engineering Company, New York.
Type: Observation amphibian.
Accommodation: Three.
Power plant: One 475 hp Packard 1A-2500.
Dimensions: Span, 45 ft; length, 35 ft 1 in; height, 12 ft 9 in; wing area, 504 sq ft.
Weights: Empty, 3,673 lb; gross, 5,316 lb.
Performance: Max speed, 122 mph; climb, 10 min to 5,500 ft; service ceiling 13,000 ft; range, 423 st miles.
Serial numbers:

OL-1: A6879–A6880.   OL-4: A7059–A7064.
OL-2: A6980–A6983, A7030.   OL-6: A7324–A7350.
OL-3: A7055–A7058.   XOL-7: A7335.

TECHNICAL DATA (OL-9)

Manufacturer: Keystone-Loening, Bristol, Pennsylvania.
Type: Observation amphibian.
Accommodation: Pilot and observer.
Power plant: One 450 hp Pratt & Whitney R-1340-4.
Dimensions: Span, 45 ft; length, 34 ft 9 in; height, 12 ft 9 in; wing area, 504 sq ft.
Weights: Empty, 3,649 lb; gross, 5,404 lb.
Performance: Max speed, 122 mph at sea level; climb, 5·7 min to 5,000 ft; service ceiling 14,300 ft; range, 625 st miles.
Serial numbers:

XOL-8: A7344.   OL-9: A8733–A8747, A8979–A8985,
OL-8: A7832–A7851.   9208–9211.
OL-8A: A8069–A8088.   XHL-1: A8275–A8276.

Two Chance Vought F8U-1Ps from VFP-62 serving aboard the USS *Independence* in June 1962. (*US Navy*)

LTV F-8 Crusader

The last Navy fighter developed by the Chance Vought company before it was absorbed into the Ling-Temco-Vought organization, the F-8 Crusader was one of eight designs submitted to a US Navy requirement for a supersonic air-superiority fighter in 1952. In May 1953 the Vought design was selected for further development, and contracts were placed for two prototypes designated XF8U-1 (later XF-8A). The XF-8A was designed around an 18,900 lb s.t. Pratt & Whitney J57-P-11 afterburning turbojet and was unusual in having a high-mounted wing, the incidence of which could be increased to reduce the landing speed, without forcing the aircraft to assume an exaggerated nose-high attitude.

The two XF-8As flew, respectively, on March 25 and September 30, 1955, and the first production aircraft appeared before the end of 1956. The production F-8As had the J57-P-12 or J57-P-14 engine and a built-in armament of four 20 mm cannon and 32 air-to-air rockets. This was supplemented by two Sidewinder missiles externally on the sides of the fuselage. Deliveries of the F-8A (to Squadron VF-32) began in March 1957, and this unit took its Crusaders aboard the USS *Saratoga* late the same year. Production totalled 318, these being followed by 130 F-8Bs (F8U-1E) with limited all-weather interception capability. Another related variant was the RF-8A (F8U-1P) carrying five cameras in place of the cannon in the forward fuselage. The first RF-8A flew on December 17, 1956, and 144 were built. During 1965 and 1966, 73 of these aircraft were converted to RF-8G standard with strengthened wings, ventral fins, fuselage reinforcement, a new navigation system and improved camera-station installations.

One F-8A airframe, after serving as the F-8E prototype, was converted to a two-seat configuration, with dual controls in tandem and two of the 20 mm cannon eliminated, but otherwise similar to the operational single-seater. First flown on February 6, 1962, it was designated YTF-8A

The first LTV F-8H remanufactured from an F-8D, in August 1967. (*LTV photo*)

(YF8U-1T). During 1961 a few F-8As were modified as directors for Regulus I and II drones and designated DF-8F, and a few others became QF-8A drones. Many eventually served as TF-8A single-seat trainers.

Succeeding the F-8A in production for the air-superiority role, the F-8C (F8U-2) differed in having a 16,900 lb s.t. J57-P-16 engine, small ventral fins under the rear fuselage and internal changes. The fire-control system was improved and the performance benefited from the extra power and a small reduction in wing span. Delivery of 187 examples was spread from January 1959 to September 1960, the first unit to use the F-8C being VF-84.

The F-8B was succeeded by the F-8D (F8U-2N) as a limited all-weather interceptor, powered by a 10,700 lb s.t. J57-P-20. This variant, first flown on February 16, 1960, had improved radar and equipment, more fuel and two more Sidewinders externally on the fuselage in lieu of the rocket pack in the fuselage. The Navy received a total of 152 with deliveries from June 1960 to January 1962.

Production of the Crusader for the Navy ended with the F-8E (F8U-2NE), a further improvement of the F-8D, incorporating APQ-94 search and fire-control radar and, on all but the first few examples, provision for up to 5,000 lb of external stores on two wing strong points to permit operation in the attack role. The first F-8E flew on June 30, 1961, and production aircraft began to appear in September 1961. A total of 286 was built, plus 42 F-8E(FN) for France.

Chance Vought F8U-2 of VF103 from the USS *Forrestal* in 1960. (*US Navy*)

290

During 1966, LTV initiated a remanufacturing programme which by mid-1968 covered 89 F-8Hs (from F-8D); 136 F-8Js (from F-8E); 87 F-8Ks (from F-8C) and 61 F-8Ls (from F-8B); the designation F-8M was reserved for modified F-8As but not used. Modifications included a boundary layer control system and strengthened landing gear. The first F-8H flew on July 17, 1967.

TECHNICAL DATA (F-8E)

Manufacturer: LTV Aerospace Corporation, Dallas, Texas (originally the Chance Vought Corporation).

Type: Carrier-based fighter.

Accommodation: Pilot only.

Power plant: One 10,700 lb s.t. Pratt & Whitney J57-P-20A turbojet.

Dimensions: Span, 35 ft 2 in; length, 54 ft 6 in; height, 15 ft 9 in; wing area, 350 sq ft.

Weights: Gross, 34,000 lb.

Performance: Max speed, 1,120 mph at 40,000 ft; cruising speed, 560 mph at 40,000 ft; climb, 6·5 min to 57,000 ft; service ceiling, 58,000 ft; range, 1,100 st miles.

Armament: Four fixed forward-firing 20 mm Colt cannon. Four Sidewinder AAM or up to 5,000 lb of bombs or rockets or ASMs under wings.

Serial numbers (allocations):

XF8U-1: 138899–138900.

F-8A: 140444–140446; 141336–141362; 142408–142415; 143677–143709; 143711–143821; 144427–144461; 145318–145353.

RF-8A: 141363; 144607–144625; 145604–145647; 146822–146901; 147078–147084.

TF-8A: 143710.

XF8U-2: 140447–140448.

F-8B: 145416–145545.

F-8C: 145546–145603; 146906–147072.

F-8D: 147035–147072; 147896–147925; 148627–148710.

F-8E: 149134–149227; 150284–150355; 150654–150683; 150843–150932.

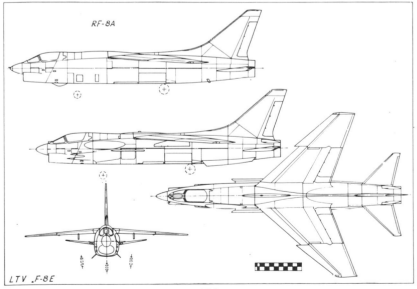

RF-8A

LTV .F-8E

291

LTV A-7A Corsair II in the markings of VA-147, the first operational squadron
to fly the type. (*LTV photo*)

LTV A-7A Corsair II

On May 17, 1963, the US Navy initiated a design competition for a
light attack aircraft (VAL) which could replace the Douglas A-4 Skyhawk.
With speed of production one of the major requirements, the in-
service target date being 1967, only four companies submitted proposals,
and that by Ling-Temco-Vought was named winner of the competition on
February 11, 1964. A key factor in this company's successful bid was use
of the F-8 Crusader design (see page 289) as the basis for the VAL submission, although there were many significant differences. Compared with
the F-8, the LTV design for VAL, subsequently designated the A-7A, had a
shorter fuselage with no afterburner, less sweepback on the wing, and no
provision for varying the wing incidence. Outboard ailerons, not needed
on the F-8, were introduced on the A-7 wing, and the structure was
strengthened to allow the wings and fuselage to carry the big load of
ordnance required by the VAL specification, totalling up to 15,000 lb.
The eight stores stations comprised two on the fuselage sides, each with
500 lb capacity; two inboard wing stations with 2,500 lb capacity each and
four outer wing points with 3,500 lb capacity each. These stations could
accommodate virtually every offensive weapon or store in the Navy's
airborne armoury when the A-7 was designed, and it was calculated that
more than 200 combinations of different stores were possible.

As the VAL requirement was subsonic, the A-7A used a Pratt & Whitney
TF30-P-6 turbofan in place of the F-8's afterburning J57. Ling-Temco-
Vought received a contract on March 19, 1964, covering seven A-7As for
flight testing and the first 35 production aircraft. A second order, for 140
more, was placed in September 1965, and a third for 17 brought total
production of the initial model to 199.

The first A-7A flew on September 27, 1965, and the remaining six test aircraft were flying by mid-1966. Two training units, VA-174 and VA-122, received their first aircraft in September and October, 1966, respectively, and began training pilots for the first Corsair II squadrons; VA-147 was commissioned as the first A-7A tactical squadron on February 1, 1967, and made the first carrier landings aboard the USS *Ranger* in June 1967. This same squadron took the A-7A into operation for the first time over Vietnam on December 4, 1967.

Following completion of the A-7A production, the Navy ordered 196 A-7Bs, differing primarily in having 12,200 lb st TF30-P-8 engines in place of the -6 model used initially. The first production A-7B flew on February 6, 1968, and production was completed with delivery of the final A-7B to the Navy on May 7, 1969. This model entered combat in Vietnam on March 4, 1969, and with the A-7A proved to be one of the most effective close support and strike aircraft used by the Navy in that conflict.

After the USAF had adopted a version of the Corsair II powered by an Allison-built Rolls-Royce Spey engine (designated TF41-A-1), the Navy decided to switch to this powerplant also, ordering the A-7E to succeed the A-7B in production. Because of production delays with the TF-41-A-2 for this model, the first 67 A-7Es were delivered with TF30-P-8 engines, and differed from the A-7B in having a completely new and greatly improved avionics installation, including head-up display, an M61 multi-barrel cannon in place of two single-barrel guns in the A-7A and A-7B, improved hydraulic system and anti-skid brakes. After delivery, these 67 aircraft were redesignated A-7C, the latter designation having earlier been reserved for a training variant that did not materialize.

The A-7E first flew on 25 November, 1968, and entered service in South-east Asia in May 1970 operating with VA-146 and VA-147 aboard

Prototype LTV YA-7H Corsair II[2] was similar to US Navy TA-7C conversions. (*LTV photo*)

the USS *America*. By FY 76, the US Navy had ordered 542 of a total planned procurement of 692 A-7Es, all of which would eventually have TRAM (Target Recognition Attack Multi-sensors) fitted. Procurement of an RA-7E reconnaissance variant, with tactical recce pods, was also planned, as a replacement for the RA-5C.

On August 29, 1972, LTV made the first flight of the YA-7H, a two-seat advanced trainer derivative with full operational capability, having longer front and rear fuselage and two seats in tandem. In 1974 the USN announced that it was allotting 40 A-7Bs and 41 A-7Cs to be converted to similar standard, with the designation TA-7C.

TECHNICAL DATA (A-7E)

Manufacturer: LTV Aerospace Corporation, Dallas, Texas.
Type: Carrier-based attack-bomber.
Accommodation: Pilot only.
Power plant: One 14,250 lb s.t. Allison TF41-A-2 turbofan.
Dimensions: Span, 38 ft 9 in; length, 46 ft 1½ in; height, 16 ft 0¾ in; wing area, 375 sq ft.
Weights: Empty, 19,490 lb; max take-off, 42,000 lb.
Performance: Max speed, 698 mph at sea level, clean; tactical radius with typical load, 700 miles; ferry range, over 2,800 miles.
Armament: One 20-mm M61-A1 multi-barrel gun in front fuselage port side with ·1,000 rounds. Two fuselage and six wing stores stations with combined maximum load of more than 15,000 lb.
Serial numbers:

A-7A: 152580–152582; 152647–152685; 153134–153273; 154344–154360.
A-7B: 154361–154556.
A-7C: 156734–156800.

A-7E: 156801–156890; 157435–157594; 158002–158028; 158652–158681; 158819–158842 (plus later contracts).

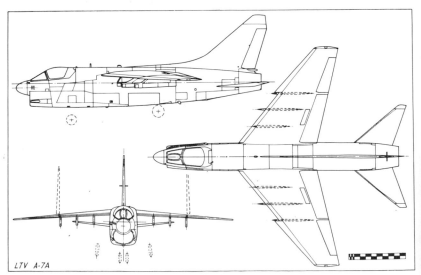

LTV A-7A

McDonnell FD-1 Phantom on a carrier deck. (*McDonnell photo*)

McDonnell FD/FH Phantom

To take advantage of the then new concept of jet propulsion, the US Navy by letter of intent on August 30, 1943, asked the McDonnell Aircraft Corporation to undertake design of a new carrier-based fighter. The McDonnell company, founded in July 1939, had not previously worked on a Navy aircraft; nor had it produced for any user any type of aircraft adopted for quantity production. Nevertheless, use of a comparatively young and inexperienced company for so radical a project as the first Navy jet fighter was advantageous in two ways: it avoided interrupting the war effort of the major companies, all of which were fully committed to design and production of aircraft needed in large quantities, and it allowed an uninhibited approach to be made to the unusual design problems involved.

This lack of inhibition by McDonnell's designers was reflected in the first project designs, which had three small Westinghouse engines in each wing. Each engine had a diameter of 9·5 in and a thrust of about 300 lb. However, development of such an engine appeared difficult and further studies showed that two larger engines, one in each wing root, offered a better estimated performance. By the end of 1943 the configuration had been finalized, with two engines, a low wing, single tail unit with dihedral tailplane mounted clear of the exhaust efflux and pilot well forward in the slender fuselage. Two prototypes had been ordered, with the designation XFD-1.

The engines were 1,165 lb s.t. Westinghouse 19 XB-2Bs, and armament comprised four 0·50-in guns in the nose. The first flight was made on

January 26, 1945, and on March 7 that year the Navy ordered 100, this contract later being cut back to 60. To avoid confusion with Douglas aircraft procured by the Navy with the D manufacturer designation, the McDonnell aircraft became FH-1 Phantoms, after delivery, which began on July 23, 1947. Engines in the FH-1s, all of which were delivered by May 27, 1948, were J30-WE-20s rated at 1,600 lb s.t. each.

On July 21, 1946, a prototype XFD-1 landed for the first time aboard USS *Franklin D. Roosevelt*, this being the first occasion on which a US pure-jet aircraft operated on an aircraft carrier. The first unit deliveries were to VF-17A, and the Phantom subsequently became the first jet fighter in operational service with Marine groups. Its operational career was limited, however, by the advent of newer jet fighters with better performance.

TECHNICAL DATA (FD-1/FH-1)

Manufacturer: McDonnell Aircraft Corporation, St Louis, Missouri.
Type: Carrier-borne fighter.
Accommodation: Pilot only.
Power plant: Two 1,600 lb s.t. Westinghouse J30-WE-20 turbojets.
Dimensions: Span, 40 ft 9 in; length, 38 ft 9 in; height, 14 ft 2 in; wing area, 276 sq ft.
Weights: Empty, 6,683 lb; gross, 12,035 lb.
Performance: Max speed, 479 mph at sea level; cruising speed, 248 mph; initial climb, 4,230 ft/min; service ceiling, 41,100 ft; range, 980 st miles.
Armament: Four fixed forward-firing 0·50-in guns.
Serial numbers:
    XFD-1: 48235–48236.   FD-1: 111749–111808.

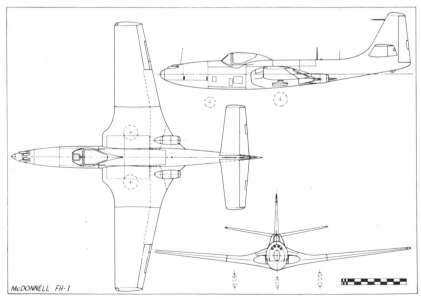

McDONNELL FH-1

McDonnell F2H-2P serving with Marine Squadron VMF-1 in Korea. (*Clay Jansson*)

McDonnell F2H Banshee

At the same time that a production contract was placed for the McDonnell FD-1 (see page 295), the US Navy authorized the company to proceed with prototypes of a larger development which was destined to be among the fighter-bombers available to the US Navy for operations in Korea. A March 2, 1945, contract covered two XF2D-1 prototypes, the designation changing to F2H soon after these were completed.

In general appearance the F2H closely resembled its progenitor, but sizes and weights were increased, and the engines were 3,000 lb s.t. Westinghouse J34-WE-22s. Armament comprised four 20 mm cannon in the nose, and the prototype, first flown on January 11, 1947, had dihedral on the tailplane like the FH-1. This latter feature was absent from the production model F2H-1 Banshees, 56 of which had been ordered in May 1947. Deliveries to Navy Squadron VF-171 began in March 1949.

The F2H-2, which followed the -1 later in 1949, had a longer fuselage and fixed wingtip tanks, both features being intended to increase the internal fuel capacity and hence the range. Engines were 3,250 lb s.t. J34-WE-34s, and all F2H-2s were eventually modified for inflight refuelling. Production, completed by September 1952, totalled 364, plus a further 14 F2H-2N night fighters with nose radomes and 58 F2H-2Ps carrying reconnaissance cameras in a lengthened nose.

The continuing search for additional range led to the third Banshee model, the F2H-3, which had a further extension in the fuselage to accommodate two more tanks. Search radar was a standard fitting, made possible by relocation of the four cannon farther aft in the fuselage sides.

The 250 F2H-3s had the same engines as the -2 model; the 150 F2H-4s delivered in 1953 had 3,600 lb s.t. J34-WE-38s and different radar. Both operated in the all-weather role.

The surviving F2H-3s and F2H-4s were redesignated F-2C and F-2D respectively in 1962. A few remained in the inventory in 1965, mostly in storage.

TECHNICAL DATA (F2H-2)

Manufacturer: McDonnell Aircraft Corporation, St Louis, Missouri.
Type: Carrier-borne fighter.
Accommodation: Pilot only.
Power plant: Two 3,250 lb s.t. Westinghouse J34-WE-34 turbojets.
Dimensions: Span, 44 ft 10 in; length, 40 ft 2 in; height, 14 ft 6 in; wing area, 294 sq ft.
Weights: Empty, 11,146 lb; gross, 22,312 lb.
Performance: Max speed, 532 mph at 10,000 ft; cruising speed, 501 mph; initial climb, 3,910 ft/min; service ceiling, 44,800 ft; range, 1,475 st miles.
Armament: Four fixed forward-firing 20 mm guns. Provision for two 500 lb bombs
Serial numbers:

XF2D-1: 99858–99860.
F2H-1: 122530–122559; 122990–123015.
F2H-2: 123204–123299; 123314–123382; 124940–125071; 125500–125505; 125649–125679; 128857–128886.

F2H-2N: 123300–123313.
F2H-2P: 125072–125079; 125680–125706; 126673–126695.
F2H-3: 126291–126350; 126354–126489; 127493–127546.
F2H-4: 126351–126353; 127547–127693.

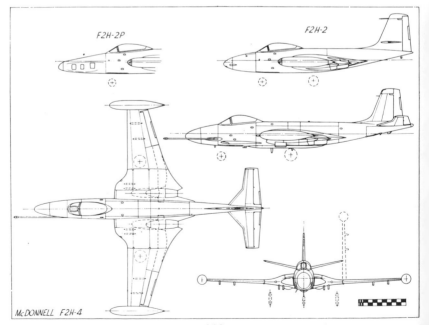

298

McDonnell F3H-2M carrying four Sparrow AAMs in 1957. (*Peter M. Bowers*)

McDonnell F3H Demon

By 1949 US Navy experience with jet fighters aboard aircraft carriers had reached the point where it was possible to plan for future operation of aircraft comparable in all respects with their land-based contemporaries. Aerodynamics, structural and propulsion advances promised to off-set the inherent limitations of carrier operation and on September 30, 1949, the Navy ordered from McDonnell two prototypes of a new design intended to prove the point. In fact, this high-performance interceptor was to be dogged by misfortune, and its operational career was seriously compromised by the costly failure of the Westinghouse J40 engine programme, around which the XF3H-1s were designed. Nevertheless, the aircraft was an essential link in the development of McDonnell Navy fighters which led eventually to the outstandingly successful F-4 Phantom II (see page 301).

The XF3H-1 Demon, first flown on August 7, 1951, was the first of the McDonnell Navy fighters to have a sweptback wing, and unlike the earlier types it had only one engine. In the prototype, this was the 7,200 lb s.t. XJ40-WE-6, but the increased weight of the production model (150 F3H-1Ns were ordered in 1951) required the more powerful J40-WE-24. This engine never materialized. Instead, the first 56 F3H-1s used the 7,200 lb s.t. J40-WE-22, with which they proved to be underpowered.

McDonnell F3H-2N, radar equipped for night and all-weather operation by VF-124 in 1956. (*Peter M. Bowers*)

299

A switch to the 9,700 lb s.t. Allison J71-A-2 was made in the F3H-2, first flown in June 1955. In addition to original production of this model, 29 of the batch of F3H-1s were re-engined to the same standard, prior to delivery of the definitive F3H-2s.

All Demons carried a built-in armament of four 20-mm cannon and were fitted with Hughes APG-51 radar. The 239 F3H-2s (F-3Bs after 1962) were strike fighters with a variety of underwing loads. The Navy also received 95 F3H-2Ms (later MF-3Bs) armed with four AIM-7C Sparrow III missiles and 125 F3H-2Ns (F-3Cs) limited all-weather fighters carrying four AIM-9C Sidewinders each. Production ended in November 1959, and the Demon remained in front-line service until 1965, when the last F-3B-equipped unit began to receive F-4Bs as replacements.

TECHNICAL DATA (F3H-2)

Manufacturer: McDonnell Aircraft Corporation, St Louis, Missouri.
Type: Carrier-borne fighter.
Accommodation: Pilot only.
Power plant: One 9,700 lb s.t. Allison J71-A-2E turbojet.
Dimensions: Span, 35 ft 4 in; length, 58 ft 11 in; height, 14 ft 7 in; wing area, 519 sq ft.
Weights: Empty, 22,133 lb; gross, 33,900 lb.
Performance: Max speed, 647 mph at 30,000 ft; initial climb, 12,795 ft/min; service
  ceiling, 42,650 ft; max range, 1,370 st miles.
Armament: Four fixed forward-firing 20 mm guns. Bomb or rocket loads underwing.
Serial numbers (allocations):

XF3H-1: 125444–125445.
F3H-1N, -2, -2N, -2M: 133489–133638.
F3H-2: 143403–143492;   145202–
  145306;   146328–146339;
  146709–146742.

F3H-2N: 136966–137032.
F3H-2M: 137033–137095.

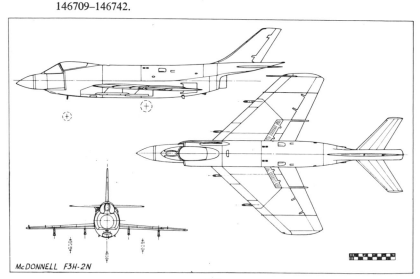

McDONNELL F3H-2N

300

McDonnell F-4B in the markings of VF-103 aboard the USS *Saratoga* in 1967.
(*Stephen P. Peltz*)

McDonnell Douglas F-4 Phantom II

Action to procure what was to become one of the finest air weapons the Navy had ever used began in 1954. On October 18 that year the Navy placed a letter of intent with McDonnell in respect of a new project which the company had started a year earlier for a twin-engine attack fighter. When first ordered, the aircraft carried an attack designation, AH-1, but early in 1955 some fundamental changes were made in the specification to make the primary role that of a long-range high-altitude interceptor. For this role, the proposed cannon armament was abandoned in favour of air-to-air missiles, and a second crew member was added to serve as a radar operator; the designation was then changed to F4H-1 and the original contract for two YAH-1s was changed to one for 23 F4H-1 test aircraft. McDonnell's earlier Navy fighter, the FH-1, being no longer in service, the new type was named Phantom II.

The F4H-1 was a large low-wing monoplane designed around two General Electric J79 turbojets, the 10,900 lb s.t. J79-GE-8 version being specified for the production aircraft. Tandem seating was provided for the pilot and observer; APQ-72 fire-control radar was located in the nose, and provision was made for six Sparrow III air-to-air missiles under the

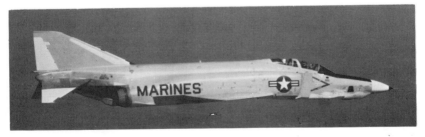

McDonnell RF-4B for the Marine Corps, showing nose-mounted cameras.
(*McDonnell photo*)

301

McDonnell F-4J on its first flight in May 1966. (*McDonnell photo*)

fuselage and wings. Powered temporarily by J79-GE-3A engines on loan from the USAF and located side by side in the fuselage behind variable-geometry intakes, the F4H-1 first flew on May 27, 1958. By the end of 1958, the Phantom II had been selected over the Chance Vought F8U-3 Crusader III as new standard equipment for US Navy squadrons, and preparations for full-scale production began.

Early flight trials led to a number of changes to the F4H-1, including dihedral on the outer wing panels and anhedral on the tailplane. A blown-flap system of boundary layer control was first tested on the sixth aircraft and was adopted for production. This same aircraft made the first carrier-suitability trials in February 1960, and at the end of the year VF-101 began to receive the first aircraft for transition training. Pending availability of the -8 engine, the trials aircraft and the first 24 production Phantom IIs had J79-GE-2 or 2A engines and were designated F4H-1F. These aircraft became F-4As in 1962, when the final production standard F4H-1 became the F-4B, and the proposed camera-equipped F4H-1P became the RF-4B.

Meanwhile, the USAF had adopted the Phantom II as an air superiority fighter and close support aircraft for Tactical Air Command—an unprecedented move. After a short period during which the USAF versions were designated F-110A and RF-110A, they became F-4C and RF-4C in the unified tri-service system.

Deliveries of the definitive F-4B to Navy and Marine squadrons proceeded during 1961, and 29 squadrons were flying the type by 1966. The primary role for which the F-4B had been purchased was interception and air superiority, but deployment of the type to Vietnam put emphasis upon the secondary strike role, carrying a wide range of external stores.

One squadron, VF-213, flew the F-4G, 12 examples of which were produced by modifying F-4Bs. These aircraft had ASW-21 two-way tactical communications for operational development of a data link system, and were operational over Vietnam from the end of 1965 onwards. First flown on March 12, 1965, the RF-4B carried a variety of cameras in the nose, and the 46 built were issued to Marine squadrons to replace the RF-8A Crusader.

Succeeding the F-4B in production after 649 had been built, the F-4J

302

incorporated a number of improvements, several of which had first been developed for USAF Phantom IIs. These changes included adoption of the AJB-7 bombing system, Westinghouse AWG-10 pulse Doppler radar fire-control, a data link system, 17,900 lb s.t. J79-GE-10 engines and larger wheels to permit a 38,000-lb landing weight. The first F-4J flew in May 1966, and this version went into production to equip both Navy and Marine squadrons. A total of 522 was built, of which 398 remained in the active inventory by January 1973.

In January 1971, the US Navy launched a programme (Project *Bee-line*) to modify and update 178 F-4Bs, pending introduction of the F-14. Designated F-4N, these aircraft were modified up to the standard of the final production F-4B, with the addition of AIMS, data link, air-to-air IFF, dogfight computer and pilot lock-on mode and other equipment, plus structural strengthening to extend the fatigue life. The first F-4N was delivered in February 1973 and the programme was completed at the end of 1975. Some F-4As were redesignated TF-4A when assigned to crew training duties, and were not carrier deployable in this guise.

During 1972, the Naval Air Development Center converted a Phantom II to QF-4B configuration to provide a supersonic manoeuvring target for use in connection with new missile development. Following successful flight testing, several more similar conversions were put in hand in 1974.

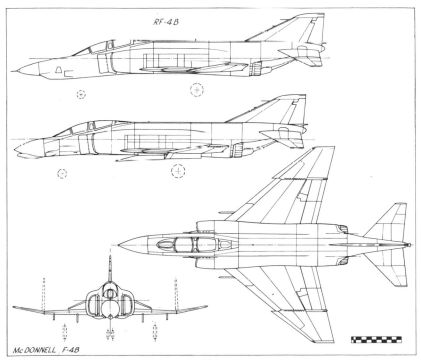

RF-4B

McDONNELL F-4B

303

A McDonnell Douglas F-4B Phantom II of US Navy tactical electronic warfare Squadron VAQ-33. (*US Navy*)

During 1975, work began to upgrade Navy F-4Js to F-4S configuration, with structural strengthening similar to that introduced in the F-4N, plus provision for the later installation of leading-edge manoeuvre slats as used on the USAF F-4E.

TECHNICAL DATA (F-4B)

Manufacturer: McDonnell Aircraft Company (Division of McDonnell Douglas Corporation), St Louis, Missouri.

Type: Carrier-borne fighter and tactical strike aircraft.

Accommodation: Pilot and radar intercept officer.

Power plant: One 10,900 lb s.t. (17,000 lb with afterburner) General Electric J79-GE-8B or -8C turbojet.

Dimensions: Span, 38 ft 4¾ in; length, 58 ft 3¾ in; height, 16 ft 3 in; wing area, 530 sq ft.

Weights: Empty, 28,000 lb; normal take-off, 46,000 lb; max take-off, 54,600 lb.

Performance: Max speed, 1,485 mph at 48,000 ft; initial climb, 28,000 ft/min; service ceiling, 62,000 ft; combat radius as intercepter, 900 miles; ferry range, 2,300 miles.

Armament: Up to six underwing AIM-7E Sparrow III missiles or four AIM-7E and four AIM-9B or -9D Sidewinder missiles. Up to 16,000 lb assorted stores on five external strong points.

Serial numbers:

F-4A: 142259–142260; 143388–143392; 145307–145317; 146817–146821; 148252–148275.

F-4B: 148363–148434; 149403–149474;* 150406–150493;*† 150624–150653;*† 150993–151021;* 151397–151519; 152207–152331; 152965–153070; 153912–153915.

RF-4B: 151975–151983; 153089–153115; 157342–157351.

F-4J: 153071–153088; 153768–153911; 154781–154788; 155504–155580; 155731–155903; 157242–157309; 158346–158379.

* 30 from these batches loaned to USAF.

† Includes 12 converted to F-4G.

A McDonnell Douglas A-4E Skyhawk of Fleet Composite Squadron One, VC-1, equipped with a tanker package. (*US Navy*)

McDonnell Douglas A-4 Skyhawk

Equipping some 30 US Navy and Marine attack squadrons during 1968, the Skyhawk was one of the primary US weapons in the Vietnam War, supporting troops in their offensive against the guerrillas in South Vietnam and attacking ground targets in North Vietnam. Close support and interdiction missions of this kind were precisely the type of operation for which the Skyhawk was designed in 1950-2, when the lessons of similar operations in Korea were still being learned. Sometimes known as 'Heinemann's Hot-Rod' after its chief designer, Ed Heinemann, the Skyhawk was an exercise in design optimization for a specific role—the US Navy purchased over 2,000 examples in five versions without ever varying the attack mission —and was an outstanding example of simple, lightweight design.

About the time the US Navy first turned its attention to a jet-powered successor to the ubiquitous AD-1 Skyraider (see page 176), the Douglas design group at El Segundo was taking a long, critical look at the growing weight and complexity of contemporary combat aircraft. Putting the results of this survey into practice, Ed Heinemann's team was able to offer the Navy a new attack aeroplane with a gross weight just half of the 30,000 lb proposed in the official specification. On June 21, 1952, a contract was placed for prototypes and pre-production aircraft, with the original A4D-1 designation, later changed to A-4.

The Skyhawk was laid out as a low-wing monoplane with a modified delta wing planform of low aspect ratio and a conventional cruciform tail unit. The wing was a three-spar structure in one piece from tip to tip, with an integral fuel tank between the spars. Additional fuel was carried in the fuselage just behind the cockpit; electronics were concentrated in the nose section, and the rear fuselage was detachable to give access to the Wright J65 (licence-built Armstrong Siddeley Sapphire) engine. Armament com-

305

prised two 20 mm guns in the wing roots and three external store stations were provided, one under the fuselage and one under each wing. These had a total capacity of 5,000 lb, carrying a wide variety of bombs, rockets, gun pods, guided missiles or fuel tanks.

The first of the two prototypes flew on June 22, 1954, and the first of the pre-production aircraft followed on August 14, 1954, powered by a 7,200 lb s.t. J65-W-2 engine. Production A-4As (A4D-1) had the 7,700 lb s.t. J65-W-4 or J65-W-4B, and deliveries to Navy Attack Squadron VA-72 began on October 26, 1956. Production of this version totalled 165.

First flown on March 26, 1956, the A-4B (A4D-2) differed from the initial production version in having numerous detail refinements, a 7,700 lb s.t. J65-W-16A engine, and provision for inflight refuelling, a probe being fitted to the starboard side of the forward fuselage. The rudder was changed to a single-surface type with external strengthening and powered operation, and the rear fuselage was strengthened. Production totalled 542, and these were widely used by Navy and Marine units until 1965, when they were phased out of front-line service, many then becoming TA-4Bs for non-combatant use.

Provision of equipment necessary to give the Skyhawk limited all-weather capability resulted in a new model flown on August 21, 1959, as the A-4C (A4D-2N). As well as the radar equipment in the nose, this model had further improvements in cockpit layout and safety provisions, with the Douglas Escapac ejection-seat. Production totalled 638. Some later became A-4Ls with -20 engine, improved avionics in a dorsal hump and provision to carry Walleye or Shrike missiles.

During 1957 the Navy ordered ten examples of the A-4D (A4D-3), a version of the A-4B with Pratt & Whitney J52-P-2 turbojet. This programme was cancelled, but the proposal to use the new engine was revised in 1960 in a more extensive revision of the basic Skyhawk which first flew on July 12, 1961, as the A-4E (A4D-5). Apart from having the 8,500 lb thrust J52-P-6A engine which, by virtue of the lower fuel consumption,

Douglas A-4C of VX-5 in February 1967, with three-tone camouflage. Oil from 'breathers' has washed off the paint from rear fuselage and the J65-powered versions of the Skyhawk were not subsequently camouflaged. (*Lawrence S. Smalley*)

306

Douglas TA-4F trainer with full underwing load. (*Douglas photo*)

extended the Skyhawk's range by some 27 per cent, the A-4E had a completely re-engineered airframe with two more underwing store stations and a total external load of 8,200 lb. Production of 499 was completed in April 1966, these aircraft being used to re-equip squadrons flying A-4Bs and A-4Cs.

Following Douglas design studies of single- and two-seat versions of the Skyhawk for export, the Navy ordered the two-seat TA-4E advanced trainer in 1964, as a replacement for the TF-9J. Two cockpits in tandem were provided in a slightly lengthened fuselage, the second cockpit restricting the size of the fuselage fuel tank; all other operational features of the A-4E were retained. The engine was the 9,300 lb s.t. J52-P-8A, and the two prototypes flew on June 30 and August 2, 1965, respectively, followed by the first production aircraft in April 1966, at which time the designation was changed to TA-4F. The first TA-4F was delivered to VA-125 at NAS Lemoore on May 19, 1966 and production eventually totalled 238. These were followed, from mid-1969 onwards, by more than 200 TA-4Js which were similar but had many items of non-essential equipment deleted, although provisions were retained.

With the same up-rated engine as the TA-4F, the A-4F first flew on August 31, 1966, and production quantities were ordered as part of the Navy procurements for Vietnam operations. New features of the A-4F

Douglas A-4F showing 'hump' over new avionics compartment. (*Douglas photo*)

included a zero-zero ejection-seat, nosewheel steering, wing spoilers and new avionics in a compartment behind the cockpit, producing a distinctive hump. Deliveries of the A-4F began on June 20, 1967, and the 146 delivered (plus one prototype, ordered as the 500th and last A-4E) were the final operational Skyhawks built for the US Navy. A new version, the A-4M, appeared in 1970, specifically for use by the US Marine Corps, however, and a programme to acquire 170 of these Skyhawk IIs began later in the same year. With an uprated J52-P-408A turbojet, the A-4M featured an enlarged windscreen and canopy, a braking parachute for short-field operation and increased ammunition capacity. The first of two prototypes flew on 10 April, 1970, and the first production delivery was made on 16 April, 1971, when four A-4Ms went to VMA-324 at Beaufort, SC. All five active USMC light attack squadrons used A-4Ms by 1976.

The US Navy also procured several Skyhawk variants for supply to overseas nations, these being the A-4G and TA-4G for Australia, A-4K and TA-4K for New Zealand and A-4H, TA-4H and A-4N for Israel; the last-mentioned also received some A-4Es from US Navy stocks during the October 1973 war. Surplus A-4Bs supplied to Argentina were re-designated A-4P and A-4Q, and those supplied to Singapore, A-4S and TA-4S. Total Skyhawk production exceeded 3,000 by 1975.

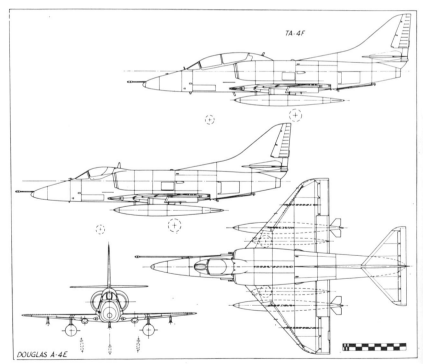

TA-4F

DOUGLAS A-4E

McDonnell Douglas TA-4J Skyhawk of Fleet Composite Squadron VC-1, with Commanding Officer's penant amidships. (*US Navy*)

TECHNICAL DATA (A-4M)

Manufacturer: Douglas Aircraft Company (Division of McDonnell Douglas Corporation), El Segundo and Long Beach, California.

Type: Carrier-based attack-bomber.

Accommodation: Pilot only.

Power plant: One 11,200 lb s.t. Pratt & Whitney J52-P-408A turbojet.

Dimensions: Span, 27 ft 6 in; length (excluding refuelling probe), 40 ft 3¾ in; height, 15 ft 0 in; wing area, 260 sq ft.

Weights: Empty, 10,465 lb; gross, 24,500 lb.

Performance: Max speed, 670 mph at sea level; initial rate of climb, 8,440 ft/min; tactical radius, with 4,000-lb bomb load, 340 miles.

Armament: Two fixed forward-firing 20-mm guns. Up to 9,155 lb of ordnance externally on five strong points.

Serial numbers:

XA4D-1: 137812.

A-4A: 137813–137831; 139919–139970; 142142–142235.

A-4B: 142082–142141; 142416–142423; 142674–142953; 144868–145061

A-4C: 145062–145146; 147669–147849; 148304–148317; 148435–148612; 149487–149646; 150581–150600.

A-4E: 148613–148614; 149647–149666; 149959–150138.

A-4F: 152101; 154172–154217; 154990–155069.

TA-4F: 152102–152103; 152846– 152878; 153459–153531; 153660–153690; 154287– 154343; 154614–154657.

TA-4J: 155070–155119; 156891– 156950; 158073–158147; 158438–158527; 158712– 158723; 159099–159104; 159546–159556; 159795– 159798 (plus later contracts).

A-4M: 158148–158196; 158412– 158435; 159470–159493; 159778-159794; 160022– 160045 (plus later contracts).

Great Lakes TG-2, the final production version of the Martin T4M airframe. (*US Navy*)

Martin T3M/T4M

From its experience in building the Navy-designed SC-1 bomber-torpedo scouts for the US Navy (see page 123), the Glenn L. Martin Company was able to offer an improved version when the Navy needed new aircraft in this category. A contract dated October 12, 1925, was placed for this new version with the designation T3M-1. It differed from the SC-1 in two major respects: the fuselage was a steel-tube structure, and the front cockpit, in which the pilot and bombardier sat side by side, was moved forward, ahead of the wing. The gunner's cockpit remained unchanged, and a large radiator was slung under the top wing between the cockpits.

In common with its contemporaries, the T3M-1 could operate on wheels or floats. The engine was a 575 hp Wright T-3B. Deliveries began in

Martin T3M-1 prototype showing the longer-span lower wing and unusual radiator position. (*Martin photo*)

September 1926, but comparatively little use was made of the 24 aircraft of this type built. They were quickly followed by the Packard-engined T3M-2, in which the span of the upper wing was increased to equal that of the lower, and the crew disposition was changed to three in individual cockpits in tandem, with the pilot in the centre cockpit and the gunner well back along the fuselage. The Navy ordered 100 T3M-2s, and by mid-1927 this version was serving with VT-1S (USS *Lexington*) and VT-2B (USS *Langley*). It quickly became the standard equipment of all Navy torpedo squadrons, whether ship- or shore-based, and it continued to serve with a number of utility units at various shore bases after its retirement from front-line units.

The first T3M-2 was re-engined by Martin with a Pratt & Whitney Hornet radial engine as the XT3M-3, and then (modified by the NAF) became the XT3M-4 with a Wright R-1750 Cyclone. As a result of these experimental installations, the Navy ordered a new prototype (Martin 74) with the Hornet engine, designated the XT4M-1 and first flown early in April 1927. Apart from the new engine, it had shorter wings and a revised rudder. Like its predecessor, it could operate on floats

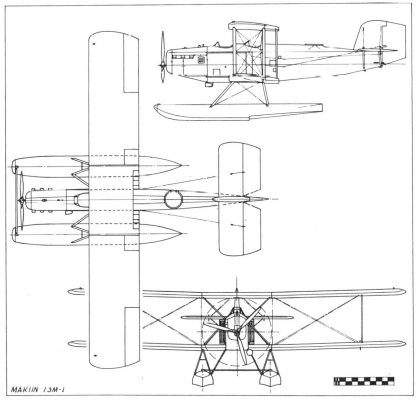

MARTIN T3M-1

311

as well as wheels, but the growing importance of the aircraft carrier during the service lifetime of the T4M-1 meant that most of the 102 built operated on wheels with carrier-based squadrons. The first production contract was placed on June 30, 1927, and operational deployment began in August 1928 with deliveries to VT-2B on the *Saratoga* and VT-1B on the *Lexington*, replacing T3M-2s in both these units.

After the Great Lakes Company had acquired Martin's Cleveland factory in October 1928, production was continued under the new company's designation of TG-1, with an R-1690-28 engine and slightly modified undercarriage. In addition to 18 TG-1s, Great Lakes built 32 TG-2s with

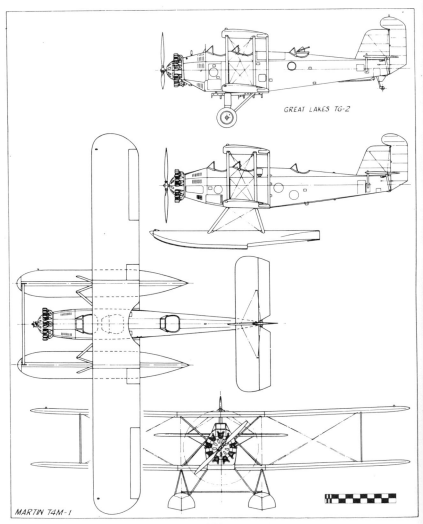

GREAT LAKES TG-2

MARTIN T4M-1

Wright R-1820-86 Cyclone engines. Deliveries by Great Lakes, by this time a subsidiary of Detroit Aircraft Corporation, began in 1930, and the TG-2 outlived the T4M-1 in operational use by several years, serving with VT-2 until 1937. Several T4M-1s went to Reserve units in 1932, where the last one remained in use also until 1937. The passing of these distinctive biplanes ended an era in Navy torpedo-bomber operations, their successors being the streamlined metal monoplanes such as the Douglas TBD-1 which opened up a whole new concept of aerial warfare at sea.

TECHNICAL DATA (T3M-2)

Manufacturer: Glenn L. Martin Company, Cleveland, Ohio.
Type: Torpedo-bomber-scout.
Accommodation: Pilot, bombardier, gunner.
Power plant: One 770 hp Packard 3A-2500.
Dimensions: Span, 56 ft 7 in; length, 41 ft 4 in; height, 15 ft 1 in; wing area, 883 sq ft.
Weights: Empty, 5,814 lb; gross, 9,503 lb.
Performance: Max speed, 109 mph at sea level; initial climb, 16·8 min to 5,000 ft; service ceiling, 7,900 ft; range, 634 st miles.
Armament: One flexible 0·30-in gun in rear cockpit.
Serial numbers:
T3M-1: A7065–A7088.   XT3M-3, XT3M-4: A7224.
T3M-2: A7224–A7323.

TECHNICAL DATA (T4M-1)

Manufacturer: Glenn L. Martin Company, Cleveland, Ohio.
Type: Torpedo-bomber-scout.
Accommodation: Pilot, bombardier, gunner.
Power plant: One 525 hp Pratt & Whitney R-1690-24.
Dimensions: Span, 53 ft; length, 35 ft 7 in; height, 14 ft 9 in; wing area, 656 sq ft.
Weight: Empty, 3,931 lb; gross, 8,071 lb.
Performance: Max speed, 114 mph at sea level; cruising speed, 98 mph; initial climb, 14 min to 5,000 ft; service ceiling, 10,150 ft; range, 363 st miles.
Armament: One flexible 0·30-in gun in rear cockpit.
Serial numbers:
XT4M-1: A7566.   T4M-1: A7596–A7649; A7852–A7899.

TECHNICAL DATA (TG-2)

Manufacturer: Great Lakes Aircraft Corporation, Cleveland, Ohio.
Type: Torpedo-bomber-scout.
Accommodation: Pilot, bombardier, gunner.
Power plant: One 620 hp Wright R-1820-86.
Dimensions: Span, 53 ft; length, 34 ft 8 in; height, 14 ft 10 in; wing area, 656 sq ft.
Weights: Empty, 4,670 lb; gross, 9,236 lb.
Performance: Max speed, 127 mph at sea level; initial climb, 11 min to 5,000 ft; service ceiling, 11,500 ft; range, 330 st miles.
Armament: One flexible 0·30-in gun in rear cockpit.
Serial numbers:
TG-1: A8458–A8475.   TG-2: A8697–A8728.

Martin BM-1, the production version of the Navy-designed XT2N-1.
(*Gordon S. Williams*)

Martin BM

Evolution of the dive-bombing technique, primarily by Navy and Marine squadrons flying two-seat fighter-bombers, led, in 1928, to a Bureau of Aeronautics specification for a special-purpose dive-bomber. BuAer design No. 77 specified an aeroplane which could carry a 1,000 lb bomb (or a torpedo—the Navy was reluctant to abandon its tradition of using dual-purpose aeroplanes) and to be stressed to pull out of a terminal velocity dive even with the bomb still attached.

Single prototypes of the BuAer design were ordered from the Naval Aircraft Factory and the Martin company, respectively designated XT2N-1 and XT5M-1 as two-seat torpedo-bombers. The contract for the Martin 125 was dated June 18, 1928, and the aircraft was delivered for trials at Anacostia Naval Air Station in 1930, powered by a 525 hp Pratt & Whitney R-1690-22 Hornet radial engine. A metal-framed biplane with fabric-covered wings and metal-covered fuselage and tail, the XT5M-1 seated pilot and rear-gunner in tandem.

Tests of the XT5M-1, completed in March 1930, demonstrated its ability to pull out of a dive as required, and a production order for 12 aircraft was placed on April 9, 1931. These were designated BM-1 in the Navy's new Bomber category, and deliveries began on September 28, 1931. They were similar to the prototype but had 625 hp R-1690-44 engines; ring cowlings for the engines and wheel fairings were designed, but were removed after the aircraft went into service with Navy Torpedo Squadron One (VT-1S) in 1932, serving aboard USS *Lexington*. In the same year, deliveries began of 16 BM-2s which had been ordered on October 17, 1931; these differed only in small details from the initial version. With four additional BM-1s ordered later, these aircraft served alongside BM-1s in VT-1S, which was redesignated VB-1B (Bombing Squadron One) in 1934. A second unit, VB-3B, was equipped with BM-1s and BM-2s in 1934 and served aboard the USS *Langley*. Subsequently, these aircraft served with a number of other units; they were withdrawn from fleet service in 1937 and continued in use at shore bases for assorted test and utility duties until 1940 when the last BM-2 was scrapped.

In addition to the 32 production BM-1s and BM-2s, Martin built a single XBM-1 in 1932. Its experimental designation indicated its use for special tests by the NACA.

TECHNICAL DATA (BM-2)

Manufacturer: Glenn L. Martin Company, Baltimore, Maryland.
Type: Dive-bomber and torpedo-bomber.
Accommodation: Pilot and gunner.
Power plant: One 625 hp Pratt & Whitney R-1690-44.
Dimensions: Span, 41 ft; length, 28 ft 9 in; height, 12 ft 4 in; wing area, 436 sq ft.
Weights: Empty, 3,662 lb; gross, 6,218 lb.
Performance: Max speed, 146 mph at 6,000 ft; climb, 5·7 min to 5,000 ft; service ceiling, 16,800 ft; range, 413 st miles with 1,000 lb bomb.
Armament: One fixed forward-firing 0·30-in gun; one flexible 0·30-in gun in rear cockpit.
Serial numbers:

XT2N-1: A8052. XBM-1: 9212.
XT5M-1: A8051. BM-2: A9170–A9185.
BM-1: A8879–A8890; 9214–9217.

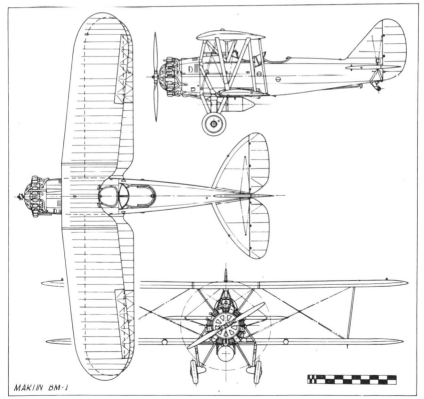

MARTIN BM-1

315

Martin P3M-2 flying-boat, the production version of the Consolidated XPY-1.
(*National Archives 80-A-3717*)

Martin P2M, P3M

The Glenn L. Martin Company's first association with flying-boats for the US Navy was as a producer of designs already completed by other organizations; first the Naval Aircraft Factory and then the Consolidated company. The foundation laid in this way allowed Martin to gain a major foothold in the flying-boat market, however, and eventually oust its principal competitor, Consolidated, to become the sole source of Navy patrol flying-boats after World War II.

The first Martin boats were versions of the NAF PN series (see page 334) ordered on May 31, 1929, and delivered in 1930–1. Only a month after this first contract, Martin successfully underbid Consolidated to produce the latter's new XPY-1. The XPY-1 was itself built to NAF designs for a 100-ft span monoplane boat—the first to break the Navy's biplane tradition. Powered by two 450 hp Pratt & Whitney R-1340-38 engines in nacelles under the wings plus a third similar engine, initially, above the centre section of the parasol wing, the XPY-1 completed its tests in the first half of 1929, and at the end of June in that year Martin received contracts for its development and production.

The development contract was for a prototype designated XP2M-1 which was to have the same wing as the XPY-1, but located closer to the fuselage and with 575 hp Wright R-1820-64 engines in nacelles on the leading edge. Like the XPY-1, it was tested with a third engine above the centre section; when this was deleted the designation became XP2M-2.

The production contract awarded Martin on June 29, 1929, was to build nine copies of the XPY-1. The first appeared in February 1931 as the P3M-1, and was powered like the XPY-1 by two R-1340s; later aircraft on the batch were P3M-2s with 525 hp Pratt & Whitney R-1690-32 Hornet engines and an enclosure over the pilots' cockpits. The P3M-1s served with VP-10S for a time in 1931–2, but these boats achieved little operational success and were used primarily for training and miscellaneous duties.

TECHNICAL DATA (P3M-2)

Manufacturer: Glenn L. Martin Company, Baltimore, Maryland.
Type: Patrol flying-boat.
Accommodation: Crew of five.
Power plant: Two 525 hp Pratt & Whitney R-1690-32s.
Dimensions: Span, 100 ft; length, 61 ft 9 in; height, 16 ft 8 in; wing area, 1,119 sq ft.
Weights: Empty, 10,032 lb; gross, 17,977 lb.
Performance: Max speed, 115 mph at sea level; climb to 5,000 ft, 10·9 min; service ceiling, 11,900 ft; range, 1,010 st miles.
Armament: Two flexible 0·30-in guns each in bows and dorsal position.
Serial numbers:

XPY-1: A8011.	P3M-1: A8412–A8414.
XP2M-1: A8358.	P3M-2: A8415–A8420.

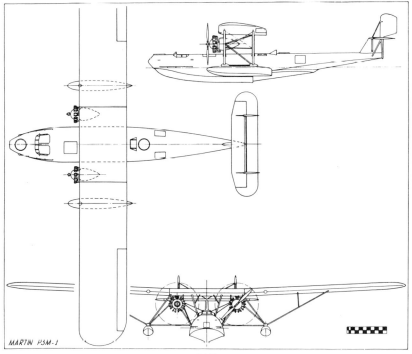

MARTIN P3M-1

317

Martin PBM-3D Mariner with search radar housing above the forward fuselage.
(*Martin photo*)

Martin PBM Mariner

Having established itself as a builder of Navy flying-boats by producing the PM, P2M and P3M series to NAF designs (see page 316), the Glenn Martin Company logically turned to the design of a new flying-boat to succeed these earlier types. Designed in 1937, the Model 162 continued the rivalry which had sprung up between Martin and Consolidated by challenging the latter company's PBY Catalina. A somewhat later design than the PBY, the Martin 162 was in due course to demonstrate a marked superiority of performance, and although it served in smaller quantities than the PBY during World War II, it continued to give important service for many years after 1945.

The Model 162 design featured a deep hull with a gull wing and two Wright Cyclone engines. To test the handling qualities of the design, Martin built a single-seat, quarter-scale model known as the Model 162A, and on June 30, 1937, the Navy placed a contract for a single full-size prototype, to be designated XPBM-1. First flown on February 18, 1939, the XPBM-1 had 1,600 hp R-2600-6 engines and provision for nose and dorsal turrets plus additional gun positions at the waist and tail positions.

Martin PBM-3 in silver finish used for training in 1942–3. (*USAF photo*)

In addition to six guns, the XPBM-1 was designed to carry 2,000 lb of bombs or depth-charges. It had retractable stabilizing floats under the wings and, as first flown, a flat tailplane with outrigged fins; later, dihedral was added to the tailplane, canting the fins inwards to give the Martin boat one of its most striking characteristics.

At the end of 1937 the Navy ordered 20 production model PBM-1s, for which the name Mariner was eventually chosen, and a single XPBM-2 with extra fuel and provision for catapult launching. All these were completed by April 1941 and were similar to the XPBM-1 with dihedral on the tailplane. Most of the PBM-1s went into service during 1941 with VP-74, a merger of the former VP-55 and VP-56 squadrons, and in the same year the prototype went to the Aircraft Armament Unit at Norfolk Naval Air Station as the XPBM-1A for armament experiments.

Orders were placed with Martin for 379 PBM-3 Mariners on November 1, 1940, but these appeared, from April 1942 onwards, in several different versions. All Mariners from the -3 model onward, however, had fixed, strut-braced wing floats and lengthened engine nacelles, the latter providing stowage for the bombs or depth-charges. The basic PBM-3 had

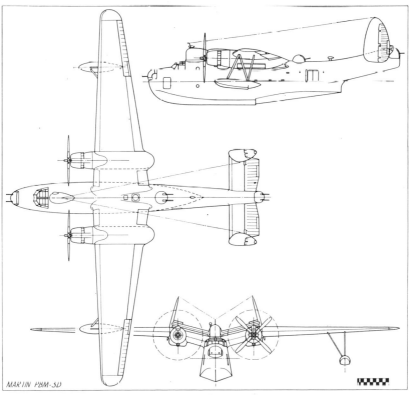

MARTIN PBM-3D

319

R-2600-12 engines, and the variants included 50 unarmed PBM-3R transports with seats for 20 passengers, 274 PBM-3Cs with standardized US/British equipment, and 201 PBM-3Ds with R-2600-22 engines and improved armament and armour protection. Many of the PBM-3Cs and -3Ds carried search radar in a large housing above and behind the cockpit, and experience with the use of this radar led to development in 1944 of a long-range anti-submarine version, the PBM-3S. A total of 156 of the latter variant was delivered, with R-2600-12 engines.

A 1941 contract for 180 Mariners specified the PBM-4 version, which would have had R-3350-8 engines, but this development was not pursued. Instead, the XPBM-5 appeared in 1943 with 2,100 hp Pratt & Whitney R-2800-22 or -34 engines, and production contracts were placed for this variant in January 1944. The PBM-5, delivered from August 1944 to the end of the war, had eight 0·50-in machine-guns and AN/APS-15 radar. The latter equipment was indicated by the PBM-5E designation, which was in practice not always used; some PBM-5S variants were also in service in the later forties, this designation indicating a change of equipment specifically for the anti-submarine role. Production totalled 631, with the last aircraft in this batch assigned for conversion to the prototype XP5M-1 Marlin (see page 323).

Final Mariner development was an amphibian version, the PBM-5A, 36 of which had been built by the time production was completed in April 1949. The PBM-5A served primarily in the air–sea rescue role with the Navy and Coast Guard, replacing PBM-5G flying-boats in the latter case.

TECHNICAL DATA (PBM-3C)

Manufacturer: Glenn L. Martin Company, Baltimore, Maryland.
Type: Patrol flying-boat.
Accommodation: Crew of nine.
Power plant: Two 1,700 hp Wright R-2600-12s.
Dimensions: Span, 118 ft; length, 80 ft; height, 27 ft 6 in; wing area, 1,408 sq ft.
Weights: Empty, 32,378 lb; gross, 58,000 lb.
Performance: Max speed, 198 mph at 13,000 ft; initial climb, 410 ft/min; service ceiling, 16,900 ft; range, 2,137 st miles.
Armament: Two flexible 0·50-in guns each in nose and dorsal turrets, single 0·50-in guns at waist and tail positions. Up to 2,000 lb bombs or depth-charges.
Serial numbers (allocations):

XPBM-1: 0796.	PBM-3S: 01674–01728; 48125–48163.
PBM-1: 1246; 1248–1266.	XPBM-5: 45275–45276.
XPBM-2: 1247.	PBM-5: 45405–45444; 59000–59348;
PBM-3C: 6505–6754; 01650–01673.	84590–84789; 85136–85160;
XPBM-3E: 6456.	98602–98616.
PBM-3R: 6455; 6457–6504.	XPBM-5A: 59349.
PBM-3D: 45205–45274; 45277–45404;	PBM-5A: 122067–122086; 122468–
48124; 48164–48223.	122471; 122602–122613.

Martin AM-1 illustrating the use of white and red only for the national markings on all-blue aircraft. (*Harold G. Martin*)

Martin AM Mauler

Throughout World War II the US Navy used a variety of carrier-based aircraft for attack duties. They were designated in either one of two categories—SB for scout (and dive) bombers and TB for torpedo-bombers. For the most part, the various designs fell into a well-defined pattern of single radial engine, low wing, two- or three-man crew, rear defensive armament and in some cases internal weapon stowage. This configuration had evolved over more than a decade, and the aircraft cast in this mould gave good service. By 1944, however, the nature of the war at sea had so changed that it was possible to consider a radical new configuration for an aeroplane which would combine both the SB and TB functions. The largest available engine would be used, to obtain high performance and big ordnance load, but internal stowage was rejected as unnecessarily complex and too restrictive in choice of weapons.

The first order for a prototype of such an attack aircraft was placed by the US Navy on the last day of 1943, for the Curtiss XBTC-1. Another similar type was ordered from Kaiser-Fleetwings on March 31, 1944; neither of these types passed the prototype stage. Martin was luckier with its XBTM-1, two prototypes of which were ordered on May 31, 1944, although this did not achieve the success of the fourth of the single-seaters, the Douglas XBT2D-1 ordered on July 21, 1944 (see page 176).

Designed as the Model 210, the Martin XBTM-1 made its first flight on August 26, 1944. Powered by a 3,000 hp Pratt & Whitney XR-4360-4 engine, it had four 20 mm cannon in the wings and 15 strong points under the wings and fuselage for bombs and rockets. Flight trials proved successful, and a production order was placed on January 15, 1945, for 750 BTM-1s, the designation changing to AM-1 Mauler before the first flight on December 16, 1946. Testing and carrier qualification trials continued during 1947, and delivery of the production aircraft to an operational unit began on March 1, 1948, the unit being VA-17A.

By the end of 1949 several other units were flying the AM-1, but production was terminated in October that year with a total of 151 built, including prototypes, so that the Navy could standardize on one attack aircraft. During 1950 all Maulers were assigned to Reserve units to allow first-line squadrons to fly the ADs. Six had been adapted for radar countermeasures with additional equipment and were designated AM-1Q.

TECHNICAL DATA (AM-1)

Manufacturer: Glenn L. Martin Company, Baltimore, Maryland.
Type: Carrier-based attack aircraft.
Accommodation: Pilot only.
Power plant: One 2,975 hp Pratt & Whitney R-3350-4.
Dimensions: Span, 50 ft; length, 41 ft 2 in; height, 16 ft 10 in; wing area, 496 sq ft.
Weights: Empty, 14,500 lb; gross, 23,386 lb.
Performance: Max speed, 367 mph at 11,600 ft; cruising speed, 189 mph; initial climb, 2,780 ft/min; service ceiling, 30,500 ft; range, 1,800 st miles.
Armament: Four fixed forward-firing 20 mm guns. Approx 4,500 lb of assorted ordnance on 15 external points. (Max demonstrated ordnance load, 10,689 lb.)
Serial numbers:
XBTM-1: 85161–85162. AM-1Q: 122388–122393.
AM-1: 22257–22355; 122394–122437.

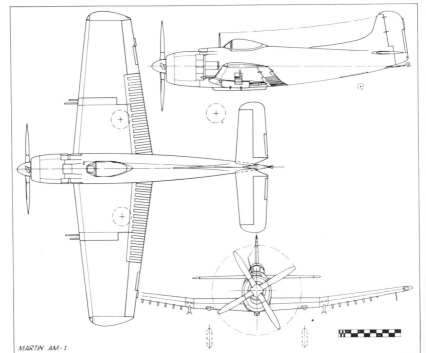

MARTIN AM-1

Martin P5M-2 Marlin in white and dark blue finish, showing the unusual
single-step hull design. (*US Navy*)

Martin P5M Marlin

Evolved from the war-time PBM Mariner (see page 318), the Marlin
was the US Navy's last operational flying-boat, entering service in 1952 and
continuing in front-line use through 1966. As Martin M-237, the design
began in 1946 with a Mariner wing and upper hull combined with a new
lower hull having a length-to-beam ratio of 8·5:1. Construction of a
prototype was authorized by a US Navy contract dated June 26, 1946, and
this prototype, designated XP5M-1, flew on May 30, 1948. Powered by
Wright R-3350 radial engines, the XP5M-1 also differed from its predecessor in having radar-operated nose and tail turrets, as well as a power-
operated dorsal turret.

A July 1950 contract put the P5M into production and the first production model flew on June 22, 1951, displaying a number of differences from
the prototype, and carrying the type name Marlin. A large radome for
APS-80 search radar replaced the nose turret, the dorsal turret was removed, the flight deck was raised for better visibility, and up-rated engines,

A silver-finished Martin P5M-1G in Coast Guard service. (*Martin photo*)

323

R-3350-30WAs, were introduced in lengthened nacelles which incorporated weapons-bays. Deliveries to the first operational unit, VP-44, began on April 23, 1952. Production of the P5M-1 series totalled 114 by 1954; at least 80 subsequently became P5M-1S when fitted with AN/ASQ-8 magnetic anomaly detection equipment, Julie active echo-sounding, Jezebel passive sono-buoy detection and other new equipment. Seven Marlins of the first series went to the Coast Guard as P5M-1G, these later becoming P5M-1T trainers when the Navy acquired them in 1961. Designations of the variants still serving in 1962 changed to P-5A (P5M-1), SP-5A (-1S) and TP-5A (-1T).

Major redesign of the Marlin produced the P5M-2 in 1953, this version having a T-tail, lower bow chine line, improved crew accommodation and 3,450 hp R-3350-32WA engines. First flown in August 1953, the P5M-2 began to reach Navy squadrons on June 23, 1954. Production continued until the end of 1960, by which time 145 of the P5M-2 series had been built, including 10 for France under MAP arrangements. A modernization

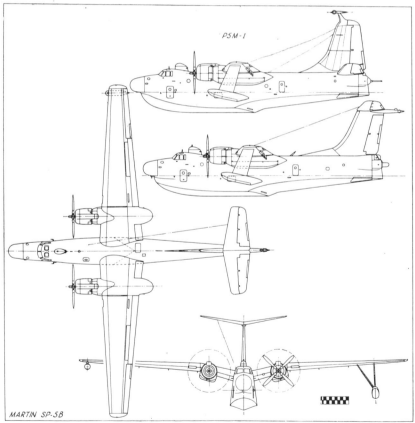

P5M-1

MARTIN SP-5B

Martin P5M-1, the initial production version of the Marlin with low-mounted tail.
(*Martin photo*)

programme to introduce Julie/Jezebel systems and other new equipment resulted in the P5M-2S variant, these two operational versions later becoming P-5B and SP-5B respectively.

Four P5M-2Gs were delivered to the Coast Guard in 1961, with ASW gear deleted and equipment for air–sea rescue added. During 1964 one SP-5B was fitted with a jet engine in the rear fuselage to boost performance, but this feature was not adopted for a retrofit programme.

TECHNICAL DATA (P5M-2)

Manufacturer. Glenn L. Martin Company, Baltimore, Maryland.

Type: Patrol flying-boat.

Accommodation: Crew of 11.

Power plant: Two 3,450 hp Wright R-3350-32WAs.

Dimensions: Span, 118 ft 2 in; length, 100 ft 7 in; height, 32 ft 8½ in; wing area, 1,406 sq ft.

Weights: Empty, 50,485 lb; gross, 85,000 lb.

Performance: Max speed, 251 mph at sea level; cruising speed, 150 mph at 1,000 ft; initial climb, 1,200 ft/min; service ceiling, 24,000 ft; range, 2,050 st miles.

Armament: Four torpedoes, four 2,000 lb bombs or mines, or combinations of smaller stores weighing up to 8,000 lb total internally in nacelles; up to eight 1,000 lb bombs or mines externally.

Serial numbers (allocations):

P5M-1: 124910–124914; 126490–126511; 127696–127719; 130265–130351; 135452–135473.

P5M-1G: CG 1284–1290.

P5M-2: 135474–135543; 137846–137848; 140140–140150; 141252–141258; 146440–146445; 147539–147542; 147926–147937; 149825–149835. 146445; 147539–147542; 147926–147937; 149825–149835.*

* 7 P5M-1G and 4 P5M-2G returned from USCG, as P5M-1T and P5M-2 respectively.

325

The N3N-3, shown here in Naval Academy service in 1960, was the last original design by the NAF built in quantity and the last biplane to serve with the US Navy. (*Howard Levy*)

Naval Aircraft Factory

SPECIAL mention must be made of the manufacturing activities of the Naval Aircraft Factory (NAF) located in the Philadelphia Navy Yard, Pennsylvania.

With the Navy actually engaged in the design and manufacture of aircraft since 1916, it was considered desirable that a permanent manufacturing and test facility be available. Consequently, the Naval Aircraft Factory was authorized in 1917 and completed early in 1918. Because of the demand for service aircraft during the war, its initial activity was the manufacture of 150 Curtiss H-16 flying-boats, the first of which was completed and flown on March 27, 1918. These were built concurrently with four Navy-designed experimental single-engine N-1 pusher seaplanes and were then followed by 140 F-5L flying-boats and the last six of the ten NC boats.

In the immediate post-war years, the NAF continued its manufacturing of outside designs in service quantities, but the practice ended in 1922. Some of the excess war-time manufacturing space was used for dead storage of aircraft. A number of experimental prototypes of Navy-designed aircraft were built in the 1920s and those designs which proved desirable as service types were turned over to the established aircraft industry for production. Large-scale overhaul and modification work continued until World War II, the most notable programmes being the modification of F-5L flying-boats and the conversion of war-time Army de Havilland Liberty Planes (see page 155) to DH-4Bs in the mid-1920s. One famous aircraft for which the NAF seldom gets credit was the rigid airship ZR-1,

326

better known as the USS *Shenandoah*. The design was adapted from the German Zeppelin L 33, which had been shot down over England in 1917. The ZR-1 parts were all fabricated at the NAF but had to be hauled to the Naval Air Station at Lakehurst, New Jersey, for assembly into a complete airship, as the only suitable hangar was located there. The ZR-1 made its first flight on September 4, 1923 (see page 522).

The NAF got back into significant manufacturing in the mid-1930s when it was decreed that the Navy would build 10 per cent of its aircraft in its own factories as a check on the state of the art and to obtain accurate manufacturing cost data. By this time the Navy had designed and built the prototype of a new primary trainer, the XN3N-1. Rather than pass this on to the industry, it was put into production in Philadelphia, with deliveries beginning in 1936. From that date until the end of World War II, the NAF was again a major aircraft producer.

Just before and during World War II, outside designs were again put into production, this time being given N-for-NAF designations to indicate their actual place of manufacture in keeping with established policy. Aircraft manufacture ended for good at the NAF at the end of World War II, and the former factory was redesignated as the Naval Air Materiel Unit (NAMU), still existing as such in 1968.

The models and numbers of aircraft that constituted the major production efforts of the NAF from 1918 through 1945 are listed below. Navy designs are described under the NAF heading in subsequent pages of this book as indicated by the page references; those designs built principally by industry appear under the actual manufacturer. Outside designs built at the NAF are shown with both the NAF and original manufacturer's designations and appear in the original manufacturer's entry in this book.

Model	Number built		Page
H-16	150	Curtiss-designed patrol flying-boat	106
F-5L	140	Improved H-16 (includes two F-6L)	114
MF	80	Small Curtiss training flying-boat	112
HD-1	10	Remanufacture of French Hanriot fighter seaplane	488
NC-5/10	6	Joint Navy/Curtiss giant flying-boat	328
M-80/81	36	Loening two-seat monoplane	283
VE-7	49	Vought all-purpose two-seat biplane	382
PT-1/2	33	Navy-designed twin-float torpedo-plane	455
TS-1/3	9	Navy-designed carrier fighter. 34 built by Curtiss	331
DT-2/4/5	6	Douglas torpedo-seaplane	160
PN-7/12	12	Navy improvements on F-5L design; production of PN-11, PN-12 variants by industry	334
N3N-1/3	817	Navy-designed primary training biplane	338
SON-1	44	Curtiss SOC-1 observation biplane	144
SBN-1	30	Production versions of Brewster XSBA-1 scout-bomber	417
OS2N-1	300	Vought OS2U observation monoplane	401
PBN-1	156	Consolidated PBY-6 flying-boat	83
TDN-1	100	Navy-designed torpedo drone, built by Interstate as TDR-1	—

Curtiss-built NC-4, designed by the NAF and used for the transatlantic flight in 1919. (*Curtiss photo*)

Navy/Curtiss NC Boats

The Navy/Curtiss boats which were to achieve undying fame as the first to complete a crossing of the Atlantic by air (albeit in stages) owed their origin to the successful German U-boat campaign of 1917. As British and American shipping losses mounted drastically, the call went out for more patrol aircraft to provide an umbrella over the Atlantic. Since many of the flying-boats then serving with the Royal Navy were built in America, that source was appealed to. In response to the urgency of the situation, the US Navy Department drew up plans in September 1917, for a new flying-boat to carry the war to the U-boat. The essential requirement was that this boat should be able to fly across the Atlantic—since shipping space was at a minimum—and be able to take on a U-boat immediately upon arrival.

Such a machine would need an endurance of 15–20 hours, which in turn meant that it would have to be very large, necessarily multi-engined, and of sufficiently rugged construction to be able to endure forced landings at sea. Alone among American manufacturers, Curtiss had the large-aircraft experience and the shop facilities that would allow it to build such a mammoth. In reply to the Navy's detailed proposal for a trans-atlantic flying-boat using a unique new Navy hull design, Curtiss quickly produced plans for a suitable three-engine 28,000 lb biplane. The principal point of departure from traditional design was the Navy's new hull. While using the established Curtiss laminated wood veneer construction, it was very short—only 45 ft—with the tail surfaces carried clear of the water on a superstructure of spruce struts. This arrangement minimized hull weight and gave, as required, an uninterrupted field of fire straight to the rear for machine gunners in the rear of the hull. Because of the joint design effort involved, the new boats were designated NCs for Navy and Curtiss. The first four, built by Curtiss at its experimental plant in Garden City, New York, and assembled at the Naval Air Station at nearby Rockaway, were given the separate designations of NC-1 through NC-4, a departure from established model designation procedure. The NC-1, which first flew on October 4, 1918, had all three Liberty engines installed as tractors, with the pilot and co-pilot located in the centre nacelle behind the engine. A ladder was provided for the pilots and other crew members to move about in flight. On November 25, the NC-1 carried 51 persons aloft to establish a new world's record for passengers carried.

Testing proved that three engines were inadequate for the transatlantic mission, so the remaining aircraft were redesigned for four. NC-2, which first flew on April 12, 1919, was completed with two tandem pairs, with the pilots in a third nacelle between the engine pairs. NC-3, which flew on April 23rd, and NC-4 reverted to the three tractor engine arrangement of NC-1 but added the fourth engine as a pusher in the middle nacelle. The pilots were removed to the hull ahead of the wing on NC-3 and NC-4, and NC-1 was modified to the same pattern.

Although the NC boats were not completed in time to see war service, it was decided to fly them across the Atlantic anyhow, as a race was developing between England and America for the honour of being the first across by air. In preparation for this mission, the NCs were redesignated NC-TA, for Transatlantic. Only three gathered at Trepassy Bay, Newfoundland, for the May 16, 1919, start. The engine arrangement of NC-2 had been declared unsatisfactory for the mission, and NC-2's wings were removed and installed on NC-1 to replace the originals which had been damaged in a storm. Under the command of Cdr (later Admiral) John H. Towers, Naval Aviator No. 3, the three boats took-off on the 1,400-mile flight to the Azores. NC-4 achieved it by air, and later completed the voyage, reaching Plymouth on May 31 via Lisbon. NC-1 and NC-3 were both forced down at sea short of the Azores. NC-1 sank after the crew was rescued by a ship, but NC-3, with Towers aboard, managed to taxi the remaining 200 miles to the Azores.

Six additional boats, NC-5 through NC-10, were built after the war at the Naval Aircraft Factory, with NC-5 and NC-6 in trimotor configuration having the centre engine installed as a pusher. The others were four-engined in the NC-3/4 pattern. After the adoption of the new designating system in 1922, the surviving NC boats were redesignated P2N for the second Navy-designed patrol model, but the term was not actually applied. Following the transatlantic flight, NC-4 made a publicity tour of eastern and southern coastal cities and flew up the Mississippi river to St Louis before being handed over to the Smithsonian Institution.

TECHNICAL DATA (NC-TA)

Manufacturer: Curtiss Aeroplane and Motor Company, Garden City, LI, NY, and Naval Aircraft Factory, Philadelphia, Pennsylvania.

Type: Long-range patrol flying-boat.

Accommodation: Two pilots; navigator/nose gunner; radio operator; two flight engineers.

Power plant: Four 400 hp Liberty 12s.

Dimensions: Span, 126 ft; length, 68 ft 3 in; height, 24 ft 6 in; wing area, 2,380 sq ft.

Weights: Empty, 15,874 lb; gross, 27,386 lb.

Performance: Max speed, 85 mph; initial climb, 1,050 ft in 5 min; service ceiling, 4,500 ft; range, 1,470 st miles.

Armament: Machine guns in bow and rear hull cockpits.

Serial numbers:

NC-1/4: A2291–A2294.   NC-5/10: A5632–A5635; A5885; A5886.

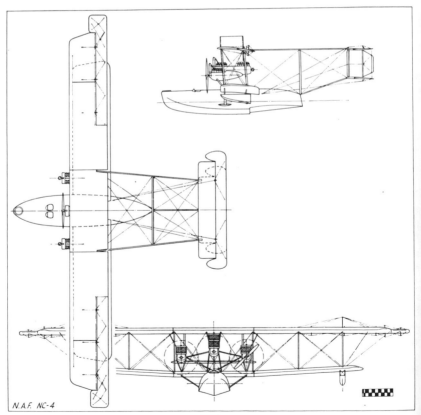

N.A.F. NC-4

330

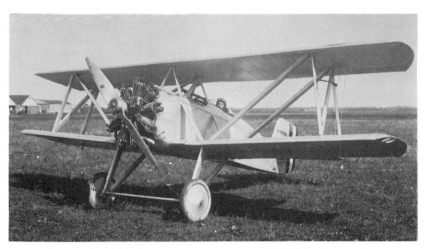

Curtiss-built F4C-1, the all-metal version of the NAF TS-1. (*Curtiss photo*)

Naval Aircraft Factory TS

The diminutive NAF TS-1 biplane fighter occupies a special place in the history of US Naval aviation, for it was the first carrier-based aeroplane specifically designed for the purpose. Designed by the Bureau of Aeronautics, it was selected for production and a contract for 34 was awarded to Curtiss Aeroplane and Motor Company under the prevailing policy of having the industry build Navy-designed aircraft. An order for five was given to the Naval Aircraft Factory to serve as a check on costs. The first TS-1 appeared in May 1922, two months after the Navy's first aircraft carrier, the USS *Langley* (CV-1) had been commissioned. It was powered by a 200 hp Lawrance J-1 air-cooled radial engine that was later to become the Wright Whirlwind. The TS-1 was designed to operate as a twin-float seaplane or on a normal wheel chassis. The fuselage was centred between the wings and fuel was carried in a thickened centre section of the lower wing. A single 0·30-in Browning gun was synchronized to fire through the propeller.

The first of the Curtiss-built TS-1s reached the *Langley* in December 1922. In addition to operating from the carrier deck, the TS-1s served for several years in floatplane configuration aboard destroyers, cruisers and battleships. No catapults were carried by these ships, the aircraft being slung over the side by cranes to operate from the water. They were retrieved the same way. Squadron VO-1 operated TS-1s this way from 1922 and VF-1 flew its float-equipped TS-1s from battleships for a time in 1925–6. VF-1 also operated TS-1s on wheels from the *Langley*.

The second NAF-built TS-1, as a floatplane. (*US Navy*)

The NAF built four improved versions; two TS-2s with 240 hp Aeromarine engines and two TS-3s with 180 hp Wright-Hispano Es. The last TS-3 was modified for the 1922 Curtiss Marine Trophy race by a change of aerofoil section on the wings and redesignated TR-2. Unsuccessful in this contest, it was then extensively modified to incorporate a more streamlined fuselage and had the upper wing lowered to the top of the fuselage to serve as a high-speed trainer for the Navy's 1923 Schneider Cup team.

The final step in the TS development came when the Navy awarded Curtiss a contract for the manufacture of two metal versions of the TS-1 for comparison of the then new metal structures with the traditional wood and wire used in the original aircraft. By this time the Navy was assigning designations according to the actual manufacturer of the aircraft, so the two metal TSs became F4C-1.

A Curtiss-built TS-1, as a landplane. (*Jack Goodwin*)

TECHNICAL DATA (TS-1)

Manufacturer: Curtiss Aeroplane and Motor Company, Buffalo, NY, and Naval Aircraft Factory, Philadelphia, Pennsylvania.

Type: Fighter seaplane (and landplane).

Accommodation: Pilot only.

Power plant: One 200 hp Wright J-4.

Dimensions: Span, 25 ft; length, 22 ft 1 in; height, 9 ft 7 in; wing area, 228 sq ft.

Weights: Empty, 1,240 lb; gross, 2,133 lb.

Performance: Max speed, 123 mph at sea level; initial climb, 5·5 min to 5,000 ft; service ceiling, 16,250 ft; range, 482 st miles.

Armament: One fixed forward-firing 0·30-in Browning gun.

Serial numbers:

TS-1 (Curtiss): A6248–A6270; A6305– A6315.

TS-1 (NAF): A6300–A6304.

TS-2: A6446–A6447.

TS-3: A6448–A6449.

TR-2: A6449.

F4C-1: A6689–A6690.

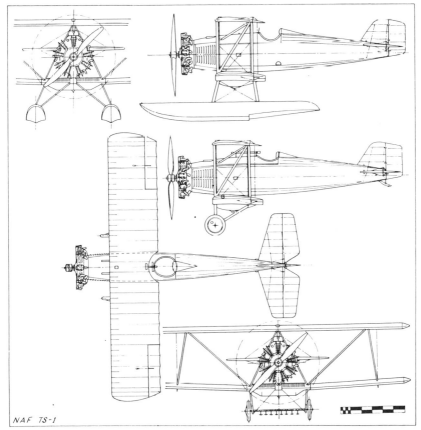

NAF TS-1

The NAF PN-7, with new wings on an F-5L wooden hull. (*US Navy*)

Naval Aircraft Factory PN

The Naval Aircraft Factory, which had built 138 F-5L flying-boats during World War I, continued to develop the design in the post-war years. When the new aircraft designation system was adopted in 1922, the F-5Ls were redesignated PN-5, for Patrol, Navy, but the term was not actually applied to the aircraft built before adoption of the system. Two modified F-5Ls were built and designated F-6L, and certain of their improved features were added to existing F-5Ls by modification.

When the Navy undertook a major redesign of the F-5L, adding entirely new wings to the established wooden F-5L hull, the two resulting aircraft, to have been called F-7L, were actually designated PN-7. The major change in appearance was in the wings, which were of fabric-covered metal frame construction with a thick USA 27 aerofoil instead of the thin RAF 6 of the NCs. The increased lift and spar depth permitted a reduction of the wing span to 95 ft with only a single bay outboard of the nacelles. The tail was the same as that of the F-6L, but experimental 525 hp Wright T-2s replaced the Liberty engines.

The redesign was highly successful, but the power plants left much to be desired, and the seaworthiness and maintenance of the wooden hull could have been improved. Consequently, two PN-8s were built, outwardly identical to the PN-7s except for metal hull construction and two 475 hp geared Packard 1A-2500 engines in streamlined nacelles fitted with large spinners. The last PN-8 was converted to PN-9 with redesigned tail surfaces and revised engine nacelles with nose radiators. This PN-9, A6878, was used for a flight from San Francisco to Hawaii. Under the command of Cdr John Rogers, Naval Aviator No. 2, it took off from San Francisco Bay on September 1, 1925, for the 2,400-mile overwater flight to Hawaii.

It was forced to alight at sea 559 miles short of its goal, but the crew, by making sails with fabric from the lower wing, managed to sail the remaining distance to the islands. In spite of the failure of the point-to-point flight, the 1,841 statute miles covered by air was recognized as a new world seaplane distance record.

The PN-9 was followed by four similar PN-10s. Since the water-cooled V-12 engines were not living up to expectations, the last two PN-10s were completed as PN-12s, one with the new air-cooled Wright Cyclone R-1750 radial engines of 525 hp and the other with Pratt & Whitney R-1850 Hornets. On May 3–5, 1928, the Cyclone-powered PN-12 set a world seaplane duration record with 1,000 kg payload and speed over 2,000 km at 80·28 mph, covering the distance of 1,243 miles in 17 hr 55 min. The Hornet-powered PN-12 set additional records on June 26 with a Class C altitude record of 15,426 ft with a 2,000 kg payload and on June 27 reached 19,593 ft with a 1,000 kg payload. The same machine set further speed, distance and duration records for the 1,000 and 2,000 kg payload categories on July 11–12.

The performance achieved by the PN-12s convinced the Navy that the radial engine was what was needed to make the optimum airframe/power plant combination desired in a patrol plane to replace the aged F-5Ls and Curtiss H-16s that were still the Navy's principal patrol aircraft. Since the Naval Aircraft Factory was not equipped for large-scale production, bids were invited from the industry for the production models. Douglas built 25 under the designation of PD-1 (see page 428), Martin built 30 as PM-1 and followed them with 25 twin-tailed PM-2s (see page 451), and Keystone built 18 twin-tailed versions as PK-1 (see page 444).

Other than the change to metal structure made in the PN-8, the first major departure from the F-5L hull design was made with four PN-11s, which featured a wider hull with entirely new lines that eliminated the sponsons that were such a noticeable feature of the F-5L hull. The other

The NAF PN-9 which was used for an epic flight across the Pacific to Hawaii, covering the last 559 miles as a sailing boat after the engines failed. (*US Navy*)

major departure on the first, which was designated as XPN-11, and fitted with geared Hornet radial engines, was the use of twin vertical tail surfaces. With slight change, these were adopted for the Martin PM-2s and the Keystone PK-1s. The last three PN-11s were fitted with Wright Cyclones and were temporarily redesignated P2N-1 to use the designation assigned to but not carried by the NC Boats (see page 328). The first of the three was to have been XP2N-1. However, this designation was dropped in favour of XP4N-1 for the first and XP4N-2 for the second and third. A modified version of the PN-11/XP4N was built by the Hall Aluminum aircraft company (see page 254) as the XPH-1, leading to production orders for PH-1s for the Navy and PH-2s and -3s for the Coast Guard, some of which served into World War II, thus resulting in the same basic flying-boat design spanning two world wars.

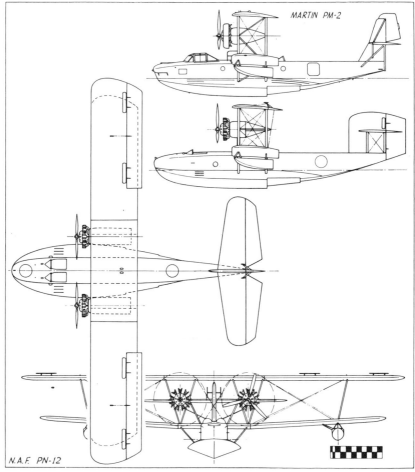

MARTIN PM-2

N.A.F. PN-12

An NAF PN-11, showing the redesigned hull. (*US Navy*)

TECHNICAL DATA (PN-7)

Manufacturer: Naval Aircraft Factory, Philadelphia, Pennsylvania.
Type: Patrol flying-boat.
Accommodation: Crew of five.
Power plant: Two 525 hp Wright T-2s.
Dimensions: Span, 72 ft 10 in; length, 49 ft 1 in; height, 15 ft 4 in; wing area, 1,217 sq ft.
Weights: Empty, 9,637 lb; gross, 14,203 lb.
Performance: Max speed, 104·5 mph at sea level; initial climb, 10·2 min to 5,000 ft; service ceiling, 9,200 ft; range, 655 st miles.
Armament: Single 0·30-in guns in bows and amidships. Four 230 lb bombs.
Serial numbers:

PN-7: A6616–A6617.	PN-10: A7028–A7029.
PN-8: A6799.	XPN-11: A8006.
PN-9: A6878.	PN-11: A7527; A8483; A8484.

TECHNICAL DATA (PN-12)

Manufacturer: Naval Aircraft Factory, Philadelphia, Pennsylvania; Douglas Aircraft Company (PD-1); Glenn L. Martin Company (PM-1, PM-2); Keystone Aircraft Corporation (PK-1).
Type: Patrol flying-boat.
Accommodation: Crew of five.
Power plant: Two 525 hp Wright R-1750Ds.
Dimensions: Span, 72 ft 10 in; length, 49 ft 2 in; height, 16 ft 9 in; wing area, 1,217 sq ft.
Weights: Empty, 7,669 lb; gross, 14,122 lb.
Performance: Max speed, 114 mph at sea level; initial climb, 16 min to 5,000 ft; service ceiling, 10,900 ft; range, 1,310 st miles.
Armament: Single 0·30-in guns in bows and amidships. Four 230 lb bombs.
Serial numbers:

PN-12: A7383–A7384.	PD-1: A7979–A8003
PM-1: A8289–A8313; A8477–A8481.	PK-1: A8507–A8524.
PM-2: A8662–A8686.	

337

The N3N-1, initial version of the widely used NAF trainer, had a cowled engine until 1941. (*William T. Larkins*)

Naval Aircraft Factory N3N

In 1934 the Navy developed a new primary trainer design. Outwardly, this was quite similar to the Consolidated NY-2s and NY-3s still in service and the Stearman NS-1 that was then under test in its civil prototype form. However, the major difference in the N3N was its structure, which featured bolted steel-tube fuselage construction with removable side panels for ease of inspection and maintenance. The wings were of all-metal construction and, like the fuselage and tail, were fabric-covered. The prototype XN3N-1, powered with a 220 hp Wright J-5 radial engine, was flown in August 1935, at the Naval Aircraft Factory. While this engine had been out of production since 1929, the Navy had quantities in storage and wanted to use them up. They were also used to power the contemporary Stearman NS-1s (see page 380).

Following the successful test of the XN3N-1 at Philadelphia and Anacostia as both a landplane and a single-float seaplane, the Navy ordered it into production. One hundred and seventy-nine N3N-1s were built, the first being delivered in June 1936. A single prototype XN3N-2 was ordered separately, and the fourth N3N-1 (0020) was modified as another prototype, XN3N-3. These two machines were fitted with Navy-built versions of the 240 hp Wright J-6-7 (R-760-96) engine, which was the same as that used in the NY-3s since 1929. Because of the suitability of this power plant and the obsolescence of the J-5, the last 20 N3N-1s were fitted with the Navy-built R-760 engine, and the other N3N-1s were eventually refitted with R-760-2s of 235 hp. When the N3N-1 order was

338

An N3N-1 in the overall yellow finish which led to the type being known as the 'Yellow Peril'. (*Peter M. Bowers*)

completed in 1938, it was followed by 816 N3N-3s, which differed outwardly from the -1s in having revised vertical tail shape and a single-strut landing gear. The N3N-1s had been delivered with a wide anti-drag ring around the engine, but the N3N-3s did not have this feature, which was also deleted from N3N-1s in 1941–2. Four N3N-3s were transferred to the Coast Guard in 1941 and acquired Coast Guard serial numbers V193—196.

The N3Ns served in primary training schools throughout World War II. The majority were declared surplus at the war's end, and most of these were fitted with more powerful engines by new owners and used as crop-dusting and spraying aircraft, work for which their biplane configuration and rugged structure suited them. A few, however, were retained by the Navy. In their seaplane configuration they were assigned to the United States Naval Academy where they fitted into the midshipmen's curriculum. The use of open cockpit biplanes, and seaplanes at that, since the

Navy N3N-3 trainer floatplane in 1959. (*Fred G. Freeman*)

fleet no longer used float seaplanes, was not quite the anachronism one might imagine. The Navy still taught its midshipmen to handle sail in the dawn of the nuclear-powered ship, so the use of obsolete propeller-driven biplanes in the jet age was understandable.

When the N3N-3s were retired in 1961, they were the last biplanes in US military service, the Air Force having disposed of all its Stearman biplane trainers by 1948.

TECHNICAL DATA (N3N-3)

Manufacturer: Naval Aircraft Factory, Philadelphia, Pennsylvania.
Type: Primary trainer.
Accommodation: Pilot and instructor in tandem.
Power plant: One 235 hp Wright R-760-2.
Dimensions: Span, 34 ft; length, 25 ft 6 in; height, 10 ft 10 in; wing area, 305 sq ft.
Weights: Empty, 2,090 lb; gross, 2,792 lb.
Performance: Max speed, 126 mph at sea level; cruising speed 90 mph; initial climb, 800 ft/min; service ceiling, 15,200 ft; range, 470 st miles.
Serial numbers:

XN3N-1: 9991.
N3N-1: 0017–0019; 0021–0101; 0644–0723; 0952–0966.
XN3N-2: 0265.

XN3N-3: 0020.
N3N-3: 1759–1808; 1908–2007; 2573–3072; 4352–4517.

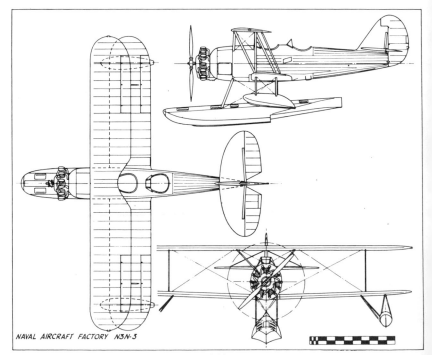

NAVAL AIRCRAFT FACTORY N3N-3

340

North American SNJ-6, the final version of the Texan procured by the Navy. (*William T. Larkins*)

North American NJ, SNJ Texan

North American's NA-16 basic trainer, built as a private venture in 1935, proved to be the progenitor of a long series of similar aircraft which remained in production for a decade and achieved wide fame and popularity as the AT-6 Texan and Harvard. By the time production of the series ended, North American had produced well over 16,000 examples of the same basic design and many hundreds more had been built in Canada, Australia, Sweden and elsewhere.

The US Navy first ordered a version of the North American trainer late in 1936, with a contract for 40 NJ-1s (designated NA-28 by the manufacturer). These were similar in most respects to the USAAC BT-9s with fixed landing gear but had the 500 hp Pratt & Whitney R-1340 Wasp engine in place of the BT-9's 400 hp R-975. The final aircraft on the order was temporarily fitted with a Ranger XV-770-4 inverted V-12 engine as NJ-2 but was reconverted to NJ-1 standard. As well as serving as trainers at Pensacola, the NJ-1s were used as command and staff transports.

Use of the R-1340 engine in the NA-16 airframe had been planned by North American, prior to the appearance of the NJ-1, in the NA-26 prototype built as a demonstrator to meet a USAAC requirement for a basic combat trainer. As well as the new engine, the NA-26 featured retractable landing gear and provision for armament. It was produced for the Army as the BC-1, and the Navy ordered 16 of an improved model in 1938. Designated SNJ-1 (NA-52), they had metal-covered rear fuselages and were assigned to the 13 Naval Reserve Air Bases but retained by the Navy. The Reserve bases received, instead, new SNJ-2s from a batch of 36 ordered in 1939; 25 more were ordered in 1940, the two batches being NA-65 and NA-79 respectively. Minor differences and an R-1340-56 engine distinguished the SNJ-2.

Also in 1940 the Navy placed its first orders for the SNJ-3, similar to

341

the Army Air Corps' AT-6A and featuring the triangular fin and rudder and blunt wingtips already introduced on the Army BC-1A. Orders totalled 368, of which 70 were purchased on a US Navy contract and the remainder diverted from US Army AT-6A production. Minor improvements were incorporated in 2,400 SNJ-4s built in Texas, and a further 1,573 SNJ-5s differed only in having a 24-volt electrical system replacing the 12-volt system. These two versions were respectively diverted from USAAF contracts for the AT-6C and AT-6D. A few SNJ-3C, -4C and -5Cs were fitted with arrester hooks and used for carrier deck-landing training.

Final Navy version of the North American trainer was the SNJ-6 of

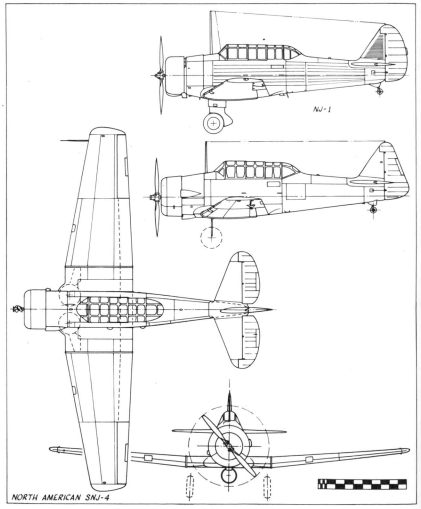

NJ-1

NORTH AMERICAN SNJ-4

342

North American NJ-1 in service at NAS Anacostia. (*Harry Thorell*)

1944 with strengthened wing panels and redesigned rear fuselage. It was identical with the AT-6F procured by the USAAF, which acted as the procuring agency for the 411 SNJ-6s (NA-121). The Navy Texans remained in service in large numbers well into the fifties as primary, basic and instrument trainers and some became SNJ-7s and armed SNJ-7Bs after modernisation at NAS Pensacola in 1952.

TECHNICAL DATA (NJ-1)

Manufacturer: North American Aviation, Inc, Inglewood, California.
Type: Basic trainer.
Accommodation: Pupil and instructor in tandem.
Power plant: One 500 hp Pratt & Whitney R-1340-6.
Dimensions: Span, 42 ft; length, 27 ft 2 in; height, 13 ft 3 in; wing area, 248 sq ft.
Weights: Empty, 3,250 lb; gross 4,440 lb.
Performance: Max speed, 167 mph at sea level; cruising speed, 140 mph; initial climb, 1,100 ft/min; service ceiling, 24,900 ft; range, 944 st miles.
Serial numbers:
  NJ-1: 0910–0949 (0949 to NJ-2).

TECHNICAL DATA (SNJ-5)

Manufacturer: North American Aviation, Inc., Inglewood, California, and Dallas, Texas.
Type: Basic trainer.
Accommodation: Pupil and instructor in tandem.
Power plant: One 550 hp Pratt & Whitney R-1340-AN-1.
Dimensions: Span, 42 ft 0¼ in; length, 29 ft 6 in; height, 11 ft 8½ in; wing area, 254 sq ft.
Weights: Empty, 4,158 lb; gross, 5,300 lb.
Performance: Max speed, 205 mph at 5,000 ft; cruising speed, 170 mph at 5,000 ft; initial climb, 1,200 ft/min; service ceiling, 21,500 ft; range, 750 st miles.
Armament: Two fixed forward-firing 0·30-in guns; provision for one 0·30-in flexible gun in rear cockpit.

Serial numbers:
  SNJ-1: 1552–1567.
  SNJ-2: 2008–2043; 2548–2572.
  SNJ-3: 6755–7024;   01771–01976;
    05435–05526.

  SNJ-4: 05527–05674;   09817–10316;
    26427–27851; 51350–51676.
  SNJ-5: 43638–44037;   51677–52049;
    84819–85093; 90582–91101.
  SNJ-6: 111949–112359.

The North American FJ-1 was used by only one Navy fighter squadron but was the first jet fighter to go to sea operationally. (*Gordon S. Williams*)

North American FJ-1 Fury

New Year's Day, 1945, marked the beginning of an era for North American, with receipt of a Navy contract for the company's first jet fighter design, the NA-134. Although only 33 examples of this particular design were destined to be built, it proved to be the progenitor of the F-86 Sabre, one of the world's greatest fighters of the fifties.

The NA-134 was one of two similar jet fighter designs developed in parallel for the Navy (see also the Vought F6U Pirate, page 473), to continue the full-scale evaluation of carrier-borne jets which had begun with the McDonnell FH Phantom (see page 295). All these three types were straight-wing, subsonic aircraft, necessary to achieve the transition between World War II's piston-engined fighters and the sweptwing fighters which were to open up new possibilities for Naval air operations.

North American chose a simple straight-through arrangement for the NA-134, which was designed around the General Electric J35 axial-flow turbojet and had a circular intake in the nose of the short and stumpy fuselage. Rather than bifurcate the air intake duct around the sides of the cockpit, the latter was mounted above the duct; this and the need to contain fuel tanks in the fuselage resulted in the decidedly squat appearance, but clean lines were achieved and external excrescences avoided, so that the aircraft had a top speed close to 550 mph.

With the Navy order on January 1, 1945, came the official designation XFJ-1. Shortly after, the USAAF ordered the similar NA-140, which had rather more slender fuselage lines and a thinner wing and was the starting-point for the F-86. On May 28, 1945, the Navy ordered 100 FJ-1s, these carrying the company designation NA-141 and the type name Fury

Powered by a 3,820 lb s.t. J35-GE-2 engine, the first XFJ-1 flew on November 27, 1946. Of conventional construction, it carried an armament of six 0·50-in guns in the sidewalls of the air intake; in the event this proved to be the final use of half-inch guns by the Navy.

Deliveries of the production FJ-1s, with Allison-built J35-A-2 engines

344

began in March 1948, by which time the contract had been cut to 30 since more promising types were rapidly overtaking the Fury. Nevertheless, it could briefly claim to be the fastest US fighter in the air when a prototype achieved Mach 0·87 in 1947. The only unit equipped was Navy Fighter Squadron VF-5A (later VF-51), and the Fury served aboard USS *Boxer* as that ship's first jet fighter. The first carrier landings were made on March 10, 1948, and VF-5A became the first jet fighter unit to serve at sea under operational conditions. Fourteen months later, the Furies were assigned to Naval Reserve units.

TECHNICAL DATA (FJ-1)

Manufacturer: North American Aviation, Inc, Inglewood, California.
Type: Carrier-borne fighter.
Accommodation: Pilot only.
Power plant: One 4,000 lb s.t. Allison J35-A-2 turbojet.
Dimensions: Span, 38 ft 2 in; length, 34 ft 5 in; height, 14 ft 10 in; wing area, 221 sq ft.
Weights: Empty, 8,843 lb; gross, 15,600 lb.
Performance: Max speed, 547 mph at 9,000 ft; initial climb, 3,300 ft/min; service ceiling, 32,000 ft; ferry range, 1,500 st miles.
Armament: Six fixed forward-firing 0·50-in guns.
Serial numbers:
   XFJ-1: 39053–39055.   FJ-1: 120342–120371.

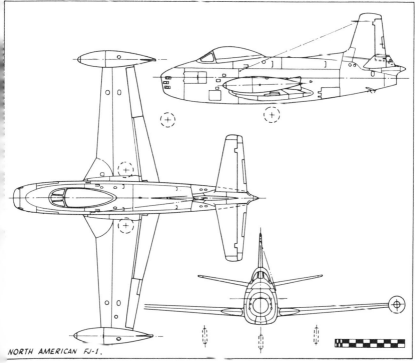

NORTH AMERICAN FJ-1.

345

The North American FJ-4 was the final development of the Fury series.
(*North American photo*)

North American FJ-2/4 Fury

The Navy's straight-winged FJ-1 fighter (see page 344) having given rise to the sweptwing F-86 Sabre, the wheel came full circle on March 8, 1951, when the Navy ordered three prototypes of the sweptwing fighter as the XFJ-2. Plans for a naval F-86 had been drawn up by North American early in 1951, largely as a result of the escalating war in Korea, and had been submitted as design NA-181, but two of the three prototypes were built as NA-179s and were essentially F-86Es with V-frame arrester hooks, catapult points and a lengthened nosewheel leg to increase the angle of attack and reduce the speed during carrier take-offs and landings. The third prototype was the NA-181, XFJ-2B, with a standard F-86 nosewheel but four 20 mm guns in place of the six fifties used on the FJ-1 and F-86; this was the first of these aircraft to fly, on December 27, 1951, followed by the first XFJ-2 on February 14, 1952. Powered by 5,200 lb s.t. General Electric J47-GE-13 engines, the prototypes were standard F-86s in all respects except the points mentioned above and the Naval blue finish. In December 1952 the two XFJ-2s went aboard the USS *Coral Sea* for carrier qualification trials.

A production contract for 300 FJ-2s had been placed on February 10, 1951, and North American assigned this task to its Columbus plant, already in operation as a second source for F-86s. The production NA-181s incorporated several new features for their Naval role, including folding wings (the outer 7 ft of each wing folding upwards at right angles), AN/APG-30 radar, increased wheel track and the 6,000 lb s.t. J47-GE-2 engine. The four-cannon armament became standard, with 600 r.p.g. The first FJ-2 Fury was completed at Columbus in the autumn of 1952 but production of the F-86 slowed down the programme, and only 2. had been completed by the end of 1953; after the Korean War ended, the contract was cut to 200.

North American FJ-2 in the markings of Marine squadron VMF-321. (*Peter M. Bowers*)

All FJ-2s were assigned to Marine units, beginning in January 1954 with VMF-122 at Cherry Point, NC. Other units to fly this version of the Fury were VMF-232 and VMF-312 in AIRLANT, and VMF-235, VMF-334 and VMF-451 in AIRPAC.

Work on a new variant of the Fury began in March 1952, the NA-194 being basically an FJ-2 with a 7,800 lb s.t. Wright J65-W-2 engine. A contract for 389 examples to be designated FJ-3 was placed on April 18, 1952, and North American initiated a trial installation in the fifth production FJ-2 as NA-196. This aircraft first flew on July 3, 1953, and was the only prototype for the FJ-3 series; it retained the standard FJ-2 nose intake but production aircraft had an enlarged intake and were powered by the 7,650 lb s.t. J65-W-4 engine. The first FJ-3 flew from Columbus on December 11, 1953. Eventually, the FJ-3 was to be redesignated F-1C in the 1962 tri-service system, as the third Fury model, but the FJ-1 and FJ-2 were by then out of service and never carried F-1 designations.

FJ-3s began to reach the Navy in September 1954 when VF-173 was re-equipped with Furies at Jacksonville, Florida. The unit took its FJ-3s aboard the USS *Bennington* on May 8, 1955, and an aircraft of this type from VF-21 was also the first to land on the USS *Forrestal* on January 4, 1956. An additional contract, initially for 214 but reduced to 149, brought the total of FJ-3s ordered to 538, the second batch being NA-215s with J65-W-4D engines. Delivery was completed in August 1956, and a total of 21 squadrons flew the type, 17 Navy units and four Marine.

North American FJ-3 in May 1956, in service with Navy squadron VF-154.
(*William T. Larkins*)

In the course of production, several significant modifications were introduced on FJ-3s, some of these changes being retrofits. The original leading-edge slats were replaced by extended leading edges with leading-edge fences, and extra fuel in the wing. All aircraft had two underwing store stations, but two more were added later, plus provision for carrying Philco AAM-N-7 Sidewinder (AIM-9A) AAMs. Entering service in 1956, the Sidewinder Furies were designated FJ-3M (MF-1C), a total of 80 being so equipped. A few FJ-3s were modified in 1957–60 to serve as drone directors: the FJ-3D (DF-1C) controlled drone versions of surplus Vought Regulus missiles and the FJ-3D2 (DF-1D) controlled F9F-6K drones and KDA targets.

Work on the final Fury version began at Columbus early in 1953 with the object of increasing the range without compromising the overall performance. A 50 per cent increase in internal fuel capacity called for a major redesign of the airframe with new fuselage contours, a new, thinner wing with mid-span ailerons and inboard high-lift flaps, wider track landing gear and thinner tail surfaces. The J65 engine was retained, the W-4 model being used in two NA-208 prototypes ordered on October 16, 1953; the first of these flew at Columbus on October 28, 1954.

Production FJ-4s (F-1Es) began to appear in February 1955 and had

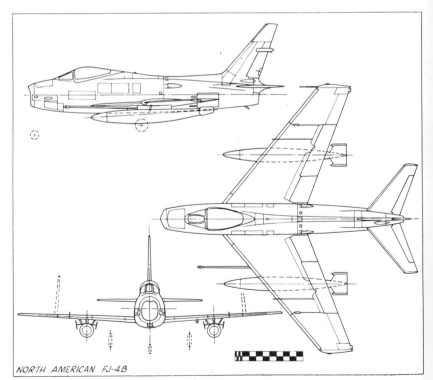

NORTH AMERICAN FJ-4B

7,700 lb s.t. J65-W-16A engines, two contracts eventually covering 152 of this version. The four-cannon armament was retained, and all four wing pylons could carry Sidewinders. Marine Squadron VMF-451 was the first to use the FJ-4, which was used almost wholly by Marine units as a replacement for the FJ-2. Deliveries were completed by March 1957, by which time Columbus was at work on the first of 222 FJ-4Bs (AF-1Es) which were to end Fury production.

The first FJ-4B flew on December 4, 1956, being an attack fighter incorporating some of the features developed for the F-86H, including a stiffened wing with six pylons and a LAB system to deliver a tactical nuclear weapon. Additional speed-brakes were fitted under the rear fuselage. For the ground attack role, the FJ-4B could carry up to five Martin ASM-N-7 Bullpup (AGM-12B) missiles, with a guidance transmitter in a pod on the sixth station. Production ended in May 1958, and FJ-4Bs were issued to nine Navy and three Marine attack squadrons, starting with VA-126 and VMA-223. The Columbus plant delivered 1,112 sweptwing Furies, only three XFJ-2s being built at Los Angeles. Two FJ-4Bs fitted experimentally with rocket motors were designated FJ-4F.

TECHNICAL DATA (FJ-2)

Manufacturer: North American Aviation, Inc, Columbus, Ohio.
Type: Carrier-borne fighter.
Accommodation: Pilot only.
Power plant: One 6,000 lb s.t. General Electric J47-GE-2 turbojet.
Dimensions: Span, 37 ft 1 in; length, 37 ft 7 in; height, 13 ft 7 in; wing area, 288 sq ft.
Weights: Empty, 11,802 lb; gross, 18,790 lb.
Performance: Max speed, 676 mph at sea level; cruising speed, 518 mph at 40,000 ft; initial climb 7,230 ft/min; combat ceiling, 41,700 ft; range, 990 st miles.
Armament: Four fixed forward-firing 20 mm guns.
Serial numbers:
XFJ-2: 133754–133755.   FJ-3: 135774–136162; 139210–139278;
XFJ-2B: 133756.   141364–141443.
FJ-2: 131927–132226.

TECHNICAL DATA (FJ-4)

Manufacturer: North American Aviation, Inc, Columbus, Ohio.
Type: Carrier-borne fighter.
Accommodation: Pilot only.
Power plant: One 7,700 lb s.t. Wright J65-W-16A turbojet.
Dimensions: Span, 39 ft 1 in; length, 36 ft 4 in; height, 13 ft 11 in; wing area, 339 sq ft.
Weights: Empty, 13,210 lb; gross, 23,700 lb.
Performance: Max speed, 680 mph at sea level; cruising speed, 534 mph; initial climb 7,660 ft/min; combat ceiling, 46,800 ft; range, 2,020 st miles.
Armament: Four fixed forward-firing 20 mm guns. Four wing pylons for 3,000 lb of bombs or up to four AIM-9A Sidewinder missiles.
Serial numbers:
FJ-4: 139279–139323; 139424–139530.
FJ-4B: 139531–139555; 141444–141489; 143493–143643.

A North American T-28B Trojan serving at NAS Ellyson Field, Pensacola. (*US Navy*)

North American T-28 Trojan

The Navy introduced North American T-28 trainers into its curriculum for new pilots in 1952 following a decision to standardize training techniques and equipment between the USAF and Navy. The T-28A had been in service with the USAF's Air Training Command since 1950, and 1,194 were delivered to that service following development of the type specifically to meet USAF requirements. As the NA-159, the design won a competition in 1948 for a new trainer combining the former primary and basic training roles; powered by an 800 hp Wright R-1300-1A radial engine, it had a conventional low wing layout with tandem seating, but was the first US primary trainer to have a tricycle undercarriage. The first flight was made on September 26, 1949.

North American T-28C in trainer colours and the markings of VA-122, in December 1966. (*Lawrence S. Smalley*)

350

For the US Navy, North American developed the T-28B version after Navy evaluation of two T-28As. The new variant had a 1,425 hp Wright R-1820-86 engine and other minor changes. A total of 489 was delivered, some of these subsequently being modified for use as drone controllers designated T-28BD (later, DT-28B). A further 299 Trojans were delivered to the Navy as T-28C, differing only in having arrester gear for use in dummy deck approach and landing training.

TECHNICAL DATA (T-28B)

Manufacturer: North American Aviation, Inc, Inglewood, California.
Type: Basic trainer.
Accommodation: Pupil and instructor in tandem.
Power plant: One 1,425 hp Wright R-1820-86.
Dimensions: Span, 40 ft 1 in; length, 33 ft; height, 12 ft 8 in; wing area, 268 sq ft.
Weights: Empty, 6,424 lb; gross, 8,500 lb.
Performance: Max speed, 343 mph; cruising speed, 310 mph at 30,000 ft; initial climb, 3,540 ft/min; service ceiling, 35,500 ft; range, 1,060 st miles.
Serial numbers:

T-28A: 137636–137637.   T-28C: 140053–140077;   140449–
T-28B: 137638–137810;   138103–   140666; 146238–146293.
138367; 140002–140052.

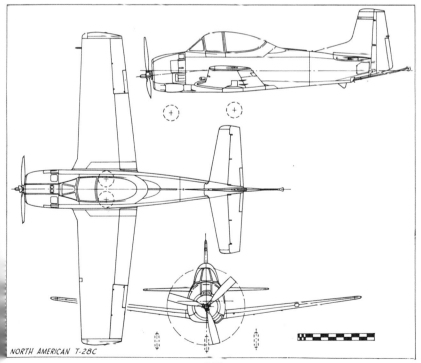

NORTH AMERICAN T-28C

351

North American A-5A in the markings of Heavy Attack Squadron VAH-3. (*US Navy*)

North American A-5 Vigilante

The heaviest aircraft accepted for service aboard US Navy aircraft carriers with the exception of the Douglas A-3 Skywarrior, the Vigilante incorporated a number of highly advanced aerodynamic concepts when it was first adopted. The design originated to meet a Navy requirement in 1955 and was intended as a high-performance attack aircraft with all-weather capability. Known for a time as the North American General-Purpose Attack Weapon (NAGPAW), the North American design was accepted in 1956, the company receiving a letter contract on June 29. Confirmation of an order for two prototypes came in August, when the designation YA3J-1 was allocated. The name Vigilante was later adopted and the designation changed from A3J to A-5 in 1962.

The A-5 had a high-wing layout, the low aspect ratio sweptback wing being of typical North American design. No ailerons were used; the wing had blown flaps for low-speed control, and a combination of spoilers in the wing surface, all-moving differential tailerons or tailplanes, and a slab fin/rudder provided control in three axes. Another advanced feature was the use of variable-geometry intakes for the two side-by-side General Electric YJ79-GE-2 engines, for the first time in a production aircraft.

North American RA-5C serving with RVAH-9 landing aboard the USS _Saratoga_. (*Stephen P. Peltz*)

Between the tailpipes of the engines North American located a linear bomb-bay, from which the A-5's primary armament—a free-falling nuclear weapon—was to be ejected rearwards. Part of the A-5's fuel load was carried in two tanks attached to this weapon; emptied before the target was reached, these tanks acted as aerodynamic stabilizers for the weapon as it fell.

The first of the prototypes flew on August 31, 1958, and production of the A-5A began. Deliveries started in 1960, and the first operational unit, VAH-7, began to receive Vigilantes in June 1961. This squadron took its A-5s aboard the USS *Enterprise* in August 1962, and VAH-1 and VAH-3 also converted on to the type. Production aircraft had J79-GE-2, J79-GE-4 or J79-GE-8 engines, REINS bombing-navigation system and provision for fuel tanks, bombs or rockets on underwing pylons; 59 were built.

North American RA-5Cs of RVAH-5 aboard USS *Ranger*, showing the folding wings and fin tips. (*North American photo*)

Considerable difficulties were encountered in clearing the linear bomb-bay for operational use and, before these could be satisfactorily overcome, a major shift in Navy policy deleted strategic bombing from that Service's role. Consequently, plans to produce an improved Vigilante attack-bomber were abandoned after design had reached an advanced stage. This projected A-5B had additional fuel in a deepened fuselage, the top line of which was humped; larger flaps, boundary layer control by air blown over the whole wing, and two additional underwing pylons, making four in all.

All these features were also incorporated in the RA-5C (originally A3J-3P), an unarmed reconnaissance version with a large array of electronic and visual reconnaissance equipment in the space previously occupied by the bomb-bay. This equipment included side-looking airborne radar in a fairing under the fuselage, vertical, oblique and split-image cameras, and active and passive ECM equipment.

The prototype RA-5C flew on June 30, 1962, and six A-5Bs in production

353

were converted to this configuration before delivery, being known as YA-5C (Limited) and the first being flown on April 29, 1962. During 1964 the Navy initiated a programme to convert 43 A-5As to RA-5C standard, and this work was put in hand after production of 91 new RA-5Cs had ended. The first squadron to receive reconnaissance Vigilantes was RVAH-5, which took its aircraft aboard the USS *Ranger* in June 1964. Additional squadrons selected to equip with the Vigilante were RVAH-1, RVAH-7, RVAH-9 and RVAH-11.

TECHNICAL DATA (RA-5C)

Manufacturer: North American Aviation, Inc, Columbus, Ohio.
Type: Carrier-borne electronic and visual reconnaissance.
Accommodation: Pilot and observer/radar operator.
Power plant: Two 10,800 lb s.t. General Electric J79-GE-8 turbojets.
Dimensions: Span, 53 ft; length, 76 ft 6 in; height, 19 ft 4¾ in; wing area, 754 sq ft.
Weight: Empty, 37,498 lb; max gross, 79,588 lb.
Performance: Max speed, Mach 2·1 (1,385 mph at 40,000 ft); cruising speed, 1,254 mph; combat ceiling, 48,400 ft; combat radius, up to 1,500 mls.
Serial numbers:

XA3J-1: 145157–145158.	YA-5C: 149300–149305.
A-5A: 146694–146702; 147850–147863;	A-5C: 149306–149317; 150823–150824;
148924–148933; 149276–	151615–151634; 151726–151728;
149299.	156608–156643.

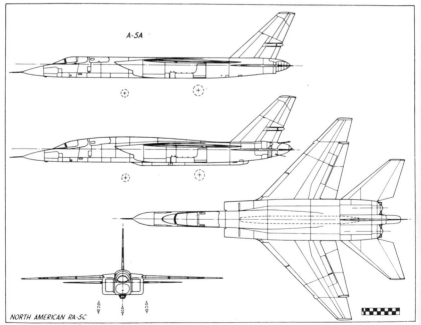

NORTH AMERICAN RA-5C

US Marine Corps North American Rockwell OV-10A. (*North American photo*)

North America Rockwell OV-10 Bronco

The first aircraft designed from scratch to meet requirements for so-called COIN (counter-insurgency) operations, the OV-10A was ordered into production for the Marine Corps and Army late in 1966 after some initial development problems with the prototypes. Although the OV-10 was developed and tested in a tri-service programme, the original specification for a LARA (Light Armed Reconnaissance Airplane) was drawn up for the Marine Corps.

A number of companies submitted design proposals to meet the LARA, and the North American NA-300 was selected as the winner of the design competition in August 1964. With the keynote on simplicity, the OV-10 had an untapered, parallel-chord wing, twin booms extending aft from the nacelles for the Garrett AiResearch T76 turboprops and a slender fuselage seating pilot and observer in tandem. For the strike mission, the OV-10 could carry up to 3,270 lb of ordnance externally on three strong points under the fuselage and two more on small stub-wings.

Seven YOV-10A prototypes were ordered in 1964, and the first of these flew on July 16, 1965, at the maker's Columbus plant, with 600 shp T-76-G-6/8 (opposite rotation) engines. Production orders were placed for both the USAF and the USMC, the versions differing in equipment details but both designated OV-10A. The production model differed from the original prototypes in several major respects, primarily because of weight growth; these changes included a 10-ft increase in span, moving the booms farther apart and using uprated T76-G-10/12 engines. The first production OV-10A flew on August 6, 1967, and deliveries (one each to USAF and USMC) began on February 23, 1968; production for the Marines totalled 114.

The Marine Corps training unit was VMO-5 at Camp Pendleton, Calif. (later HML-267) and the first operational unit was VMO-2, operating out of Da Nang in Vietnam, where the Bronco became operational in July 1968. Subsequently, VMO-6 also became operational in Vietnam and VMO-1, VMO-4 and VMO-8 used the Bronço in USA. Eighteen OV-10As were acquired from the Marine Corps by the US Navy to equip VA(L)-4 operating riverine patrols in the Mekong Delta.

During 1976, USMC was evaluating the YOV-10D equipped as a night observation surveillance aircraft with FLIR, uprated T76-G-420/421 engines and other special equipment. Conversion of about 18 earlier models to OV-10D configuration was planned.

TECHNICAL DATA (OV-10A)

Manufacturer: North American Rockwell Corp., Columbus, Ohio.

Type: Light attack and forward air control.

Accommodation: Pilot and observer in tandem.

Power plant: Two 715 shp Garrett AiResearch T76-G-10/12 turboprops.

Dimensions: Span, 40 ft; length, 39 ft 9 in; height, 15 ft 1 in; wing area, 291 sq ft.

Weights: Empty, 7,190 lb; normal loaded, 12,500 lb; max take-off, 14,444 lb.

Performance: Max speed, 281 mph at 5,000 ft; cruising speed, 220 mph at 18,000 ft; initial climb, 2,320 ft/min; service ceiling, 29,000 ft; mission radius with full internal fuel and 2,800-lb ordnance load, 190 miles plus 1-hour loiter.

Armament: Four fixed forward-firing 7·62 mm guns in sponsons; four sponson stations for rockets, gun pods or other stores; fuselage centre-line station for 20 mm gun pod; two underwing stations for AIM-9 missiles. Max eternal stores 4,600 lb.

Serial numbers:

YOV-10A: 152879–152885. OV-10A: 155390–155503.

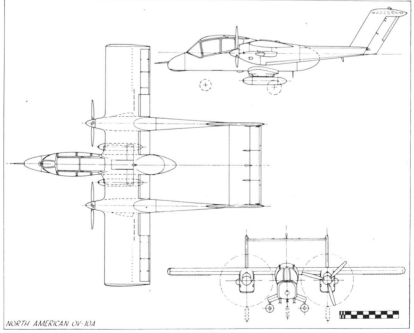

NORTH AMERICAN OV-10A

356

The Northrop BT-1, photographed in May 1938. (*Douglas photo*)

Northrop BT

John K. Northrop, famed as the designer of the Lockheed Vega, formed his own company, Northrop Aircraft Corporation, at Burbank, California, in 1929 and began producing new all-metal designs with advanced structural features. This firm soon became a division of United Aircraft and was merged with Stearman in 1931. Northrop then formed another company, the Northrop Corporation, at Inglewood, California, which was backed by and eventually became a wholly-owned subsidiary of the Douglas Aircraft Company. Northrop continued to develop military and civil aircraft to his established structural concepts, notably the Delta transport (US Coast Guard RT-1, see page 458) and the Gamma high-speed mailplane. The Gamma was quickly developed into an export attack model. One, tested by the Army as the YA-13, was converted to XA-16 and became the prototype of the A-17 series. A scaled-down version was tested by the Navy in 1935 as XFT-1, later redesignated XFT-2.

A parallel development, fitted with a semi-retractable undercarriage and split trailing-edge flaps, was submitted to the Navy as XBT-1. After minor refinements, 54 production BT-1s were ordered and deliveries began to VB-5 in April 1938. The last aircraft on the contract was held back and fitted with a non-retractable tricycle landing gear to test the suitability of this type of gear, which was beginning to reappear on aircraft after having been abandoned at the beginning of World War I, for use on aircraft carriers.

One BT-1 was diverted to a design improvement programme and re-designated XBT-2. The initial change was replacement of the rearward-retracting landing gear with an inward-folding type that retracted completely as on the Army A-17A. The next change was replacement of the 825 hp Pratt & Whitney R-1535-94 Twin Wasp Jr two-row engine with an

800 hp single-row Wright XR-1820-32 Cyclone. Then followed gradual revisions to canopy and vertical tail shape.

The Northrop Corporation, meanwhile, had become the El Segundo Division of Douglas, but the original T-for-Northrop designation was retained for the Naval aircraft on order. By the time the XBT-2 configuration was finalized and production aircraft were desired it was decided that they should carry D-for-Douglas designations, and these aircraft became the famous SBD Dauntless series of World War II (see page 167).

TECHNICAL DATA (BT-1)

Manufacturer: Northrop Corporation, Inglewood, California.
Type: Carrier-based scout/dive-bomber.
Accommodation: Pilot and observer/rear gunner.
Power plant: One 825 hp Pratt & Whitney R-1535-94.
Dimensions: Span, 41 ft 6 in; length, 31 ft 8 in; height, 9 ft 11 in; wing area, 319 sq ft.
Weights: Empty, 4,606 lb; gross, 7,197 lb.
Performance: Max speed, 222 mph at 9,500 ft; cruising speed, 192 mph; initial climb, 1,270 ft/min; service ceiling, 25,300 ft; range, 1,150 st miles.
Armament: One fixed forward-firing 0·50-in gun and one flexible dorsal 0·30-in gun. One 1,000 lb bomb under fuselage.
`Serial numbers:`
   XBT-1: 9745.   XBT-2: 0627.
   BT-1: 0590–0626; 0628–0643.

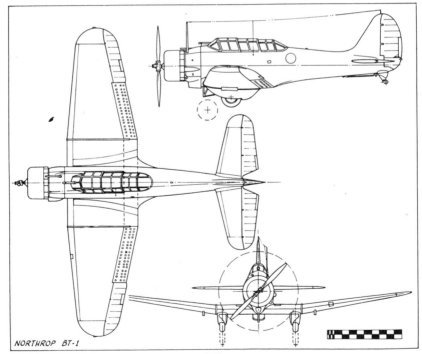

NORTHROP BT-1

A North American T2J-1 Buckeye, prior to redesignation as T-2A. (*North American photo*)

Rockwell T-2 Buckeye

The US Navy drew up requirements for an all-purpose jet trainer during 1956, seeking an aeroplane which could be used by students from the time they had completed *ab initio* training right up to carrier qualification, advanced training and fighter tactics. North American, already in production with the T-28 for Navy basic training, won the contract to develop the new aircraft, in the design of which proven components and equipment were used wherever possible. Thus, the wing design was derived from that of the FJ-1 Fury, and the control system was similar to that of the T-28C.

A single 3,400 lb s.t. Westinghouse J34-WE-36 was located in the under-fuselage with bifurcated intakes under the nose and a single tailpipe. The tandem cockpit arrangement placed the instructor well above the level of the student in the front, for good forward visibility; both occupants had North American zero-level ejection-seats. For armament training the aircraft had underwing strong points to carry bombs, rockets or machine-gun pods, and an arrester hook was fitted for real or simulated carrier deck-landings.

The North American trainer was ordered straight into production when the NA-241 design was accepted by the Navy, with an initial contract for six followed by one for 121 in 1956 and a third and final batch of 90 ordered in February 1959. First flight of the new trainer, designated T2J-1, was made at Columbus on 31 January, 1958, and after preliminary evaluation at the NATC Patuxent River, deliveries began in July 1959, shortly after the name Buckeye had been adopted. The first unit to receive Buckeyes (which were redesignated T-2As in 1962) was Basic Training Group Seven (BTG-7) which eventually became Training Squadron VT-7 at NAAS Meridian. A second unit, VT-9, was formed at Meridian later, each unit having a minimum of 60 aircraft on strength. The type was also used by VT-4 and VT-19 at Pensacola for gunnery and carrier qualification training.

359

North American T-2B prototype. (*North American photo*)

Production of the T2J-1 ended in 1961, but on January 26, 1962, the Columbus Aircraft Division (of what became North American Rockwell and eventually Rockwell International) received a contract to modify two T2J-1s to a twin-engined configuration, initially known as YT2J-2 but soon changed to YT-2B. The modification comprised fitting two 3,000 lb thrust Pratt & Whitney J60-P-6 engines side-by-side in the fuselage in place of the single J34; few other changes were necessary but performance was significantly improved. First flight of the YT-2B was made on August 30, 1962, and an initial production contract was placed in March 1964, subsequent orders bringing the total buy to 97. The first production T-2B flew on May 21, 1965, and deliveries to VT-4 began in December the same year.

Another change of power plant was made in 1968, when General

North American Rockwell T-2B Buckeye of Training Squadron VT-4.
(*North American photo*)

Electric J85-GE-4s were fitted in place of the J60s. This resulted in a change of designation to T-2C, first flown on April 17, 1968, followed by the first production model on December 10. Subsequent purchases up to FY 74 brought total T-2C production for the US Navy to 231, with others procured through Navy channels for export (including 12 T-2D for Venezuela and 30 T-2Es for Greece). The last T-2As were retired from active service in mid-1973, leaving the T-2B and T-2C to continue in service as the Navy's basic jet trainer into the 'eighties.

TECHNICAL DATA (T-2C)

Manufacturer: Rockwell International Corporation, Columbus Aircraft Division, Columbus, Ohio.
Type: All-purpose jet trainer
Accommodation: Pupil and instructor in tandem.
Power Plant: Two 2,950 lb s.t. General Electric 085-GE-4 turbojets
Dimensions: Span, 38 ft 2 in; length, 38 ft 8 in; height, 14 ft 9½ in; wing area, 255 sq ft.
Weights: Empty, 8,115 lb; gross, 13,180 lb.
Performance: Max speed, 521 mph at 25,000 ft; initial climb, 6,200 ft/min; service ceiling, 44,400 ft; range, 910 st miles.
Armament: Provision for gun pods, bombs or rockets under wings.

Serial numbers:		T-2B:	152382–152391; 152440–152475; 153538–153555; 155206–155238.
XT2J-1: 144217–144218.			
T-2A: 144219–144222;	145996–		
146015; (145997 to XT2J-2);		T-2C:	155239–155241; 156686–
147430–147530; 148150–148239			156733; 157030–157065;
			158310–158333; 158575–
			158610; 158876–158911;
			159150–159173

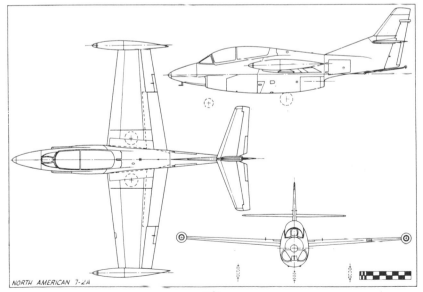

NORTH AMERICAN T-2A

361

A Ryan FR-1 in Navy service in 1946. (*Peter M. Bowers*)

Ryan FR Fireball

The US Navy's first experience of jet aircraft operation was obtained with an aeroplane which retained, so far as was possible, all the characteristics and attributes of the conventional piston-engined fighter. This highly conservative approach to the application of the jet engine was dictated by the overriding requirements of aircraft-carrier operation. Long take-off and landing runs, and short endurance—characteristics of the early jet fighters—were unacceptable to the Navy. Consequently, late in 1942, a proposal for a composite aircraft with both piston engine and jet, appeared to have considerable merit.

Of nine manufacturers invited to submit designs, Ryan showed the best appreciation of the problems, and a letter of intent dated February 11, 1943, authorized Ryan to proceed with three prototypes of its Model 28 design. Ryan had no previous experience of building aircraft for the Navy and had never built a combat aeroplane; nevertheless, work proceeded rapidly to achieve a first flight on June 25, 1944. In layout, the XFR-1 was a simple low-wing monoplane with a Wright R-1820-72W radial engine in the nose, a tricycle landing gear, a laminar-flow wing-section, a flush-riveted skin and metal-covered control surfaces. Several of these features were novel in a Navy fighter, but it was the General Electric I-16 jet engine in the rear fuselage which singled the XFR-1 out from its contemporaries.

The first production contract for Ryan's composite fighter, which was named the Fireball, was placed by the Navy on December 2, 1943, calling for 100 FR-1s. A further 600 were ordered on January 31, 1945, but 634 of the total on order were cancelled on V-J Day. Delivery of the 66 FR-1s began in January 1945 and was completed in November the same year. The production aircraft were similar to the prototypes, but single-slotted flaps replaced the original double-slotted type after the first 14 aircraft.

In March 1945 the first FR-1 was delivered to Navy Squadron VF-66 at San Diego, this unit having been specially formed for the purpose. Carrier qualification trials were conducted with three aircraft aboard USS *Ranger* on May 1, 1945, but the squadron was decommissioned on October 18, 1945, without having put to sea. Its aircraft and personnel were transferred to VF-41 (later VF-1E) which conducted a number of exercises aboard the USS *Wake Island* (November 1945), USS *Bairoko* (March 1946) and USS *Badoeng Strait* (March 1947 and June 1947). The FR-1s were withdrawn from Naval service shortly after the final tour aboard the USS *Badoeng Strait*.

TECHNICAL DATA (FR-1)

Manufacturer: Ryan Aeronautical Corporation, San Diego, California.

Type: Carrier-borne fighter.

Accommodation: Pilot only.

Power plant: One 1,350 hp Wright R-1820-72W radial and one 1,600 lb s.t. General Electric J31 turbojet.

Dimensions: Span, 40 ft; length, 32 ft 4 in; height, 13 ft 11 in; wing area, 275 sq ft.

Weights: Empty, 7,689 lb; gross, 11,652 lb.

Performance: Max speed, 404 mph at 17,800 ft; cruising speed, 152 mph; service ceiling, 43,100 ft; range, 1,620 st miles.

Armament: Four fixed forward-firing 0·50-in guns.

Serial numbers:
XFR-1: 48232–48234. FR-1: 39647–39712.

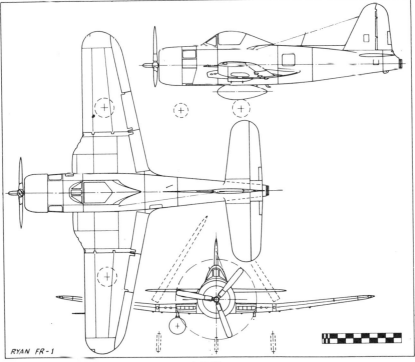

RYAN FR-1

363

A Sikorsky RS-3 serving with Utility Squadron Five (VJ-5). 'D11' in the fuselage marking indicated the Eleventh Naval District. (*US Navy*)

Sikorsky PS, RS

Distinguished by their sesquiplane wings and twin-boom tail arrangement, the series of Sikorsky flying-boat amphibians from the S-36 to the S-41 became well known in the thirties in commercial, military and naval service. The single S-36 prototype, lost in an emergency landing during a test flight, was followed by five examples of the S-38, and one of these was purchased by the US Navy. Designated XPS-1, it was evaluated as a potential new patrol aircraft, with a gunner in the bow, but was relegated to transport and utility duties soon after delivery in 1927. This version had Wright J-5 engines.

Two more Sikorsky amphibians were ordered in the patrol category on October 13, 1928, these being S-38As designated XPS-2. Delivered in 1928, they were transferred to Utility Squadron One, VJ-1B, in 1929 at Aroostook. Four similar aircraft were delivered to the Navy between 1929 and 1932, being versions of the S-38B and designated PS-3. These also had nose- and rear-gunners' cockpits when delivered but were later removed.

Late in 1930 the Navy ordered three examples of the similar Sikorsky S-41, and, having finally abandoned the pretext that these aircraft served in the patrol category, designated them RS-1s. At the same time, the XPS-2s became XRS-2s, and the PS-3s took the designation RS-3. Three more S-38Bs purchased by the Navy on two contracts in 1931 were designated RS-3 from the start.

The Sikorsky amphibians saw service with a number of Navy units, and RS-1s and RS-3s served also with the Marine Corps at home and overseas. All had been withdrawn from service by the end of 1934, but four S-38s acquired from Pan American early in World War II became RS-4s and RS-5s.

Although only a few S-38s saw service with the Navy—a few others went to the Army Air Corps as C-6s—the amphibian proved a great success commercially and gained a sound reputation for its performance and ruggedness. It was this success, in fact, which proved a turning-point in the fortunes of the Sikorsky Manufacturing Corporation and provided the foundation for its later achievements, first in the development of larger flying-boats, and then with helicopters.

TECHNICAL DATA (RS-3)

Manufacturer: Sikorsky Manufacturing Corporation, College Point, LI, NY.
Type: Utility transport.
Accommodation: Crew of four and four passengers.
Power plant: Two 450 hp Pratt & Whitney R-1340Cs.
Dimensions: Span, 71 ft 8 in; length, 40 ft 3 in; height, 13 ft 10 in; wing area, 720 sq ft.
Weights: Empty, 6,740 lb; gross, 10,323 lb.
Performance: Max speed, 124 mph at sea level; cruising speed, 110 mph; initial climb, 5,000 ft in 8 min; service ceiling, 15,300 ft; range, 594 st miles.
Serial numbers:

XPS-1: A8005.
XPS-2 (XRS-2): A8089–A8090.
PS-3 (RS-3): A8284–A8287;   A8922
A8923; A9055.

RS-1: A8842–A8844.
RS-4: 37854–37855.
RS-5: 37852–37853.

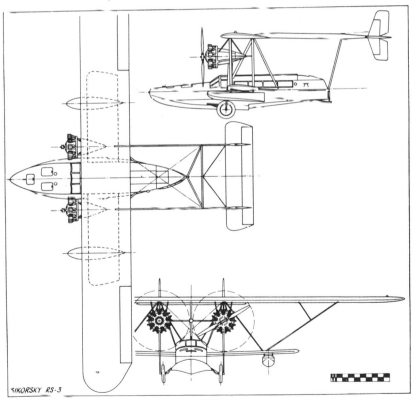

SIKORSKY RS-3

365

A Coast Guard Sikorsky HO4S-3G demonstrating sea rescue techniques.
(*US Coast Guard*)

Sikorsky HO4S, HRS

Versions of the Sikorsky S-55 were used by all the US Services during the fifties and played an important part in developing helicopter roles and techniques. Of classic helicopter configuration, with a single main rotor and anti-torque tail rotor, the S-55 was the first of the Sikorsky helicopters with adequate cabin space and lifting ability to permit satisfactory operation in such roles as troop transport and air–sea rescue.

The first Navy contract for an S-55 variant was placed on April 28, 1950, for the general purpose and anti-submarine observation HO4S-1, equivalent to the Army's H-19A. Deliveries began on December 27, 1950, to Utility Squadron HU-2. Only 10 were built, with 550 hp Pratt & Whitney R-1340 engines. They were followed by 61 HO4S-3s with 700 hp Wright R-1300s. The HO4S-2 was a proposed version of the HO4S-1 for Coast Guard use in the air–sea rescue role, but this was not produced; instead, 30 of the more powerful Wright-engined model were assigned to the USCG as HO4S-3Gs. These were redesignated HH-19G in 1962; most of the helicopters of this type which remained with the Navy operated as HO4S-3s before becoming UH-19Fs in the utility role in 1962.

A second series of S-55s was purchased for Navy and Marine use as troop transports, carrying eight troops plus the pilot. These were designated HRS, Sikorsky receiving the go-ahead for production of this variant on August 2, 1950. The HRS-1 was similar to the HO4S-1 apart from troop seating and self-sealing fuel tanks. Deliveries to Marine Squadron HMX-1 began on April 2, 1951, and 60 were built. The 91 HRS-2s differed only in detail, and these two versions equipped nine Marine transport (HMR) squadrons in time to serve operationally in Korea. Final production version was the HRS-3 with the Wright R-1300-3 engine; 89 were built to give a grand total of 235 HRS helicopters produced for the USN/UMC. In addition, some HRS-2s were converted

366

to HRS-3 with the change of engine. Surviving examples were designated CH-19E in 1962.

TECHNICAL DATA (HRS-2)

Manufacturer: Sikorsky Aircraft Division of United Aircraft Corporation, Stratford, Connecticut.

Type: Transport helicopter.

Accommodation: Pilot and eight troops.

Power plant: One 550 hp Pratt & Whitney R-1340.

Dimensions: Rotor diameter; 53 ft; length, 42 ft 2 in; height, 13 ft 4 in; rotor disc area, 2,210 sq ft.

Weights: Empty, 4,590 lb; gross, 7,900 lb.

Performance: Max speed, 101 mph at sea level; cruising speed, 85 mph at 1,000 ft; initial climb, 700 ft/min; service ceiling, 10,500 ft; range, 370 st miles.

Serial numbers (allocations):

HO4S-1: 125506–125515.

HO4S-3: 133739–133753,* 133777–133779;* 138494–138529; 138577–138601; 150193–150194.

HRS-1: 127783–127842.

* MDAP supply.

HRS-2: 129017–129049; 130138–130181; 130182–130191;* 130192–130205.

HRS-3: 130206–130264; 137836–137845; 140958–140961;* 141029;* 141230;* 142430–142436;* 144244–144258; 144268–144270;* 146298–146302.

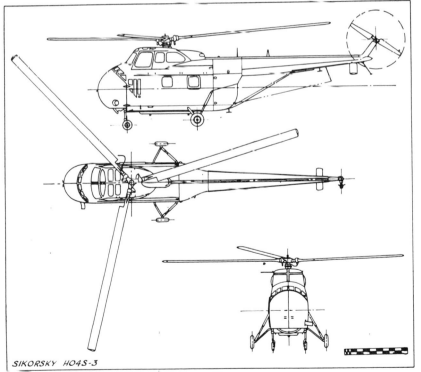

SIKORSKY HO4S-3

367

Sikorsky HR2S-1 transport. Note the 'eye' decoration on front of nacelle. (*US Navy*)

Sikorsky HR2S-1

A Marine Corps requirement in 1950 led to development of the first of a series of very large helicopters by Sikorsky. For more than 10 years after its first flight this design, Sikorsky S-56, was the largest helicopter flying anywhere outside the USSR, and from the experience gained with this type, Sikorsky went on to evolve the S-60 crane, the S-64 and the S-65.

The Marine Corps requirement was for an assault helicopter which would carry about 26 troops or a variety of vehicles and other military loads. The classic Sikorsky single-rotor layout was retained, but two engines were essential for a helicopter of this size, and, to leave the fuselage clear for load-carrying, these were located in external nacelles on stub-wings which sprouted from the fuselage. The main legs of the landing gear retracted into the engine nacelles.

The emphasis upon the transport role was reflected in the clam-shell nose-opening doors, permitting straight-in loading beneath the cockpit. A rail along the cabin ceiling and a winch with 2,000 lb capacity aided loading of cargo through the nose doors or the rear side door.

A contract for a prototype XHR2S-1 was placed by the Navy on May 9, 1951, and this helicopter first flew on December 18, 1953. A number of changes were introduced in the production model HR2S-1 which flew on October 25, 1955, and had a dorsal fin, changes in the nacelles, and smaller twin wheels in place of the single wheel on the main legs.

Delivery of HR2S-1s began on July 26, 1956, with Squadron HMX-1 the first unit to receive this type. Production of 55 for the Marines was completed in February 1959, and these helicopters were redesignated CH-37C in 1962. Two examples were modified in 1957 to HR2S-1Ws for early-warning duties with the Navy. The large-diameter scanner for the AN/APS-20E radar was located in the nose, producing a distinctive chin, and the size and location of the tail trimming surfaces was changed.

TECHNICAL DATA (HR2S-1)

Manufacturer: Sikorsky Aircraft Division of United Aircraft Corp. Stratford, Conn.
Type: Transport helicopter.
Accommodation: Two pilots and 20 troops or 24 litters.
Power plant: Two 1,900 hp Pratt & Whitney R-2800-54s.
Dimensions: Rotor diameter, 72 ft; length, 64 ft 3 in; height, 22 ft; rotor disc area, 4,080 sq ft.
Weights: Empty, 20,831 lb; gross, 31,000 lb.
Performance: Max speed, 130 mph at sea level; cruising speed, 115 mph; initial climb, 910 ft/min; service ceiling, 8,700 ft; range, 145 st miles.
Serial numbers:

    XHR2S-1: 133732–133735.   HRS-1W: 141646–141647.
    HR2S-1: 138418-138424;   140314–140325;   141603–141617; 145855–145875.

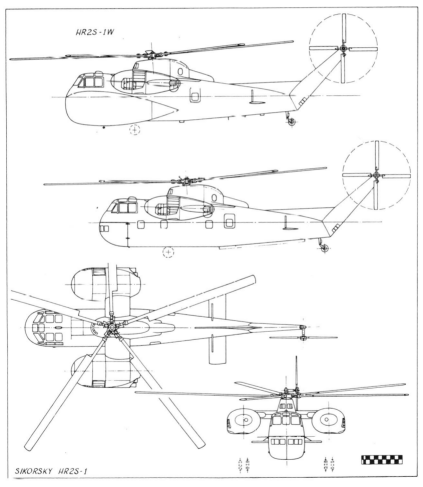

HR2S-1W

SIKORSKY HR2S-1

369

Sikorsky HSS-1N was equipped for all-weather anti-submarine search role.
(*Sikorsky photo*)

Sikorsky HSS Seabat, HUS Seahorse

Although the Sikorsky HO4S (see page 366) was operated by the US Navy nominally in an anti-submarine role, its lack of range and the small useful load which could be carried severely restricted its usefulness in operation. The helicopter's potential value in this offensive role was clearly shown with the HO4S, however, and Sikorsky set about the design of a similar but larger helicopter, the Model S-58. A prototype was ordered by the Navy on June 30, 1952, with the designation XHSS-1.

The new helicopter made its first flight on March 8, 1954. Apart from its size, the S-58 closely resembled the S-55. The engine was a 1,525 hp Wright R-1820 radial, located obliquely in the nose so that the transmission shaft ran at right angles to the engine straight into the gearbox beneath the rotor hub. Unlike the S-55 with its nosewheel landing gear, the S-58 had a tailwheel, and, to facilitate ship-board stowage, the main rotor blades could be folded aft and the entire rear fuselage and tail rotor folded forward.

Production orders for the HSS-1 were placed by the Navy before the first flight of the prototype on March 8, 1954. Subsequently, a transport and utility version was ordered by the Marine Corps with the designation HUS-1, and the two types were officially named Seabat and Seahorse, respectively. In 1962, while versions of the S-58 were still in production, all Navy and Marine variants were redesignated in the H-34 series as indicated below.

First flown on September 20, 1954, the HSS-1 (SH-34G) was delivered to a number of Navy anti-submarine squadrons, starting in August 1955 with HS-3. Since the load-carrying ability of the HSS-1 was still somewhat limited, these helicopters operated either as hunters, with dipping

ASDIC as the primary search aid, or as killers carrying homing torpedoes externally on the fuselage. Paired hunter/killer operations were possible, but it was more usual for the HSS-1s to search and to call upon destroyers to make the attack when a target was located. For night operations, Sikorsky developed the HSS-1N (SH-34J), incorporating Sikorsky automatic stabilization equipment, Ryan AN/APN-97 Doppler navigation equipment and an automatic hover coupler. Production of these Navy versions totalled 350, including 135 of the HSS-1N model. After being replaced in service by SH-3As, many Seabats were stripped of ASW gear and used in the utility role as UH-34G and UH-34J. Others were built for the Mutual Aid Program.

On January 30, 1957, the single HSS-1F (SH-34H) made its first flight. This had two General Electric T58 turboshaft engines replacing the piston engine and was used as a test-bed for these engines and also to evaluate the operational advantage of turboshafts.

The first Marine Corps order for S-58s was placed with Sikorsky on October 15, 1954, for the HUS-1 (UH-34D) version. This had the ASW gear deleted and provision for carrying up to 12 passengers in the cabin. Deliveries, to HMRL-363, began in February 1957, and these helicopters were widely used in the support role and in the further development of helicopter armament for attacking ground targets. Four were modified for Arctic operation with VX-6 as HUS-1L (LH-34D) and seven HSS-1Z (VH-34D) aircraft were assigned in 1960 to the joint Army–Marine Corps

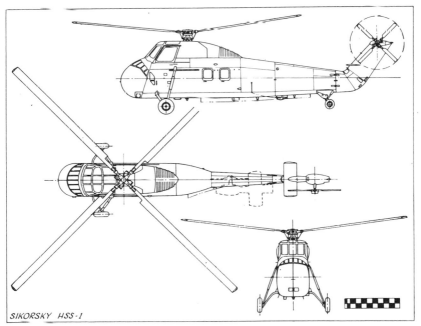

SIKORSKY HSS-1

371

Sikorsky UH-34 Seahorse serving with Helicopter Training Squadron Eight, HT-8. (*US Navy*)

Executive Flight Detachment providing a VIP service for the President and others.

Operating on amphibious pontoons, 40 of these Marine helicopters were designated HUS-1A (UH-34E), and six others went to the Coast Guard for search and rescue duties, as HUS-1G (HH-34F). Production for the Marine Corps totalled 603, including the variants noted above and at least one for MAP.

TECHNICAL DATA (UH-34D)

Manufacturer: Sikorsky Aircraft Division of United Aircraft Corporation, Stratford Connecticut.

Type: Utility helicopter.

Accommodation: Two pilots and 12–18 passengers:

Power plant: One 1,525 hp Wright R-1820-84.

Dimensions: Rotor diameter, 56 ft; length, 46 ft 9 in; height, 15 ft 11 in; rotor disc area, 2,460 sq ft.

Weights: Empty, 7,900 lb; gross, 14,000 lb.

Performance: Max speed, 123 mph at sea level; cruising speed, 98 mph; initial climb, 1,100 ft/min; service ceiling, 9,500 ft; range, 182 st miles.

Serial numbers (allocations):

XHSS-1: 134668–134670.

HSS-1: 137849–137858; 138460–138493; 139017–139029; 140121–140139; 141571–141602; 143864–143960; 145660–145669.

HSS-1N: 145670–145712; 147631–147635;* 147984–148032; 148934–148963; 149082–149087;* 149131–149133;* 149840–149842;* 150730–150732;* 150808–150819;* 150821–150822.

* Procured for MAP supply.

HUS-1: 143961–143983; 144630–144662; 145713–145812; 147147–147201; 148053–148122; 148753–148822; 149318–149402; 149840–149842; 150195–150264; 150552–150580; 150691;* 150717–150729.

SH-34J: 151729–151731; 152380–152381; 153617–153622.

UH-34D: 152686;* 153116–153133; 153556–153558; 153695–153704; 154045; 154889–154902; 156592–156598.

Sikorsky SH-3A serving with HS-5 as the standard Navy anti-submarine helicopter in 1967. (*Stephen P. Peltz*)

Sikorsky H-3 Sea King

Based on experience with the HSS-1 helicopters operating in anti-submarine hunter/killer pairs, the US Navy contracted on December 24, 1957, for development of a new helicopter by Sikorsky in which the two roles could be combined. The ability to carry both search equipment (dipping sonar) and 840 lb of weapons, as well as instrumentation for all-weather operations, meant that a considerably larger helicopter was needed than the S-58 series already in service (see page 370). In the design of this new type, with the company designation S-61, conventional Sikorsky practice was followed in overall layout but a number of new features were introduced.

A watertight hull design was used since most of the operational life would be spent over water, and stabilizing floats provided a convenient housing for the retractable wheels, making the S-61 a true amphibian. Two General Electric T58 turboshaft engines side by side above the main fuselage were located close to the gearbox driving the five-blade main rotor. Primary search equipment was the Bendix AQS-10 or AQS-13 sonar, with a coupler to hold hovering altitude automatically in conjunction with the Ryan APN-130 Doppler and radar altimeter. A Hamilton Standard auto-stabilization system was incorporated for all-weather capability.

An initial batch of seven trials' aircraft carried the designation YHSS-2, but the Sea King was subsequently redesignated SH-3 in the unified tri-service system of September 1962. First flight was made on March 11, 1959, and Navy BIS (Board of Inspection and Survey) service acceptance trials began in February 1961; fleet deliveries began in September 1961 when HSS-2s were delivered to VHS-10 and VHS-3, respectively at Ream

The Sikorsky RH-3A was specially equipped to tow mine-counter-measures equipment. (*Sikorsky photo*)

Field, near San Diego, and Norfolk, Virginia. A total of 255 SH-3As were delivered before introduction of 1,400 shp T58-GE-10 engines in place of the original 1,250 shp GE-8Bs led to a designation change, to SH-3D, on helicopters in production during 1966. The SH-3D, produced for export under MAP as well as for the US Navy, which acquired 73, also had increased fuel capacity.

During 1965, nine SH-3As were converted to RH-3As for mine counter-measures (MCM) duties, with anti-submarine equipment removed and provision for towing MCM gear. They operated from shore bases or MCM ships. Eight specially equipped VH-3As (originally HSS-2Z) were supplied to the Executive Flight Detachment which provided a VIP transport and emergency evacuation service for key personnel in Washington,

Delivered to the Coast Guard in 1967, the HH-3F was based on the USAF CH-3C rather than the SH-3A. (*Sikorsky photo*)

374

including the President. Operation of these aircraft was shared between the USMC and the USAF. In 1975, 11 VH-3Ds (with up-rated engines) were procured as replacements for the VH-3As, after plans to procure larger VH-53Fs for this purpose had been dropped as an economy measure. In another conversion programme, the Navy procured 12 HH-3A Sea Kings for search and rescue duties in combat areas, including Vietnam; these had -8F engines and a TAT-102 turret mounting a 7·62 mm Mini-gun in the rear of each sponson, remotely controlled by a crewman in the cabin.

The HH-3A, which was operated by Squadron HC-7, had a number of other new features, such as a 'Highdrink' refuelling system, high-speed refuelling hoist, fuel dumping system and provision for long-range fuel tanks. All these features, less the armament and armour, were introduced in 105 SH-3As converted to SH-3Gs from 1970 onwards. A further improvement in the ASW capability of the Sea King was brought about in the SH-3H, which had six new items of equipment specifically for better ASW performance, an LN66HP radar and other equipment for anti-ship-

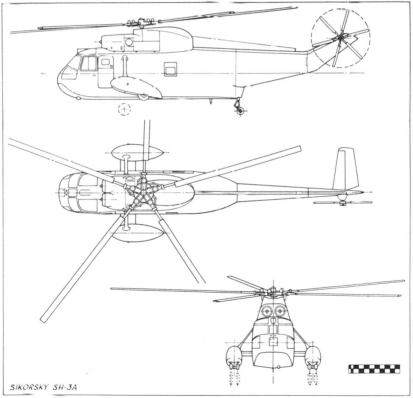

SIKORSKY SH-3A

375

Sikorsky SH-3G Sea King of Helicopter Anti-Submarine Squadron HS-15, with spray guard ahead of engine intakes. (*US Navy*)

missile detection and some general aircraft improvements permitting a gross weight of 21,000 lb. Flight testing of a converted SH-3G began in 1972 and a programme of conversions was begun, including most SH-3Gs and some SH-3As and SH-3Ds, to allow the US Navy to equip 10 squadrons, plus four USN Reserve squadrons, with eight SH-3Hs each.

TECHNICAL DATA (SH-3D)

Manufacturer: Sikorsky Aircraft Division of United Aircraft Corporation, Stratford, Connecticut.
Type: Anti-submarine helicopter.
Accommodation: Two pilots and two radar operators.
Power Plant: Two 1,500 shp General Electric T58-GE-10 shaft turbines.
Dimensions: Rotor diameter, 62 ft overall; length, 72 ft 8 in; height, 16 ft 10 in; rotor disc area, 3,019 sq ft.
Weights: Empty, 12,087 lb; ASW mission gross weight, 18,897 lb; max take-off, 20,500 lb.
Performance: Max speed, 166 mph at sea level; cruising speed, 133 mph; initial climb, 2,200 ft/min; service ceiling, 14,700 ft; range, 625 st miles.
Armament: 840 lb of homing torpedoes, depth-bombs or other stores.
Serial numbers:

VH-3A: 150610–150617.	SH-3A: 147137–147146; 148033–
SH-3D: 152139; 152690–152713;	148052; 148964–149012;
153532–153537;* 154100–	149679–149738; 149893–
154123; 1568483–156506.	149934; 150618–150620;
	15122–151557; 152104–
	152138.

* For MAP supply.

The Sikorsky CH-53A in the overall olive finish of the Marine Corps in 1965.
(*Sikorsky photo*)

Sikorsky H-53 Sea Stallion

Ordered in August 1962 for service with the Marine Corps, the CH-53A
Sea Stallion was, at the time of its development, the largest helicopter in
the family of designs·by the Sikorsky company. Work on large, heavy-lift
helicopter designs at Sikorsky had begun a decade earlier when the S-56
assault transport had been produced for the Navy and Marine Corps.
This type went into service eventually with the Army also, providing
Sikorsky with a wide spread of experience in large helicopter design and
operation.

Out of this experience two new, related designs began to take shape in
the early sixties. One, Sikorsky model S-64, was a flying crane, adopted by
the Army as the CH-54A; the other, using many of the same components
but having a conventional fuselage, was the S-65 and was designed
primarily to the Marine Corps specification for a new assault transport.
Plans to purchase the S-65 were announced on August 27, 1962, and an
initial contract was placed for two prototypes, a static test airframe and a
mock-up. Production contracts were placed subsequently.

The CH-53A was designed to have good load-carrying ability in a
fuselage comparable to conventional fixed-wing designs, the helicopter's
primary mission being cargo transport. Rear-loading doors were incorporated beneath the boom carrying the tail rotor, with remotely controlled winches at the forward end of the hold to facilitate loading and
unloading. Among the items which the specification required to be
carried were a 1½-ton truck and trailer, the Hawk missile system, an
Honest John missile on its trailer and a 105 mm Howitzer or a ½-ton jeep
with a ½-ton two-wheeled trailer. For the secondary missions of troop
transport and casualty evacuation, the CH-53A accommodated 38 combat-
equipped troops or 24 stretchers. The fuselage included a watertight lower

377

A US Marine Corps CH-53D Sea Stallion in the mine-sweeping role, with magnetic detector in tow. (*US Navy*)

section to give emergency water-landing capability, aided by the two sponsons into which the main units of the undercarriage retracted.

The first CH-53A made its first flight on October 14, 1964, and deliveries began in mid-1966, followed by deployment to Vietnam from the beginning of 1967. At the end of 1968, production switched to the CH-53D version, the first example of which was delivered on March 3, 1969. Whereas the CH-53A was powered by 2,850 shp General Electric T64-GE-6, 3,080 shp -1 or 3,435 shp -16 (mod) engines, the CH-53D had either 3,695 shp T64-GE-412 or 3,925 shp -413 engines. Internal changes made it possible for the CH-53D to accommodate up to 55 troops, compared with 38 in the CH-53A, and provision was made for automatic folding of the main and tail rotors for stowage aboard aircraft carriers. Production of the Sea Stallion for the USMC ended in January 1972 with a combined total of 265 A and D models delivered.

All but the first 34 CH-53As were fitted with hardpoints so that mine-sweeping equipment could be towed, and in 1971 the US Navy borrowed 15 CH-53As from the Marines to equip its first Helicopter Mine Counter-measures Squadron, HM-12. Following successful operation of these helicopters, which were re-engined with T64-GE-413s and designated RH-53As, the Navy ordered 30 specially-equipped versions of the Sea Stallion as RH-53Ds, and the first of these flew on October 27, 1972, deliveries to HM-12 beginning in September 1973. Initially fitted with T64-GE-413A engines, the RH-53Ds, production of which was completed in 1974, were to be retrofitted in due course with 4,380 shp T64-GE-415 engines. Features of the RH-53D included a flight refuelling probe in the nose and a 500-US gal fuel tank on each sponson. Max take-off weight was increased to 50,000 lb and provision was made to carry two 0·50-in machine guns to detonate surface mines.

On March 1, 1974, Sikorsky flew the first of two prototypes of the

YCH-53E, a three-engined development (with T64-GE-415s) of the Sea Stallion developed to meet USN and USMC requirements for a heavy-duty multi-purpose helicopter. Production of 70 was planned to meet USN and USMC needs; a pre-production CH-53E first flew on December 8, 1975 and the first batch of 10 was ordered in 1976.

TECHNICAL DATA (CH-53D)

Manufacturer: Sikorsky Aircraft Division of United Aircraft Corporation, Stratford, Connecticut.

Type: Assault transport helicopter.

Accommodation: Two pilots and a crew chief; up to 55 troops in cabin.

Power plant: Two 3,925 shp General Electric T64-GE-413 shaft turbines.

Dimensions: Rotor diameter, 72 ft 2¾ in; length, 88 ft 2½ in; height, 17 ft 1½ in; rotor disc area, 4,070 sq ft.

Weights: Empty, 22,444 lb; normal take-off, 35,000 lb; overload, 42,000 lb.

Performance: Max speed, 196 mph at sea level; cruising speed, 173 mph; initial climb, 2,180 ft/min; service ceiling, 21,000 ft; range, 257 st miles.

Serial numbers:
CH-53A/CH-53D:

151613–151614; 151686–151701;	154887–154888; 156654–156677;
152392–152415; 153274–153313;	156951–156970; 157127–157176;
153705–153739; 154863–154884;	157727–157756.

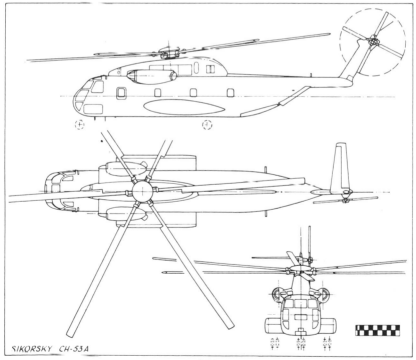

SIKORSKY CH-53A

379

The Stearman N2S-5/PT-13D was a standardized Navy/Air Force model; this example has a USAF designation and serial number on the fuselage, plus Navy markings on the tail. (*Boeing photo*)

Stearman NS, N2S Kaydet

This famous series of biplane trainers—perhaps the world's most-produced biplane—began as a Stearman private venture in 1934. In that year, when Stearman also became a Boeing subsidiary, the Model 70 was built and flown as a potential replacement for the PT-1, PT-3 and PT-11 trainers used by the Army Air Corps. The new design was a conventional biplane with a 225 hp Wright or 215 hp Lycoming radial engine.

The Model 70 won an Army Primary Trainer contest in 1934, but the first production orders came from the Navy, which specified installation of the older Wright J-5 (R-790-8) engine in order to use up existing stocks. In this guise it became Model 73 and carried the Navy designation NS-1; 61 were delivered in 1935–36.

Improvements were incorporated in the Stearman Model 75 which appeared in 1936 with a Lycoming engine and was the first of the series ordered by the Army, as PT-13. In the ensuing 10 years, well over 8,000 Model 75s were built, the majority for the Army Air Force and the Navy, which eventually adopted a standardized model, fully interchangeable between the two Services. The first Navy order for Model 75s (actually A75N1s) was for 250 N2S-1s, similar to the PT-17 and powered by the Continental R-670-14 engine. A further 125 with Lycoming R-680-8 engines were Stearman Model B-75 and designated N2S-2 by the Navy.

Another engine change, to the Continental R-670-4, distinguished the N2S-3 (Model A75N-1), of which the Navy acquired 1,875 examples. The 577 N2S-4s were virtually identical, and included 99 diverted Army PT-17s with R-670-5 engines. Full interchangeability with the Army model was achieved in 1942 with the N2S-5 and PT-13D Kaydet (Model E-75) with

Lycoming R-680-17 engine and overall silver finish in place of the earlier Naval all-yellow. Of 1,768 E-75s built, the Navy received 1,430, including a few fitted with a cockpit canopy similar to that on the lend–lease PT-27s for the RCAF.

TECHNICAL DATA (N2S-5)

Manufacturer: Stearman Aircraft Division, Boeing Aircraft Company, Wichita, Kansas.
Type: Primary trainer.
Accommodation: Student and instructor in tandem.
Power plant: One 220 hp Lycoming R-680-17.
Dimensions: Span, 32 ft 2 in; length, 25 ft 0¼ in; height, 9 ft 2 in; wing area, 297 sq ft.
Weights: Empty, 1,936 lb; gross, 2,717 lb.
Performance: Max speed, 124 mph at sea level; cruising speed, 106 mph at sea level; initial climb, 840 ft/min; service ceiling, 11,200 ft; range, 505 st miles.
Serial numbers:

NS-1: 9677–9717; 0191–0210.
N2S-1: 3145–3394.
N2S-2: 3520–3644.
N2S-3: 3395–3519; 4252–4351; 05235–05434; 07005–08004; 37988–38437.

N2S-4: 27960–28058;   29923–30146; 34097–34101;   34107–34111; 37856–37967;   37978–37987; 55650–55771.
N2S-5: 38438–38610;   43138 43637; 52550–52626;   61037–61097; 61105–61120;   61137; 61143–61155;   61176–61190; • 61224; 61232–61240;   61261–61267.

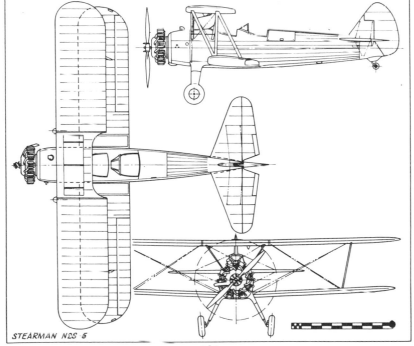

STEARMAN N2S-5

381

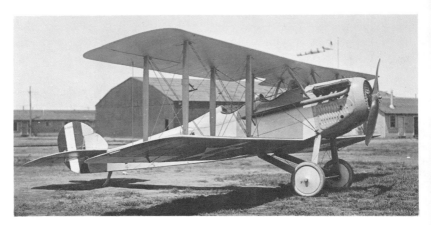

A Vought VE-7, as produced for the Navy after the end of World War I. (*Navy photo*)

Vought VE-7, VE-9

The VE-7, product of the new Lewis & Vought Corporation, appeared in the summer of 1918 after the Aircraft Production Board urged American industry to turn out new original designs for the war programme instead of trying to adapt European designs, as had been the previous policy. Vought was asked to develop an advanced trainer to use the 150 hp French Hispano-Suiza Model A engine, then in production by the Simplex Automobile Division of the Wright-Martin Company. (When Wright-Martin broke up, Wright became the Wright Aeronautical Corporation and concentrated its major efforts on engines until merging with Curtiss in 1929 to form Curtiss-Wright. After the war, Lewis & Vought was renamed Chance Vought Corporation after its founder, the famed aircraft designer Chance M. Vought.)

The VE-7 proved to be an excellent design that generally resembled a slightly scaled-down British de Havilland D.H.4 with the nose of a French Spad. Large orders were soon placed, and additional manufacturers were lined up to assist in production. However, the VE-7 did not go into production during the war. This was due to an economy and time-saving measure that resulted in the same 150 hp Wright-Hispano engine being installed in the existing 90 hp Curtiss JN-4D primary trainer, making it the JN-4H for advanced training duties (see page 100). After the war, the Navy became interested in a version of the VE-7 fitted with the 180 hp Wright-Hispano E engine. The naval versions were built by Vought and by the Naval Aircraft Factory to a total of 128, which was really large-scale production for the early 1920s.

Vought VE-7 used by NACA at Langley Field. (*NACA photo*)

Procured originally for training purposes, the performance of the VE-7 was such that it was used for a great variety of work under a number of sub designations:

VE-7—Standard two-seat trainer.

VE-7G—Armed VE-7 with flexible 0·30-in Lewis machine gun in the rear cockpit and a synchronized Vickers gun forward.

VE-7GF—VE 7G with emergency flotation gear.

VE-7H—Trainer or unarmed observation hydroplane (seaplane).

VE-7S—Single-seat fighter with one synchronized Vickers (later Browning) gun.

VE-7SF—VE-7S with flotation gear.

VE-7SH—VE-7S with VE-7H floats.

A Vought VE-7GF, built by the NAF, showing flotation gear and hydrovanes for emergency use. (*US Navy*)

383

In the observation models, the observer rode in the forward cockpit, a reversion to early World War I practice when the disposable load of light-weight aircraft was carried right at the centre of gravity. Service experience proved the VE-7s (and later UO-1s) to be notoriously tail-heavy. In spite of its relatively large wingspan, however, the VE-7 made a nimble single-seat fighter, a role in which it served as first-line equipment until 1926, with the pilot occupying the former rear cockpit.

VE-7 landplanes operating over water were frequently fitted with emergency flotation gear of a design developed at the RAF Experimental Station on the Isle of Grain during World War I. Although this feature had been tested on earlier experimental US Navy aircraft, the VE-7s were

A Vought VE-9H of Navy squadron VO-6 in December 1924. (*US Navy*)

the first US service models to be so equipped. To prevent nosing over when alighting on water, a Grain-developed hydrovane was installed ahead of the wheels. The seaplane versions were the standard observation and scouting aircraft of the fleet in the early post-war years, being carried aboard battleships and cruisers and launched by catapult. A larger vertical fin was frequently installed on the seaplane versions and was sometimes left in place when the aeroplane was temporarily converted to a landplane.

The VE-9 was identical with the VE-7 except for minor details and the improved E-3 version of the 180 hp Wright-Hispano engine. The Navy ordered 21 as observation models, most being the unarmed VE-9H for use with battleships and cruisers. Two experimental VE-9Ws, to have been fitted with the new 200 hp Lawrance J-1 air-cooled radial engine, were cancelled after it was decided to fit the new Vought UO-1 then on order with the radial instead of the higher-powered water-cooled engine for which it had been designed.

TECHNICAL DATA (VE-7SF)

Manufacturer: Lewis & Vought Corporation, Long Island City, NY, and Naval Aircraft Factory, Philadelphia, Pennsylvania.

Type: Fighter.

Accommodation: Pilot only.

Power plant: One 180 hp Wright E-2.

Dimensions: Span, 34 ft 1⅜ in; length, 24 ft 5⅛ in; height, 8 ft 7 in; wing area, 284·5 sq ft.

Weights: Empty, 1,505 lb; gross, 2,100 lb.

Performance: Max speed, 117 mph at sea level; climb, 5·5 min to 5,000 ft; service ceiling, 15,000 ft; range, 291 st miles.

Armament: Two fixed forward-firing Vickers 0·303-in or Browning 0·30-in guns.

Serial numbers:

VE-7 (Vought): A5661–A5700; A5912–A5941; A6021–A6030.

VE-7 (NAF): A5942–A5969; A5971; A6011–A6020; A6436–A6444.

VE-9 (Vought): A6461–A6481.

VE-9 (NAF): A5970.

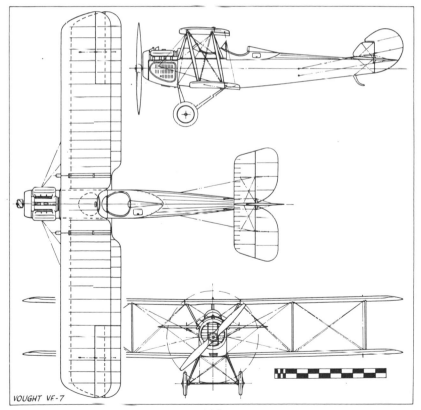

VOUGHT VF-7

385

A Vought UO-1 attached to the USS *Tennessee*. (*US Navy*)

Vought UO, FU

The Vought UO-1, of which 141 were built, was an improved VE-7/9 given a standardized Naval designation after the new designation system was adopted in 1921. The new design was originally intended for the fighter role then filled by the VE-7SF and was given the designation of UF-1 (U for Vought in place of the logical letter V since the latter was assigned to all aeroplanes as a class). However, since more specialized fighters were then under development, the new Voughts were reclassified and became unarmed observation types.

The UO-1 airframe was identical with that of the VE. Wings and horizontal tail surfaces were interchangeable. Streamlining was improved by rounding out the fuselage with formers and stringers, and the space for the observer in the front cockpit was increased by installing cheek-type fuel tanks on each side of the cockpit, outside the basic fuselage structure, where they matched the contours of the added formers. Access to the front cockpit was greatly improved by a revision of the centre-section strut arrangement to eliminate the cross-bracing wires that blocked each side of the cockpit and made it necessary to enter and leave by the rear instead of the side. Other than the rounded fuselage and a different power plant, the principal departure from the VE was the use of an entirely new shape for the vertical tail surfaces. The added area made it unnecessary to use a different fin for the seaplane versions. The various modifications to the VE airframe that resulted in the UO-1 design were worked out and tested on VE-9 A6478.

The UO-1 was originally intended to use the 250 hp Aeromarine U-873 engine, but a Navy decision to use only air-cooled engines for subsequent fleet aircraft up to 300 hp resulted in a decision to re-engine the UO-1 with the new 200 hp Lawrance J-1, a nine-cylinder air-cooled radial that became

A Vought FU-1, single-seat fighter adaptation of the UO-1. (*US Navy*)

the Wright J-1 when the Lawrance Aero Engine Corporation became part of the Wright Aeronautical Corporation. The loss of performance resulting from the decrease in power was largely compensated by the lighter weight of the air-cooled unit. From 1927 on, many UO-1s had their original Wright/Lawrance J-1s or J-3s replaced by the new 220 hp Wright J-5, which, as the Whirlwind, was destined to become one of the most famous air-cooled radials ever built.

The 19th UO, which had been completed with the original Aeromarine engine, was stripped down by the Navy to the basic flat-sided VE fuselage and was used as a single-seat racer under the designation of UO-2. French Lamblin radiators, widely used on racers of the period, were used in place of the traditional Spad nose radiator of the VEs. After the UO-2 was outclassed as a racer, it was reconverted to a two-seater with the 180 hp Wright-Hispano E and was assigned to utility work. It retained the Lamblin radiators, which were moved up from between the undercarriage struts to a position above the lower wing on each side of the front cockpit.

Float-equipped UO-1s replaced the VEs on the ships of the fleet. Those which were specially reinforced for catapult launching were designated UO-1C, but under the new standardized designation system, no distinction was made between landplane and seaplane versions of the same design. Other UO-1s were fitted with arrester hooks for operation from aircraft carriers. Following their replacement as first-line fleet equipment, the UO-1s were used for a great variety of utility, training and experimental duties long beyond the service life of their contemporaries. One was fitted with a skyhook above the centre section and was used for the Navy's first experiments in hooking an aeroplane to and releasing it from an airship in flight (see page 524).

To meet an interim Navy requirement for a catapult-launched floatplane fighter to serve aboard battleships, the original UF concept was revived, and 20 improved single-seaters were ordered as UO-3 on June 30, 1926.

Before delivery, they were reclassified as fighters with the designation of FU-1, the letters being turned around to comply with the revised designation procedures of 1923. It is a tribute to the superior flight characteristics of the basic 1918 design that repeat orders should have been placed for such an obsolescent 220 hp model in a day when 400 to 500 hp fighters were standard. Also, the FU-1 is notable for being the last wood-and-wire design of the World War I concept ordered by any US military service. The few wooden designs of World War II were monoplanes of entirely different structural concept.

The only outward difference from the UO-1 was in the shape of the vertical tail, which used a much smaller fin, and the use of an improved

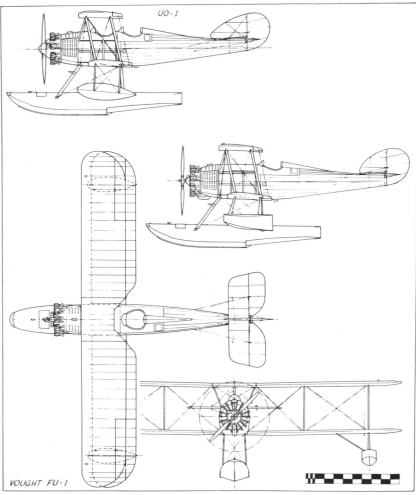

VOUGHT FU-1

wing with rounded instead of raked wing tips and a Navy N-9 aerofoil instead of the original RAF 15. Thin steel-tube struts were substituted for the original wooden ones. Armament was a pair of 0·30-in calibre Browning guns, and performance was enhanced by the fitting of a Rootes supercharger to the 220 hp J-5 engine. All of the FU-1s were assigned to Fighting Squadron Two (VF-2B), which had its aircraft scattered among 12 capital ships of the Battle Fleet. They served in the fighter role for a year until September 1928 and were then withdrawn and converted to two-seat FU-2s for assignment to utility and training duties. They could be distinguished from the near-duplicate UO-5s only by their smaller vertical tail surfaces. All were out of service by the middle of 1932.

In 1926, two modified UO-1s were built for the US Coast Guard with the designation of UO-4. These were seaplanes using the new J-5 engine and were fitted with the improved UO-3/FU-1 wings. The UO-4s were given separate Coast Guard serial numbers 4 and 5, which were later changed to 404 and 405, and the former was still in service in 1935. A few Navy UO-1s were fitted with UO-3/FU-1 wings and J-5 engines and were redesignated UO-5.

TECHNICAL DATA (UO-1)

Manufacturer: Chance Vought Corporation, Long Island City, NY.
Type: Observation biplane, ship or shore-based.
Accommodation: Pilot and observer.
Power plant: One 200 hp Wright J-3.
Dimensions: Span, 34 ft 3½ in; length, 24 ft 5½ in; height, 8 ft 9 in; wing area, 289·8 sq ft.
Weights: Empty, 1,494 lb; gross, 2,305 lb.
Performance: Max speed, 124 mph at sea level; climb, 4·9 min to 5,000 ft; service ceiling, 18,800 ft; range, 398 st miles.
Serial numbers:

UO-1: A6482–A6499; A6546–A6551;
A6603–A6615; A6706–A6729;
A6858–A6877; A6984–A7023;
A7031–A7050.
UO-2: A6546.

UO-4: 4–5 (later 404–405; V104–V105).
UO-5: A6729; A6860; A6866; A6988;
A6997; A6999; A7005; A7010–
A7012; A7016; A7035; A7038
(plus others).

TECHNICAL DATA (FU-1)

Manufacturer: Chance Vought Corporation, Long Island City, NY.
Type: Catapult fighter seaplane, ship-based.
Accommodation: Pilot only.
Power plant: One 220 hp Wright J-5.
Dimensions: Span, 34 ft 4 in; length, 28 ft 4½ in; height, 10 ft 2 in; wing area, 270 sq ft.
Weights: Empty, 2,074 lb; gross, 2,774 lb.
Performance: Max speed, 122 mph at sea level; initial climb, 5 min to 5,000 ft; service ceiling, 26,500 ft; range, 410 st miles.
Armament: Two fixed forward-firing 0·30-in guns.
Serial numbers:
FU-1: A7361–A7380.

A Vought O2U-1 Corsair serving with VO-3S in 1928 and attached to the USS *Raleigh*.

Vought O2U Corsair

During 1926, the Chance Vought Corporation designed a new observation biplane for the Navy, combining experience gained from the UO and FU series (see page 386) with the newly available Pratt & Whitney Wasp engine. As well as being the first Navy aeroplane designed around this famous power unit, the new Vought aeroplane was one of the first to have an all-steel-tube fuselage. Features of the earlier Vought designs which were retained included the cheek tank and the method of fuselage streamlining. Two prototypes ordered in 1926 were designated O2U-1 (the Navy had not then adopted X designations) and were followed by 130 similarly designated production models which began to reach the Navy in 1927 and were the first to carry the famous name Corsair.

The Corsair carried a crew of two in tandem open cockpits, with an armament of one fixed forward-firing gun and a Scarff ring for one or two guns in the rear cockpit. Bombs could be carried under the wings. As well as operating from land or aircraft carriers with a wheel chassis, the Corsair was designed for use from battleships and cruisers as a floatplane with a single central float and wingtip stabilizers. To permit greater interchangeability of role an amphibious float was developed for the O2U-1 in 1928, consisting of wheels each side of the main float which swung up to lie partially submerged in the top of the float when the aircraft was operating as a floatplane.

By mid-1928, the Corsair was standard equipment with the observation sqaudrons VO-3B, VO-4B and VO-5B, which provided aircraft for the battleship divisions, and was serving with VS-1B aboard USS *Langley* as well as other units. In this year, four Corsairs flown by Marine Squadron VO-7M were in action against rebel strongholds in Nicaragua—an action which first brought the type public acclaim. During 1927 four world class records were set by Corsairs in Class C2 for seaplanes—three for altitude and speed with a 500 kg load and one for speed with no payload; the latter was achieved with a flight over a 1,000 km circuit at an average speed of 130 mph.

Production of the O2U series continued in 1928 with the O2U-2 model; 37 were built and could be distinguished by minor alterations which in-

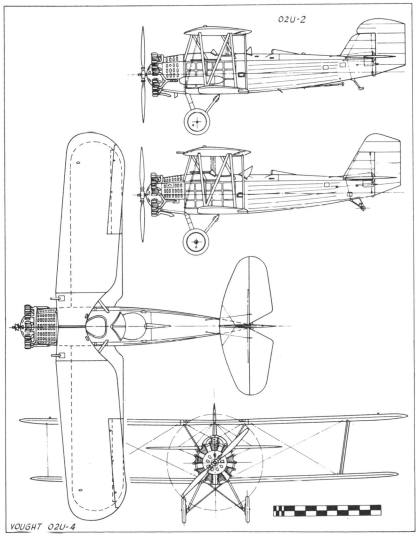

O2U-2

VOUGHT O2U-4

391

A Vought O2U-2 showing lower wing dihedral and revised rudder.
(*Chance Vought photo*)

cluded dihedral on the lower wing, a different cut-out shape in the upper centre section and a larger rudder. Six of this batch were allocated to the Coast Guard.

Further revisions to the vertical and horizontal tail shape distinguished the O2U-3, which also had pronounced dihedral on the upper wing. Eighty were built for the Navy as well as a single example for Army evaluation as O-28. Finally in the initial Corsair series came 42 O2U-4s with only minor differences 'from the O2U-3. During 1930 one O2U-4 was fitted experimentally with the new Grumman amphibious float which allowed full retraction of the wheels and legs into the float. A similar float was adopted in production for the O3U-1 Corsair which began to supersede the O2U in 1930 (see page 393).

TECHNICAL DATA (O2U-1)

Manufacturer: Chance Vought Corporation, Long Island City, NY.
Type: Observation.
Accommodation: Pilot and observer.
Power plant: One 450 hp Pratt & Whitney R-1340-88.
Dimensions: Span, 34 ft 6 in; length, 24 ft 5¾ in; height, 10 ft 1¼ in; wing area, 320 sq ft.
Weights: Empty, 2,342 lb; gross, 3,635 lb.
Performance: Max speed, 150 mph at sea level; climb, 3·6 min to 5,000 ft; service ceiling, 18,700 ft; range, 608 st miles.
Armament: One fixed forward-firing 0·30-in gun; two flexible 0·30-in guns on Scarff ring.
Serial numbers:

O2U-1: A7221–A7222; A7528–A7560; A7567–A7586; A7796–A7831; A7900–A7940.

O2U-2: A8091–A8127 (USCG: 301–306, later V117–V122).
O2U-3: A8193–A8272.
O2U-4: A8315–8356.

Vought O3U-3 attached to the USS *Colorado* in 1938. (*William T. Larkins*)

Vought O3U, SU Corsair

Soon after moving into a new factory at East Hartford, Connecticut, in association with United Aircraft Corporation, the Chance Vought company continued its highly successful series of Corsair biplanes (see page 390) with a new model for the Navy. Designated O3U-1 and delivered in 1930, this new Corsair was but a small improvement on the O2U-4, distinguished primarily by an increase in sweepback and dihedral on the lower wing, which now matched the upper wing. All other characteristics, including the R-1340C Wasp engine, were the same, but the newly developed Grumman amphibious float was adopted for this model. Pro-

A Vought O3U-6 converted for radio-controlled operation in 1939, with tricycle undercarriage. (*National Archives 72-AF-217394*)

393

duction totalled 87, all going to Navy observation units aboard carriers, battleships and cruisers.

Major changes occurred in the next model, the O3U-2, which was powered by the 600 hp Pratt & Whitney R-1690 Hornet engine with a low-drag cowling. The fin shape was changed to incorporate a small dorsal fin, and the tripod-type landing gear of the O2U series was dropped in favour of the simpler straight-axle type which had been featured on the two O2U-1 prototypes. The Scarff ring in the rear cockpit was deleted and the cockpit details changed. Production of the O3U-2 totalled 29, but all were redesignated SU-1s in the Scout category almost as soon as they reached Navy units. In this role they equipped Marine Squadrons VS-14M and VS-15M from 1932 to 1934 aboard the USS *Saratoga* and *Lexington* as the only Marine units to serve on aircraft carriers prior to World War II.

Vought SU-4, scout version of the Corsair based on the O3U-4.
(*Chance Vought photo*)

Reverting to the Pratt & Whitney Wasp, but in its 550 hp R-1340-12 version, the O3U-3 had a new, more rounded rudder which increased the overall length by 20 in. Seventy-six were built; delivery began in June 1933, and all but two of the total were in service in 1935 aboard battleships and with utility units attached to carriers and shore stations.

Another batch of the Hornet-powered Corsairs appeared with the designation O3U-4, and all but one of the 65 were redesignated in the Scout category. Forty-three became SU-2s and 20 became SU-3s with minor equipment changes. Another became the XSU-4 with a long-chord engine cowling, a full canopy over the two cockpits, enlarged tailplane and rudder shape as on the O3U-3. Forty production models designated SU-4 were the only Corsairs built from scratch in the Scout category. They reverted to the narrow-chord cowling of the O3U-3 but had larger tails. The SU-4s replaced SU-1s in the Marine unit aboard USS *Lexington* until 1934, and served in various utility roles with Navy units.

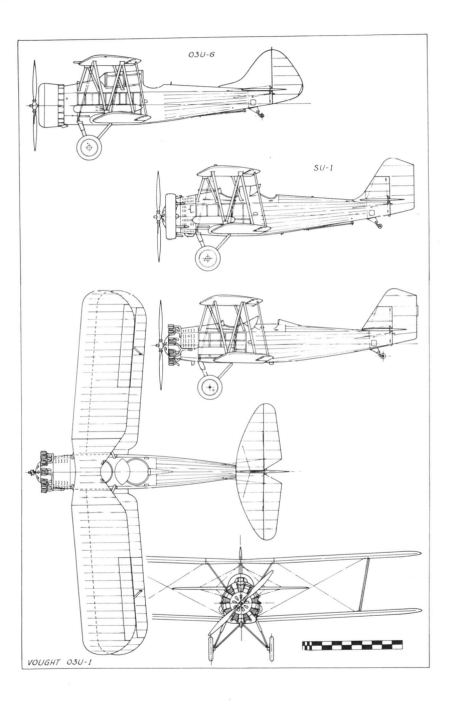

O3U-6

SU-1

VOUGHT O3U-1

One O3U-4 became the XO3U-5 to test new ideas for the observation role, and the final O3U-3 became the XO3U-6 with a full-chord cowling for the Wasp engine and a partial canopy between the cockpits. These prototypes led to a final order for 32 O3U-6s, delivered to the Marines and divided equally between VO-8M and VO-7M in 1935. The first O3U-6 was tested as a single-float seaplane, but the Marine units did not operate them in this configuration. Another experiment was made on the last O3U-6 in January 1937, when it was tested with full-span ailerons and larger fin as the XOSU-1.

By the end of 1941 the Corsair biplanes had been replaced in operational units, but 33 O3U-1s, 35 O3U-3s, 6 O3U-6s and 67 assorted SU models were still on strength and distributed around Naval air stations. Shortly before the war the Naval Aircraft Factory at Philadelphia converted a small number of O3U-6s to drone configuration for flight testing under extreme conditions where a pilot might be endangered. The conversion involved use of a fixed nosewheel undercarriage.

TECHNICAL DATA (O3U-3)

Manufacturer: Chance Vought Corporation, East Hartford, Connecticut.
Type: Observation biplane.
Accommodation: Pilot and observer.
Power plant: One 550 hp Pratt & Whitney R-1340-12.
Dimensions: Span, 36 ft; length, 27 ft 3 in; height, 11 ft 6 in; wing area, 337 sq ft.
Weights: Empty, 2,938 lb; gross, 4,451 lb.
Performance: Max speed, 164 mph at sea level; climb, 4·2 min to 5,000 ft; service ceiling, 18,000 ft; range, 650 st miles.
Armament: One fixed forward-firing 0·30-in gun; two 0·30-in guns in rear cockpit.
Serial numbers:
    O3U-1: A8547–A8582; A8810–A8839;   XO3U-5: A9078.
        A8851–A8871.   XO3U-6: 9330.
    O3U-3: A9142–A9169; 9283–9330.   O3U-6: 9729–9744; 0001–0016.

TECHNICAL DATA (SU-4)

Manufacturer: Chance Vought Corporation, East Hartford, Connecticut.
Type: Scout.
Accommodation: Pilot and observer.
Power plant: One 600 hp Pratt & Whitney R-1690-42.
Dimensions: Span, 36 ft; length, 27 ft 5½ in; height, 11 ft 4 in; wing area, 337 sq ft.
Weights: Empty, 3,312 lb; gross, 4,765 lb.
Performance: Max speed, 167 mph at sea level; climb, 4·4 min to 5,000 ft; service ceiling, 18,600 ft; range, 680 st miles.
Armament: One fixed forward-firing 0·30-in machine gun; two 0·30-in guns in rear cockpit.
Serial numbers:
    SU-1: A8872–A8875; A8928–A8937;   SU-3: A9122–A9141.
        A9062–A9076.   XSU-4: A9109.
    SU-2: A9077; A9079–A9108; A9110–   SU-4: 9379–9398; 9414–9433.
        A9121.

A Vought SBU-1 in February 1940 with Neutrality Patrol star on the nose. (*US Navy*)

Vought SBU

Vought's last biplane for the US Navy originated in 1932 in response to a specification for a two-seat fighter, but it was produced and put into service only in the scout-bomber role. The specification, Bureau of Aeronautics Design 113, was written in the light of somewhat unsatisfactory operations by the Curtiss F8C-4 Helldivers, and specified higher engine power and better slow-speed performance for carrier deck operations. The Chance Vought company, which had produced the XF2U-1 prototype in competition with the Curtiss F8C, was one of seven manufacturers submitting proposals for Design 113, and on June 30, 1932, received a contract for a prototype designated XF3U-1.

First flown in May 1933, the XF3U-1 was a biplane of conventional metal construction with fabric covering and powered by the small-diameter two-row Pratt & Whitney R-1535-64 engine, fully-cowled. In November 1933 the Navy asked Vought to modify the XF3U-1 into a prototype scout-bomber designated XSBU-1, and indicated a likely requirement for at least 27 production models. Modifications included greater internal fuel capacity, stronger and larger wings, provision for a 500 lb bomb under the fuselage and other smaller items. The XSBU-1 was, in fact, a new airframe with the XF3U-1's original engine and equipment, but when delivered to the Navy for trials in June 1934 it retained the same serial number. A year later, the original XF3U-1 airframe was resurrected as a test-bed for Pratt & Whitney engines and received a new serial number.

A production order for 84 SBU-1s was placed with Vought in January 1935. They were similar to the prototype scout-bomber version, with an armament of one fixed and one movable machine gun plus the external bomb. Deliveries began on November 20, 1935, with Scouting Squadron VS-3B the first to be equipped, followed by VS-2B and VS-1B.

397

A second batch of 40 of the Vought scout-bombers were built as SBU-2s with R-1535-98 engines and minor changes. Late in 1937, 24 of these SBU-2s were delivered, new, to the Naval Reserves, serving until 1941.

TECHNICAL DATA (SBU-1)

Manufacturer: Chance Vought Division of United Aircraft and Transport Corporation, East Hartford, Connecticut.
Type: Scout bomber.
Accommodation: Pilot and observer/gunner.
Power plant: One 700 hp Pratt & Whitney R-1535-80.
Dimensions: Span, 33 ft 3 in; length, 27 ft 10 in; height, 11 ft 11 in; wing area, 327 sq ft.
Weights: Empty, 3,645 lb; gross, 5,520 lb.
Performance: Max speed, 205 mph at 8,900 ft; cruising speed, 122 mph; initial climb, 1,180 ft/min; service ceiling, 23,700 ft; range, 548 st miles.
Armament: One fixed forward-firing 0·30-in Browning gun and one 0·30-in flexibly mounted in rear cockpit.
Serial numbers:

XF3U-1: 9222 (later 9746).   SBU-1: 9750–9833.
XSBU-1: 9222.   SBU-2: 0802–0841.

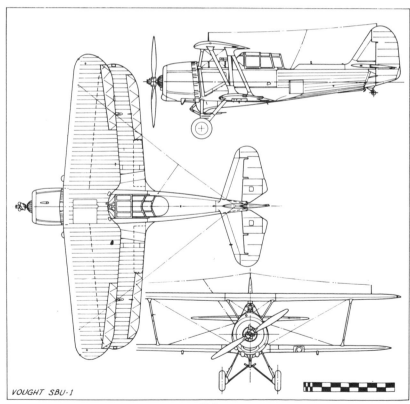

VOUGHT SBU-1

398

Vought-Sikorsky SB2U-3 of VS-41 at Midway in June 1942.

Vought SB2U Vindicator

An all-metal, low-wing monoplane with two crew seated in tandem beneath a long glasshouse, simple radial engine and retractable undercarriage: the definition was typical of the US Navy scout-bombers and torpedo-bombers of the late thirties —the new generation which followed the classic biplanes and laid the foundation for the combat types of World War II. An aircraft fitting this description in every respect was the XSB2U-1 ordered from Vought on October 11, 1934, as the Navy's first monoplane scout-bomber. Suitability of a monoplane was by no means accepted without question at that time, however, and four months later Vought received a contract for a new biplane scout-bomber prototype, the XSB3U-1, as well.

Both the new Vought prototypes reached Anacostia in April 1936, the XSB2U-1 having first flown on January 4, 1936; comparative trials clearly established the superiority of the monoplane. A production order for 54 SB2U-1s was placed on October 26, 1936, and further work on the biplane was dropped. Completed in July 1937, the SB2U-1 was powered by an 825 hp Pratt & Whitney R-1535-96 engine, improving the top speed performance by some 20 mph.

Delivery of the SB2U-1s began on December 20, 1937, to Navy Squadron VB-3. In January 1938 the Navy ordered 58 SB2U-2s with the same engines, differing from the earlier model in having a small increase in gross weight due to new equipment. These were delivered during the latter months of 1938, being followed at the end of 1940 by the first of 57 SB2U-3s ordered on September 25, 1939. These were the last variants in the series and the first to carry the name Vindicator (aircraft supplied to Britain had previously been named Chesapeake). They had the R-1535-02 engine, increased standard fuel capacity plus provision for long-range tanks for ferry flights, 0·50-in guns fore and aft, increased armour protection and higher operating weights. One SB2U-1 was converted to the XSB2U-3 as a floatplane in 1939, but the production SB2U-3s all served as landplanes.

By 1940, SB2U-1s and SB2U-2s were serving with VB-2 (USS *Lexington*), VB-3 (*Saratoga*), VB-4, VS-41 and VS-42 (*Ranger*) and VS-71 and VS-72 (*Wasp*). Most of the SB2U-3s were issued to Marine squadrons, primarily equipping VMSB-131 and VMSB-231. These squadrons saw some action against Japanese forces in the Pacific during 1942, including the Battle of Midway, but were soon replaced by later types.

TECHNICAL DATA (SB2U-3)

Manufacturer: Vought-Sikorsky Division, United Aircraft Corporation, Stratford, Connecticut.
Type: Carrier-based scout and dive-bomber.
Accommodation: Pilot and observer/gunner.
Power plant: One 825 hp Pratt & Whitney R-1535-02.
Dimensions: Span, 42 ft; length, 34 ft; height, 10 ft 3 in; wing area, 305 sq ft.
Weights: Empty, 5,634 lb; gross, 9,421 lb.
Performance: Max speed, 243 mph at 9,500 ft; cruising speed, 152 mph; initial climb, 1,070 ft/min; service ceiling, 23,600 ft; range, 1,120 st miles.
Armament: One fixed forward-firing and one flexible rear 0·50-in gun.
Serial numbers:

XSB2U-1: 9725.
SB2U-1: 0726–0778.
SB2U-2: 1326–1383.

XSB2U-3: 0779.
SB2U-3: 2044–2100.

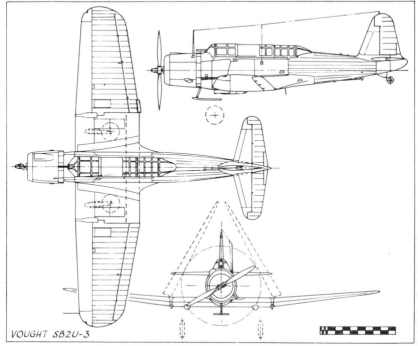

VOUGHT SB2U-3

A Vought-Sikorsky OS2N-1, built by the NAF, in late-1942 markings. (*US Navy*)

Vought OS2U Kingfisher

Of the several observation and scouting aeroplanes available to the Navy when the US entered World War II, the VS-310 Kingfisher was to prove the most useful and the most used. Incorporating some revolutionary structural techniques including spot welding, the Kingfisher was based upon the Vought company's considerable experience of observation aircraft and was designed to replace earlier Vought biplanes in a similar role. Layout of the fuselage was similar to that employed in the biplane scouts, as also was the use of a large single float plus underwing stabilizing floats. Alternatively, a conventional tailwheel undercarriage could be fitted.

A contract to build a single prototype XOS2U-1 was placed by the Navy on March 22, 1937, and this duly appeared in 1938 with a 450 hp Pratt & Whitney R-985-4 engine, making its first flight on July 20. Some small changes in the float attachments and an R-985-48 engine distinguished the production model OS2U-1. The first of 54 examples reached the Fleet in August 1940, and six had been assigned to the Pearl Harbor Battle Force before the end of the year.

Small changes of equipment and an R-985-50 engine were introduced in the OS2U-2, 158 examples of which were delivered in 1940–1. Many of these early Kingfishers were attached to Pensacola Naval Air Station but 53 OS2U-2s were assigned to equip Inshore Patrol squadrons based at Jacksonville Naval Air Station. Nine more Inshore Patrol squadrons newly formed in 1942 received exclusively OS2N-1 Kingfishers, these being versions of the OS2U-3 built by the Naval Aircraft Factory. The OS2U-3 itself differed from earlier models in having extra fuel tanks in the wings and better armour protection for the pilot and observer. Starting in 1941, Vought delivered 1,006 OS2U-3s before production ended in 1942.

TECHNICAL DATA (OS2U-3 floatplane)

Manufacturer: Vought-Sikorsky Division, United Aircraft Corporation, Stratford, Connecticut, and Naval Aircraft Factory, Philadelphia, Pennsylvania.

Type: Observation-scout.

Accommodation: Pilot and observer/gunner.

Power plant: One 450 hp Pratt & Whitney R-985-AN-2 or -8.

Dimensions: Span, 35 ft $10\frac{7}{8}$ in; length, 33 ft 10 in; height, 15 ft $1\frac{1}{2}$ in; wing area, 262 sq ft.

Weights: Empty, 4,123 lb; gross, 6,000 lb.

Performance: Max speed, 164 mph at 5,500 ft; cruising speed, 119 mph at 5,000 ft; climb to 5,000 ft, 12·1 min; service ceiling, 13,000 ft; range, 805 st miles.

Armament: One fixed forward-firing and one flexibly-mounted rear 0·30-in guns.

Serial numbers:

XOS2U-1: 0951.
OS2U-1: 1681–1734.
OS2U-2: 2189–2288; 3073–3130.

OS2U-3: 5284–5989; 09393–09692.
OS2N-1: 01216–01515.

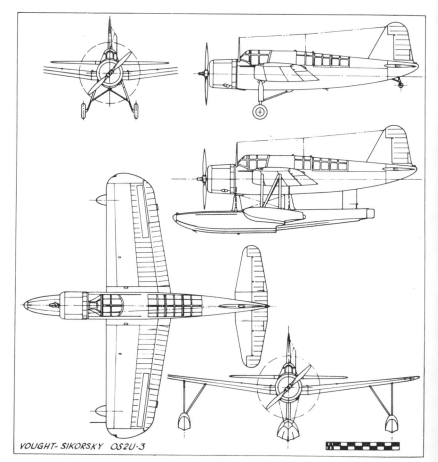

VOUGHT- SIKORSKY OS2U-3

402

An early production model Vought-Sikorsky F4U-1 Corsair. (*Vought-Sikorsky photo*)

Vought F4U Corsair

In production longer than any other US fighter of World War II, the Corsair had several claims to fame. It was credited with an 11 : 1 ratio of kills to losses in action against Japanese aircraft and was the last piston-engined fighter in production for any of the US services. Its greatest attribute, though, was the excellence of its overall performance, making it certainly the finest carrier-based fighter of any used by the combatants in World War II, and perhaps the best of any US fighters in that conflict.

Operational requirements for a new single-seat shipboard fighter were circulated to manufacturers by the US Navy early in 1938, and the Vought V-166B design, by a team led by Rex B. Beisel, was one of the proposals which resulted. To obtain the very high performance specified—matching that of contemporary land-based fighters—Beisel designed the smallest possible airframe around the most powerful available engine, the Pratt & Whitney XR-2800 Double Wasp. A characteristic feature of the Vought V-166B design was the inverted gull wing with the backward-retracting main legs of the landing gear located at the wing knuckles. This arrangement kept the legs short despite the height of the fuselage from the ground dictated by the large-diameter propeller. The wings folded upwards.

On June 30, 1938, the Vought company received a contract to build a single prototype of its Model V-166B, and this aircraft, designated XF4U-1, made its first flight on May 29, 1940. Powered by a 2,000 hp XR-2800-4 engine, the XF4U-1 had a 0·30-in and a 0·50-in gun in the forward fuselage, one 0·50-in in each wing and compartments in the wings for 10 small bombs for use against bomber formations. Before the end of 1940 the XF4U-1 had flown at 404 mph, faster than any US fighter then in the air, and on June 30, 1941, the Navy ordered production of 584 F4U-1s.

Deliveries of the F4U-1s began on October 3, 1942, four months after the first flight of the production Corsair, with the initial aircraft going to VF-12. Production aircraft had the R-2800-8 engine, two more guns in

403

the wings with extra ammunition, self-sealing fuel tanks and armour protection; the cockpit was located 3 ft further aft to allow additional fuel to be carried in the fuselage. This last-mentioned change adversely affected the pilot's view, and carrier landing trials aboard the USS *Sangamon* in September 1942 cast doubts about the aircraft's suitability for the carrier role. Consequently, F4U-1s were issued primarily to land-based Marine units, starting with VMF-124. This unit took the Corsair into operation for the first time on February 13, 1943, at Bougainville, and seven more Marine units were flying the Corsair by August 1943. A month later, Navy Squadron VF-17, also land-based, became operational in New Georgia.

To speed production of the Corsair, contracts were placed with the Brewster and Goodyear companies for versions similar to the F4U-1, designated F3A-1 and FG-1 respectively and both incorporating a raised cockpit hood, which was introduced on the 689th F4U-1 built by Vought; the FG-1A version had fixed wings. Some Corsairs were fitted with four 20 mm cannon in the wings in place of the machine guns and were designated F4U-1C, and others for use as fighter-bombers had fittings for a long-range tank under the fuselage and two 1,000 lb bombs or eight 5-in rockets under the wings, and the R-2800-8W engines with water injection; these were designated F4U-1D, FG-1D or F3A-1D according to source.

Production of these initial Corsair versions totalled 4,699 F4U-1s (including 2,066 F4U-1A, 1,675 F4U-1D and 200 F4U-1C), 738 F3A-1s, 1,704 FG-1s and -1As and 2,303 FG-1Ds. Of these totals, 2,012 were supplied to Britain's Royal Navy under lend–lease, and another 370 went

Vought-Sikorsky F4U-5P with oblique camera in rear fuselage. (*Vought-Sikorsky photo*)

Goodyear-built F2G-1 Corsair showing new engine cowling and cockpit hood.
(*Goodyear photo*)

This low-level attack version of the Corsair was used as the AU-1 in Korea by Marine squadrons. (*US Navy*)

to the RNZAF. Operating from HMS *Victorious*, Corsair IIs of No. 1834 Squadron of the Fleet Air Arm went into operation on April 3, 1944, during the attacks on the *Tirpitz*, this being the first time Corsairs had operated from aircraft carriers. The US Navy was still reluctant to commit its F4Us to carrier operation, but in April 1944 a further series of trials by VF-301 on the USS *Gambier Bay* showed that no serious problems remained and approval was finally given for the Navy squadrons to take their aircraft to sea.

Work on a night-fighter version of the Corsair had begun as early as January 1942 when a prototype XF4U-2 was ordered. Design features included AI radar on the starboard wingtip and an autopilot. The original XF4U-2 was not completed but the NAF at Philadelphia modified 32 F4U-1s as F4U-2 night fighters with the same features, and examples were issued to VFN-75 at Munda, New Guinea, and VFN-101 which operated in turn from the USS *Essex*, *Hornet* and *Intrepid*. Another service modification was designated F4U-1P, with cameras for reconnaissance duties.

Use of a turbosupercharged version of the Double Wasp engine was projected during 1941, and in March 1942 Vought received a contract for three XF4U-3s, all being F4U-1 conversions. This work proceeded on low priority throughout the war, however, and the first XF4U-3 did not appear until 1946, with an R-2800-16 engine. Twenty-seven similar FG-3s were ordered from Goodyear, but only one was completed. The second major production version of the Corsair, therefore, was the F4U-4, the prototype of which first flew on April 19, 1944, with a 2,100 hp R-2800-18W engine. The additional power from this engine increased the maximum speed of the F4U-4 to 446 mph, and other small changes were made to improve the operational characteristics of this version.

Production of the F4U-4 variants totalled 2,357 including 297 F4U-4Bs with four-cannon armament, nine F4U-4Ps and one F4U-4N, all by Vought. The similar FG-4 was to have been built by Goodyear.

Despite large-scale cancellation of contracts following V-J Day, production of the F4U-4 by Vought continued to 1947. Goodyear production was stopped, and the Goodyear-developed low-altitude versions, the

F2G-1 and F2G-2 (with folding wings), were cancelled after five of each had been completed.

In 1946 Vought produced a new Corsair variant, the XF4U-5, by fitting a 2,300 hp two-stage R-2800-32W engine and four 20 mm wing guns in an F4U-4. To meet immediate requirements for a carrier-based fighter-bomber and night fighter, the Navy purchased 223 F4U-5s, 315 F4U-5Ns and -5NLs and 30 F4U-5Ps during 1947 and 1948. Then came the low altitude XF4U-6 with a single-stage R-2800-83W engine, additional armour protection and increased underwing load-carrying ability. Re-designated AU-1, this type went into production for use by Marine squadrons operating in Korea, 111 being built.

The Corsair line ended with 94 F4U-7s, similar to the AU-1 but with the R-2800-18W engine. Procured by the Navy, they were supplied through MAP to the French *Aéronavale* and used in Indochina. Production of the

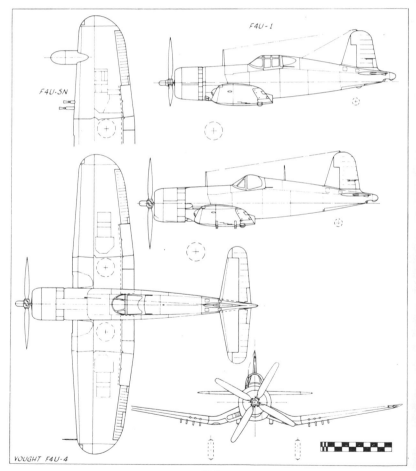

later Corsair versions was centred at Dallas where Chance Vought operated as a Division of United Aircraft and later as an independent company. The last Corsair left the Dallas plant late in December 1952, ending more than a decade of F4U production.

TECHNICAL DATA (F4U)

Manufacturer: Vought-Sikorsky Division, United Aircraft Corporation, Stratford, Connecticut; Chance Vought Division, United Aircraft Corporation (later Chance Vought Aircraft Inc), Dallas, Texas; Brewster Aeronautical Corporation, Long Island City, NY; Goodyear Aircraft Corporation, Akron, Ohio.
Type: Carrier-based fighter.
Accommodation: Pilot only.

	F4U-1	F4U-4	F4U-5N	AU-1
Power plant:	2,000 hp R-2800-8	2,100 hp R-2800-18W	2,300 hp R-2800-32W	2,300 hp R-2800-83W
Dimensions:				
Span, ft in	41 0	41 0	41 0	41 0
Length, ft in	33 4	33 8	33 6	34 1
Height, ft in	16 1	14 9	14 9	14 10
Wing area, sq ft	314	314	314	314
Weights:				
Empty, lb	8,982	9,205	9,683	9,835
Gross, lb	14,000	14,670	14,106	19,398
Performance:				
Max speed, mph/ft	417/19,900	446/26,200	470/26,800	238/9,500
Cruising speed, mph	182	215	227	184
Climb, ft/min	2,890	3,870	3,780	920
Service ceiling, ft	36,900	41,500	41,400	19,500
Range, miles	1,015	1,005	1,120	484
Armament:		6 × 0·50-in	4 × 20 mm	4 × 20 mm
	6 × 0·50-in	2 × 1,000 lb	2 × 1,000 lb	4 × 1,000 lb

Serial numbers:
XF4U-1: 1443.
F4U-1: 02153–02736;   03802–03841; 17392–17455; 18122–18191.
F4U-1A: 17456–18121;   49660–50359 55784–56483.
F4U-1C, -1D: 50360–50659;   57084– 57983; 82178–82852.
(XF4U-3): 02157; 17516; 49664.
(XF4U-4): 49763; 50301; 80759–80763.
F4U-4: 62915–63071;   80764–82177; 96752–97531.
(XF4U-5): 97296; 97364; 97415.
F4U-5: 121793–122066;   122153– 122206;   123144–123203; 124441–124560;   124665– 124724.

(XAU-1): 124665.
AU-1: 129318–129417; 133833–133843.
F4U-7:* 133652–133731;   133819– 133832.
F3A-1: 04515–04774;   08550–08797; 11067–11293.
FG-1, -1A: 12992–14685; 76139–76148.
FG-1D: 14686–14991;   67055–67099; 76149–76739;   87788–88453; 92007–92701.
(FG-3): 76450.
(XF2G-1): 12992;   13471;   13472; 14691–14695.
F2G-1: 88454–88458.
F2G-2: 88459–88463.
* MAP supply to France.

A Chance Vought F7U-3M, equipped to carry four Sparrow missiles. (*Vought photo*)

Vought F7U Cutlass

Among the large quantities of German aeronautical research data which began to reach the USA in the latter part of 1945 were details of some work on tailless designs done by the Arado company. Elaboration of these designs by the Chance Vought company led to production of the highly unconventional F7U Cutlass. The wing, with a sweepback of 38 degrees, was of very low aspect ratio, 3 : 1, and almost parallel chord. Pitch and roll controls were combined in elevons on the wing; fins and rudders were located on the wing at the ends of the centre section.

Laid out as a carrier-based fighter, this design offered a high rate of climb and high top speed combined with comparatively small size when the outer wing panels were folded up for carrier stowage. The US Navy ordered three XF7U-1 prototypes on June 25, 1946, specifying Westinghouse J34-WE-32 engines with afterburners. The first flight was made on September 29, 1948, by which time a production contract for 14 F7U-1s had been placed while development of the F7U-2 with J34-WE-42 engines and the F7U-3 with J46-WE-8As was initiated.

The first F7U-1 flew on March 1, 1950, and the entire batch of this model was assigned to the Advanced Training Command at Corpus Christi Naval Air Station during 1952. Difficulties with the Westinghouse J34 programme resulted in cancellation of the F7U-2 before completion, while early experience with the F7U-1 airframe led to an extensive redesign for the F7U-3. First flown on December 20, 1951, the latter model had a new nose shape, redesigned fins and other changes. Four squadrons were equipped—VF-81, VF-83, VF-122 and VF-124—and production of the definitive version totalled 180.

Basic armament of the F7U-3 comprised four 20 mm cannon in the upper lips of the intake fairing, with provision for underwing rocket pods or various other stores. Subsequently, provision was made for the Cutlass to carry four Sperry Sparrow I beam-riding missiles in the F7U-3M version, of which 98 were built, and 12 examples of a camera-equipped variant, the F7U-3P, also went into service. Production ended in December 1955 when 290 F7U-3 Cutlass variants had been delivered.

TECHNICAL DATA (F7U-3)

Manufacturer: Chance Vought Division of United Aircraft Corporation (later Chance Vought Aircraft Inc), Dallas, Texas.

Type: Carrier-based fighter.

Accommodation: Pilot only.

Power plant: Two 4,600 lb s.t. Westinghouse J46-WE-8A turbojets with afterburners.

Dimensions: Span, 38 ft 8 in; length, 44 ft 3 in; height, 14 ft 7½ in; wing area, 496 sq ft.

Weights: Empty, 18,210 lb; gross, 31,642 lb.

Performance: Max speed, 680 mph at 10,000 ft; initial climb, 13,000 ft/min; service ceiling, 40,000 ft; range, 660 st miles.

Armament: Four fixed forward-firing 20 mm guns; provision for four Sparrow I AAMs.

Serial numbers:

XF7U-1: 122472–122474.	F7U-3, -3M, -3P: 128451–128478;
F7U-1: 124415–124434.	129545–129756;
F7U-2: 125322–125409 (cancelled).	139868–139917.

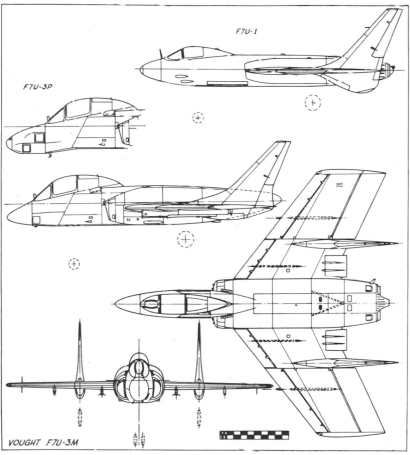

409

APPENDIX A

MINOR TYPES

Illustrated and described in this section are those aircraft of American design and manufacture which were procured for service by the US Navy, Marines or Coast Guard, but were of less significance than the types given lengthier treatment in the main portion of this volume. Purely experimental types are not included, but aircraft of which only single examples were procured are illustrated if they served in an operational role. Foreign types are separately described in Appendix B, gliders in Appendix C and balloons and airships in Appendix D.

The Aeromarine 700 used for early torpedo-dropping tests in the US. (*Aeromarine photo*)

AEROMARINE 700

The Navy ordered four Model 700 seaplanes from the Aeromarine Plane & Motor Company during 1917, but accepted delivery of only the first two (A142, A143). Powered by a 90 hp Aeromarine six-cylinder engine, one of these was used for the first US experiments in the dropping of torpedoes from aircraft. The twin floats were carried on independent structures to allow the torpedo to be dropped between them from the belly rack. Because of the limited payload of the Model 700 (700 lb) only a lightweight dummy torpedo could be used.

An Aeromarine 40. (*US Navy*)

AEROMARINE 40

In 1918 the Navy ordered 200 Aeromarine Model 40F flying-boats to supplement the ageing Curtiss Fs and the improved MFs then on order. The Aeromarines, powered with 100 hp Curtiss OXX-6 engines, were conventional two-seat training flying-boats of the period, with the pilot and student seated side by side in a single open cockpit in the wooden hull. The Armistice caused cancellation of the majority of the order, and only the first 50 (A5040–A5089) were delivered. Since most deliveries did not take place until after the Armistice, the 40Fs saw little service in the training schools. Span, 48 ft 6 in; length, 28 ft 11 in; gross weight 2,592 lb; max speed, 180 mph.

The first of two Aeromarine AS-2 scouts used by the Navy. (*US Navy*)

AEROMARINE AS-1, 2

The AS-1 (A5612) and two improved AS-2s (A5613, A5614) grew out of the Navy's interest in the two-seat German Brandenburg seaplane scouts used along the Belgian coast in 1918. A unique design feature of each was the low position of the vertical tail to give the gunner a clear field of fire to the rear. Power plant, 300 hp Wright-Hispano E. Span, 39 ft 4 in; length, 25 ft 5 in; gross weight, 2,910 lb; max speed, 114 mph.

411

A Beech GB-1 delivered to the Navy as a personnel transport in June 1939. (*Beech photo*)

BEECH GB, JB TRAVELLER

The first example of the Beech 'staggerwing' Model 17 to serve with the Navy was purchased in 1937 and was a civil C-17R with a Wright R-975-26 engine. Designated JB-1 (0801), it flew as a staff transport for two years. In 1939 the Navy purchased its first GB-1s; these were Model D-17S with 400 hp Pratt & Whitney R-985-48. Ten (1589–1595, 1898–1900) were delivered, followed by eight impressed civil D-17s similarly designated and a total of 342 GB-2s with 450 hp R-985-50s; the latter model, named Traveller, was procured both for the Navy and for lend–lease delivery to Britain. Span, 32 ft; length, 26 ft; gross weight, 4,250 lb; max speed, 189 mph.

A white-finished Beech T-34B trainer in November 1954. (*Beech photo*)

BEECH T-34 MENTOR

Seeking a new primary trainer, the US Navy began evaluation of the Beech Model 45 in July 1953. Four months earlier, the USAF had adopted the same aircraft, following protracted testing, as its first post-war primary trainer, designated the T-34A. The Navy followed suit on June 17, 1954, with an initial order of 290 examples designated T-34B, and subsequent contracts brought the total delivered by the end of 1957 to 423 (140667–140956, 143984–144116). The engine was the 225 hp Continental O-470-4. Span, 32 ft 10 in; length, 25 ft 11 in; gross weight, 3,000 lb; max speed, 188 mph.

One of the two Beechcraft YT-34C Mentor prototypes. (*Beech photo*)

BEECH T-34C

During 1973, the US Navy initiated a research and development pro-gramme to improve the T-34 Mentor in order to meet its requirement for a new primary flight trainer. Two T-34Bs were modified by Beech to YT-34C configuration, the principal change being installation of a 715 shp Pratt & Whitney (UACL) PT6A-25 turboprop, with some structural strengthening for higher operating weights. The first YT-34C flew on September 21, 1973, and after successful evaluation, the US Navy planned procurement of 252 T-34Cs. Deliveries were to start in 1976. Span, 33 ft 6 in; length, 28 ft 8½ in; gross weight, 4,000 lb; max speed, 257 mph.

A Bell HSL-1 during its short-lived Navy service. (*Peter M. Bowers*)

BELL HSL

The Bell company won a Navy design competition in June 1950 for a helicopter intended specifically for anti-submarine warfare. This design, Bell Model 61, was the only Bell helicopter using the tandem-rotor layout; it was powered by a 2,400 hp Pratt & Whitney R-2800-50 engine and was intended to carry air-to-surface missiles such as the Fairchild Petrel, as well as dipping ASDIC. Three XHSL-1s were ordered (129133–129135) the first of these flying on March 4, 1953, followed by a production contract for 78, including 18 destined for Britain's Fleet Air Arm. Deliveries to Squadron HU-1 began in January 1957 but production ended after a total of 50 (129154–129168, 129843–129877) had been built. Rotor diameter, 51 ft 6 in each; gross weight, 26,500 lb; range, 350 miles at 100 mph.

413

A Bell TH-57A Sea Ranger in US Navy training colours. (*Bell photo*)

BELL H-57 SEA RANGER

In January 1968 the Navy announced selection of the Bell 206A Jet Ranger as its new primary light turbine training helicopter. Virtually an off-the-shelf purchase, the single contract was for 40 dual-control TH-57As (157355–157394) to replace 36 Bell TH-13Ms at Pensacola. Power plant, one 317 shp Allison 250-C-18. Rotor diameter, 33 ft 4 in; overall length, 38 ft 9½ in; gross weight, 2,900 lb; max speed, 135 mph.

The Bellanca JE-1 in service at Anacostia. (*Bellanca photo*)

BELLANCA RE, JE

During 1932 three examples of the civil Bellanca CH-400 Skyrocket were acquired by the Navy; all had minor differences and were designated XRE-1 (8938), XRE-2 (9207) and XRE-3 (9341). The XRE-1 was used at NAS Anacostia for radio research, XRE-2 also served at Anacostia and XRE-3 went to the Marines as a two-stretcher ambulance. All had the Pratt & Whitney R-1340 engine, in different sub-series. Span, 46 ft 4 in; length, 27 ft 10 in; gross weight, 4,600–4,710 lb; max speed, 148–161 mph.

A single example of the commercial Bellanca Model 31-42 Senior Skyrocket was acquired by the Navy in 1938 as the JE-1 (0795). Power plant, 570 hp Pratt & Whitney R-1340-27. Span, 50 ft 6 in; length, 28 ft 6 in; gross weight, 5,595 lb; max speed, 185 mph.

414

Berliner Joyce OJ-2 attached to the cruiser USS *Memphis*. (*William T. Larkins*)

BERLINER JOYCE OJ

Bureau of Aeronautics Design No. 86 in 1930 was for a lightweight observation biplane suitable for use from the catapults of light cruisers. Keystone and Berliner Joyce built prototypes to this requirement, around the 400 hp Pratt & Whitney R-985-A Wasp Junior engine, as the XOK-1 and XOJ-1 respectively. The latter was built in series as the OJ-2 and 39 were built (A9187–A9204, A9403–A9411, A9572–A9583) and entered service in 1933. Squadrons VS-5B and VS-6B operated the OJ-2 as two-plane detachments on cruisers until 1935. Span, 33 ft 8 in; length, 25 ft 8 in; gross weight, 3,629 lb; max speed, 151 mph.

Boeing TB-1 torpedo-bomber in 1927. (*Boeing photo*)

BOEING TB

The Boeing Airplane Company built three of these big single-engined biplanes to Navy designs, the first flight being made on May 4, 1927. All three (A7024–A7026) were delivered before mid-1927 and were the last non-Boeing designs built by the company until World War II. Related to the Martin T3M design (see page 310), the three-seat TB-1s had 730 hp Packard 3A-2500 engines and could operate as seaplanes or landplanes. Span, 53 ft; length, 40 ft 10 in; gross weight, 9,786 lb; max speed, 115 mph.

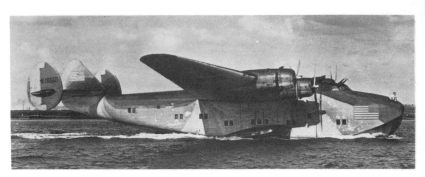

One of five Boeing B-314s used by the Navy in World War II.
(*Pan American Airways photo*)

BOEING B-314

Following the lead of the Army, the Navy obtained some transport-type aircraft from the airlines early in World War II. Among these were five Boeing Model 314A flying-boat airliners (48224–48228). Two were obtained directly from Pan American Airways in 1942 and three others came from the Army, which had impressed them into service as C-98 and then found little use for them. The Boeings were given Navy warpaint and were assigned serial numbers, but were merely called B (for Boeing) 314. They were operated with civil registrations by PAA crews. Power plant, four 1,600 hp Wright R-2600. Span, 152 ft; length, 106 ft; gross weight, 82,500 lb; accommodation, 84; max speed, 193 mph.

Boeing P2B-1 Superfortress carrying the Douglas D-558-II Skystreak. (*James C. Fahey*)

BOEING P2B SUPERFORTRESS

Four B-29s were acquired by the Navy from USAF stock on April 14, 1947, for use as test-beds in anti-submarine projects and as mother ships for the Douglas D-558-II research aircraft. Two (84028–84029) were designated P2B-1S, the second of these becoming the D-558 carrier; the other two (84030–84031) were designated P2B-2S. Small ventral radomes indicated their role as equipment test-beds. Span, 141 ft 3 in; length, 99 ft 9 in; gross weight, 133,500 lb; max speed, over 330 mph; range, 4,000 miles. Power plant, four 2,200 hp Wright R-3350-23s.

The Brewster XSBA-1 modified for use by NACA. (*NACA photo*)

BREWSTER SBA (NAF SBN)

First product of Brewster Aeronautical Corp, the two-seat XSBA-1 was designed during 1934 to meet expanding Navy requirements for scout-bombers to serve aboard new carriers under construction. A single proto-type (9726) was ordered on October 15, 1934, and appeared on April 15, 1936, with a Wright R-1820-4 engine. It was modified in 1937 with 950 hp XR-1820-22 engine and demonstrated a 263 mph speed, aided by internal stowage for the 500 lb bomb. Thirty production examples ordered on September 29, 1938, with 950 hp R-1820-38 engines were built by the NAF as SBN-1s (1522–1551), and were delivered between November 1940 and March 1942 to VB-3. They later served as trainers for VT-8 aboard USS *Hornet*. Span, 39 ft; length, 27 ft 8 in; gross weight, 6,759 lb; max speed, 254 mph; range, 1,015 miles.

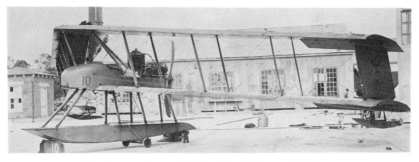

The Burgess-Dunne AH-10 tailless biplane. (*US Navy*)

BURGESS-DUNNE AH-7, 10

Manufacturing rights to the British Dunne tailless aircraft were licensed to the Burgess Company of Marblehead, Mass, which built two seaplane versions (AH-7, AH-10) in 1916. Longitudinal stability was achieved by extreme sweepback (30 degrees). The Navy's first experiments in aerial gunnery were conducted with these pushers. Power plant, 100 hp Curtiss OXX-2. Span, 46 ft 6 in; max speed 69 mph.

417

A Burgess S seaplane in October 1916. (*US Navy*)

BURGESS L

The Burgess Company, sometimes called Burgess & Curtis because of the prominence of Greely S. Curtis (no relation to Glenn L.) in its affairs and also called Burgess-Wright because of its manufacture of aircraft under the Wright's patents, produced several training designs for the Navy before becoming a Division of Curtiss Aeroplane & Motor Company in 1917. These included six Model S seaplanes (A70–A75), two HT-B (A155, A156), six HT-2 (A374–A379) and six U-2 (A380–A385). The 125 hp Hall-Scott powered Model S illustrated was AH-25 number before being re-serialled A70.

A Navy Cessna JRC-1 attached to NAS Alameda. (*William T. Larkins*)

CESSNA JRC

War-time expansion of Navy ferry squadrons and aircraft delivery units brought a need for small reliable transports to carry ferry pilots to and from their home bases at the end of delivery flights. Typical of the commercial types acquired in 1942–43 to fill this need was the four/five-seat Cessna T-50. Used by the USAAF as the UC-78 Bobcat, it bore the Navy designation JRC-1 and was powered by two 450 hp Jacobs R-755 engines; 67 were acquired by the Navy (55772–55783 and 64442–64496). Span, 41 ft 11 in; length, 32 ft 9 in; gross weight, 5,100 lb; max speed, 200 mph.

One of 60 Cessna OE-1s operated by the Marine Corps. (*Peter M. Bowers*)

CESSNA OE-1 BIRD DOG

Following adoption of the Cessna Model 305A by the US Army as the L-19, the Navy procured an identical version for use by the Marine Corps. Orders totalled 60 (133782–133816, 136887–136911), delivered in 1951–53. Some served in Korea. In 1959, the Navy acquired two L-19E Bird Dogs (144663–144664), retaining the original OE-1 designation. Powered by the 213 hp Continental O-470-11 engine, all were redesignated O-1B in 1962. Eight O-1Gs (156678–156685) were obtained in 1968. Span, 36 ft; length, 25 ft 9 in; gross weight, 2,430 lb; max speed, 130 mph.

The third production model Cessna OE-2. (*Peter M. Bowers*)

CESSNA OE-2

Developed to meet a specific Marine Corps requirement, the Cessna Model 321 was similar to the OE-1/L-19 series, but used Model 180 wings, a new fuselage and a 260 hp Continental O-470-2 engine. A batch of 25 (140078–140102) went to the Marines as OE-2s in 1955 and two in 1956 (148250–148251), but the higher cost of this model prevented its adoption on a wider scale. The designation changed to O-1C in 1962. Span, 36 ft; length, 26 ft 3 in; gross weight, 2,650 lb; max speed, 185 mph.

The first production N2Y-1 in overall yellow finish, under test at Anacostia.
(*US Navy*)

CONSOLIDATED N2Y

In 1929 the Navy tested a civilian Fleet I tandem-seat trainer made by the Fleet Aircraft division of Consolidated as the XN2Y-1 (A8019), powered by a 110 hp Warner Scarab engine. Six N2Y-1s (A8600–A8605) were purchased, with 115 hp Kinner K-5 engines, to serve as familiarization trainers for the 'skyhook' pilots with the USS *Akron* and *Macon* (see page 132). Span, 28 ft; length, 21 ft 5 in; gross weight, 1,637 lb; max speed, 108 mph.

The Coast Guard's single Consolidated 21-A, later designated N4Y-1.
(*Fred E. Bamberger, Jr.*)

CONSOLIDATED 21-A/N4Y

The Consolidated Model 21-A was a commercial development of the Army PT-3 and the Navy NY trainer series and was produced for the Army as PT-11. The Coast Guard bought one (CG-10; later 310, V110) with 165 hp Wright J-6-5 engine, which was soon changed to a 220 hp Lycoming R-680. The Navy also bought three Lycoming models (21-C) in 1934 as XN4Y-1 (9456–9458) and the Coast Guard model was then redesignated N4Y-1. Span, 31 ft 6 in; length, 26 ft 11 in; gross weight, 2,544 lb; max speed, 118 mph.

A production model Consolidated TBY-2 in dark blue finish, 1945. (*US Navy*)

CONSOLIDATED TBY SEA WOLF

Two weeks after Grumman received the contract initiating prototype construction of the Avenger (page 212), a similar three-seat torpedc-bomber was ordered from Vought as the XTBU-1. Powered by a 2,000 hp Pratt & Whitney R-2800-20 engine it was first flown on December 22, 1941, and delivered to Anacostia in March 1942. As Vought lacked production capacity, a contract for 1,100 TBY-2s was placed with Consolidated in September 1943. Production was terminated with 180 built (30299–30367, 30369, 30371–30480) and the Sea Wolf never became operational. Span, 57 ft 2 in; length, 39 ft; gross weight, 16,247 lb; max speed, 311 mph at 14,700 ft.

Convair OY-1 in Navy service at Moffett Field. (*William T. Larkins*)

CONVAIR OY SENTINEL

The Marine Corps acquired 306 Convair OY-1s during World War II for use on general liaison and light transport duties, artillery spotting and casualty evacuation in forward areas. The OY-1 originated as the Stinson Model V-76, adopted by the Army in 1942 as the L-5 after the Stinson company had become a division of Vultee Aircraft. By 1943, when Marine acquisition of the type began, Vultee had merged with Consolidated, hence the use of Consolidated's letter Y in the designation. Serials were 02747–02788, 03862–04020, 60460–60507, 75159–75182 and 120442–120474. Power plant, 185 hp Lycoming O-435. Span, 34 ft; length, 24 ft 1 in; gross weight, 2,050 lb; max speed, 129 mph.

One of six Convair R3Y-2s with nose-loading doors. (*Convair photo*)

CONVAIR R3Y TRADEWIND

Only Navy flying-boat to use turboprop engines, the Convair XP5Y-1 (121455–121456) was a 160,000 lb patrol boat powered by four 5,500 ehp Allison XT40-A-4 engines and was first flown on April 18, 1950. The design was adapted as a long-range transport and the Navy put into service five R3Y-1s (128445–128449) the first of which flew on February 25, 1954; and six R3Y-2s (128450, 131720–131724), first flown on October 22, 1954. The R3Y-2 incorporated a nose-loading door to discharge vehicles straight on to landing beaches, and was later adapted as a four-point tanker. Power plant, 5,850 ehp T40-A-10 turboprops. Span, 145 ft 9 in; length, 139 ft 8 in; gross weight, 160,000 lb; cruising speed, 300 mph.

A Convair R4Y-1 VIP transport assigned to NAS Anacostia, with blue nacelles. (*Peter M. Bowers*)

CONVAIR R4Y

Marine and Navy fleet support units received a total of 36 R4Y-1s (140993–141028) twin-engined transports similar to the USAF's C-131D and commercial CV-340, with 2,500 hp Pratt & Whitney R-2800-52W engines. They carried 44 passengers and were delivered from 1952 onwards; the designation changed to C-131F in 1962. A single R4Y-1Z (140378) had a VIP interior with 24 seats and became the VC-131F. Two R4Y-2s (C-131G) delivered in 1957 (145962–145963) were similar to the commercial CV-440. Span, 105 ft 4 in; length, 79 ft 2 in; gross weight, 47,000 lb; max speed, 275 mph.

First of the six metal Martin MS-1 scout seaplanes built to NAF designs.
(*Glenn L. Martin photo*)

COX-KLEMIN XS (MARTIN MS)

Demonstrations with the tiny M-1 Messenger, built by Sperry to designs by the Army Engineering Division, aroused Navy interest in a similar design for a single-seat scouting seaplane. Such a type, it was believed, would be a practicable proposition for stowage aboard submarines. Consequently, the Bureau of Aeronautics designed a single-seat seaplane scout that could be disassembled and reassembled quickly, with struts substituted for the traditional interplane wires to avoid tedious rigging procedures. Power plant was the same 60 hp Lawrence used by the Army in the M-1.

Instead of building the new model in its own workshops, the Navy let two contracts to industry. The Cox-Klemin Aircraft Corporation, College Point, LI, received an order for six under the designation of XS-1 (A6515–A6520) using wood construction for the airframe and the floats. The Glenn L. Martin Company, Cleveland, Ohio, received an order for another six, designated MS-1 (A6521–A6526) to be built of aluminium. The airframe was fabric covered, but the floats were metal skinned. After delivery in 1923 one XS-1 (A6519) was redesignated XS-2 when fitted with an experimental three-cylinder Kinner engine. Data for XS-1: span, 18 ft; length, 18 ft 2 in.; gross weight, 1,030 lb; max speed, 103 mph.

The Culver TD2C-1 target drone in all-red finish. (*E. M. Sommerich Collection*)

CULVER TDC, TD2C

The TDC-2 and TD2C-1 drones were Navy duplicates of the USAF PQ-8A and PQ-14A targets from Culver Aircraft Corp of Wichita, Kansas. The PQ-8 was a radio-controlled version of the commercial Culver LCA sportplane. After evaluation of one improved PQ-8A with 125 hp Lycoming O-290-1 engine, the Navy ordered 200 in 1942 as TDC-2 (44355–44554). In 1943 Culver introduced a new target powered by a 150 hp Franklin O-300-11 engine and with a retractable undercarriage. Of a total of 1,348 PQ-14As built for the USAF, 1,201 were transferred to the Navy as TD2C-1 (69539–69739, 75739–76138, 83752–83991, 119979–120338). TD2C-1 data: span, 30 ft; length, 19 ft 6 in; gross weight, 1,820 lb; max speed, 180 mph.

The Curtiss HA-1 single-float fighter. (*US Navy*)

CURTISS HA DUNKIRK FIGHTER

The HA was designed to serve as a two-seat escort and air superiority fighter for the Dunkirk–Calais area. Powered with the 400 hp Liberty, the prototype HA (A2278) flew on March 21, 1918, but soon crashed. Two modified versions were then ordered as HA-1 (A4110) which included salvageable parts of A2278, and the longer-winged HA-2 (A4111). Following the war's end, both were used for miscellaneous testing, and a landplane variant of HA-2 was developed for the Post Office department as a mailplane. Span, 36 ft (HA-2 span, 42 ft); length, 30 ft 9 in; gross weight, 4,012 lb; max speed, 132 mph.

424

Lt Al Williams of the Marine Corps with the Curtiss—Cox *Cactus Kitten* in 1922. (*Curtiss photo*)

CURTISS-COX CACTUS KITTEN

One of the Navy's most unusual aircraft acquisitions, the *Cactus Kitten* was built by Curtiss for Texas millionaire S. E. J. Cox as one of his two entries in the 1920 Gordon Bennett Cup Race. Powered with the new 435 hp Curtiss C-12 engine, the Kitten first appeared as a monoplane but proved so unmanageable that it was not raced. Rebuilt in 1921 as a triplane was placed second in the Pulitzer Trophy Race and was then sold to the Navy for $1.00 to be used as a trainer by the Navy's 1922 Pulitzer racing team. Span, 20 ft; max speed, over 170 mph.

The single Curtiss CT-1 torpedo-plane. (*Curtiss photo*)

CURTISS CT

The Curtiss Model 24 was a radical innovation as a twin-float twin-engine torpedo-plane. The Navy ordered nine as CT-1 for Curtiss Torpedo (A5890–A5898) in 1923. Using a Fokker-type cantilever wooden wing, the prototype was fitted with two 300 hp Wright-Hispano engines. These were soon changed to the new 435 hp Curtiss D-12 engines introduced on the CR racers. The other eight CT-1s were cancelled. Span, 65 ft; length, 52 ft; gross weight, 11,208 lb; max speed, 107 mph.

The prototype Curtiss F7C-1 after purchase by the Navy. (*US Navy*)

CURTISS F7C

Following the Navy decision to use air-cooled radial engines in place of the liquid-cooled inline power plant which proved difficult to maintain at sea, Curtiss produced a new fighter designed around the 450 hp Pratt & Whitney R-1340B Wasp. First flown on February 28, 1927, the XF7C-1 became part of an 18-plane production order (A7653–A7670). The F7C-1s were delivered between August 1927 and January 1929, and served operationally only with Marine Corps Squadron VF-5M at Quantico. Span, 32 ft 8 in; length, 22 ft 2 in; gross weight, 2,782 lb; max speed, 151 mph.

This Curtiss RC-1 served with the Marine Corps for six years. (*US Navy*)

CURTISS RC

A single example of the unusual Curtiss Kingbird was purchased by the Navy and assigned for transport duties with the Marine Corps. Powered by two 300 hp Wright R-975 engines, the RC-1 (A8846) was delivered in March 1931 and went into service with Squadron VF-9M at Quantico. Towards the end of 1933 it was transferred to VJ-7M at San Diego, where it remained until 1936. Span, 54 ft 6 in; length, 34 ft 10 in; gross weight, 6,115 lb; max speed, 137 mph.

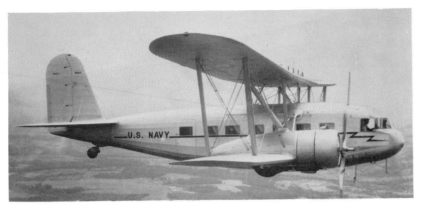

One of the two Curtiss R4C-1s purchased in 1934. (*US Navy*)

CURTISS R4C

Two examples of the famous Curtiss-Wright T-32 Condor II were purchased by the Navy in 1934 for general transport duties. They were similar to the civilian models, which had a 3,200 lb payload (hence the T-32 designation), and were powered by two 700 hp Wright R-1820-12 radial engines. Both R4C-1s (9584–9585) went to Marine Utility Squadron Seven in 1935 and were then attached to the U.S. Antarctic expedition in 1940–41, being subsequently abandoned there. Span, 82 ft; length, 49 ft 1 in; gross weight, 17,500 lb; max speed, 181 mph.

A Curtiss SNC-1 in service early in 1942. (*Fred E. Bamberger, Jr.*)

CURTISS SNC

To help meet the expanding need for trainers of all types in 1940, the Navy ordered a variant of the Curtiss CW-21 design as a combat trainer designated SNC-1 and powered by a 420 hp Wright R-974 engine. An initial contract for 150 placed in November 1940 was followed by others to a total of 305 (6290–6439, 05085–05234, 32987–32991). Span, 35 ft; length, 28 ft 6 in; gross weight, 3,200 lb; max speed, 215 mph.

427

A Marine Corps Curtiss R5C-1 in overall dark blue finish, 1947. (*Peter M. Bowers*)

CURTISS R5C COMMANDO

Of the more than 3,000 C-46 transports built by the St Louis, Louisville, and Buffalo factories of Curtiss, 160 went to the Marine Corps, which used them primarily to support the island-hopping campaign against Japanese forces in the South Pacific. Powered by two Pratt & Whitney R-2800 engines, the R5C-1s (39492–39611, 50690–50729) carried 50 troops and were equivalent to the USAAF's C-46A model. A number remained in service for several years after the war ended. Span, 108 ft 1 in; length, 76 ft 4 in; gross weight, 56,000 lb; max speed, 269 mph.

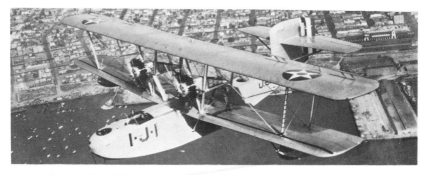

Douglas PD-1 flying-boat photographed in January 1929. (*Douglas photo*)

DOUGLAS PD

The 25 PD-1s (A7979–A8003) were Douglas-built versions of the NAF PN-12. Principal recognition feature was the unique shape of the engine nacelles, which were flattened to horizontal knife-edges at the rear. Powered by two 575 hp Wright R-1750s, these were the first flying-boats produced for the Navy in quantity by a West Coast manufacturer since World War I, delivery starting in 1929. Span, 72 ft 10 in; length, 49 ft 1 in; gross weight, 14,837 lb; max speed, 121 mph.

One of two Douglas OD-1 biplanes, in the markings of VJ-7M at NAS San Diego.
(*US Navy*)

DOUGLAS OD-1

Two examples of the Douglas O-2, a widely-used Army Air Corps observation biplane, were acquired by the Marine Corps in 1929 (serials A7203–A7204). They served only briefly in the observation role before being relegated to the utility role with Squadron VJ-7, and no further examples of this Douglas biplane family served with the Navy or Marines. Power plant, one 400 hp Liberty. Span, 39 ft 8 in; length, 29 ft 6 in; gross weight, 4,706 lb; max speed, 126 mph.

The Coast Guard operated this single example of the Douglas O-38C with its original AAC designation. (*Fred E. Bamberger, Jr.*)

DOUGLAS O-38C

In 1931 the Army Air Corps ordered one standard O-38B aeroplane from the Douglas Aircraft Company for delivery to the US Coast Guard. Minor differences in equipment resulted in a designation change to O-38C. Air Corps serial number was 32-394. This was the Coast Guard's ninth aeroplane at the time and was merely identified as CG9 prior to receiving a new serial number, V108, in 1936. Power plant, 525 hp Pratt & Whitney R-1690-5 Hornet. Span, 40 ft; length, 32 ft; gross weight, 4,458 lb; max speed, 149 mph.

The third Douglas R2D-1 at Santa Monica in December 1934. (*Douglas photo*)

DOUGLAS R2D

Dramatic advances in civil air transport in the US in the early 1930s had an inevitable impact upon the armed forces. Limited budgets prevented over-rapid replacement of older transports like the Ford Trimotor or Curtiss Condor, but in 1934 the Navy purchased its first Douglas DC-2, later acquiring four more with the designation R2D-1 (9620–9622, 9993–9994). Like their civil counterparts, the R2D-1s had 710 hp Wright R-1820-12 engines. Span, 85 ft; length, 61 ft 9 in; gross weight, 18,200 lb; max speed, 210 mph.

One of three Douglas R3D-2s assigned to the Marine Corps in 1940. (*Douglas photo*)

DOUGLAS R3D

First flown in February 1939, the Douglas DC-5 commercial transport was overtaken by World War II and of the 12 built, seven were on a US Navy contract placed in 1939. This was for three R3D-1s (1901–1903) for the Navy and four R3D-2s (1904–1907) assigned to the Marine Corps in 1940 as paratroop trainers. The first R3D-1 crashed before delivery but one privately owned DC-5 was later acquired by the Navy as the single R3D-3 (08005). Engines were two 1,000 hp Wright R-1820-44s. Span, 78 ft; length, 62 ft 2 in; gross weight, 19,582 lb; max speed, 221 mph.

430

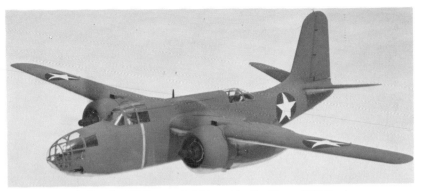

A Douglas BD-2, one of eight used by the Navy. (*US Navy*)

DOUGLAS BD

In 1941 a single example of the Douglas A-20A attack aircraft was acquired from the Army for evaluation, with the designation BD-1 (4251). Eight more followed in 1942 with the designation BD-2 (7035-7042), being converted from the USAAF A-20B. Powered by two 1,600 hp Wright R-2600-11 Cyclone engines, the BD-2s were used by Marine Corps units as target tugs and for general utility duties. Span, 61 ft 4 in; length, 48 ft; gross weight, 21,000 lb; max speed, 350 mph at 12,000 ft.

An early production model Douglas BTD-1 at El Segundo in December 1943. (*Douglas photo*)

DOUGLAS BTD DESTROYER

Six months before Pearl Harbor, the Navy gave Douglas contracts to build two prototypes of a new two-seat scout bomber, the XSB2D-1 (03551-03552). First flown April 8, 1943, the XSB2D-1 was not produced but was modified into the BTD-1, first of a new class of single-seat torpedo-bombers. A contract dated August 31, 1943, increased earlier orders to a total of 358 but only 28 BTD-1s (04959-04971, 09048-09062) were completed, deliveries starting in June 1944 and ending in October 1945. The BTD-1 was powered by a 2,300 hp Wright R-3350-14 engine. Span, 45 ft; length, 38 ft 7 in; gross weight, 19,000 lb; max speed, 344 mph.

A Douglas JD-1D carrying a Ryan Firebee beneath the starboard wing. (*US Navy*)

DOUGLAS JD INVADER

A single example of the USAF A-26B Invader was converted for the Navy in 1945 as the XJD-1, followed by a total of 140 JD-1s (57990–57991, 77139–77224, 140326–140377) converted from A-26Cs as target tugs. They were operated by Utility Squadrons VU-3, VU-4, VU-7 and VU-10 in the Navy training areas. Some were later designated JD-1D when equipped to launch and control KD target drones. The designations were changed to UB-26J and DB-26J respectively in 1962. The JD-1 was powered by two 2,100 hp Pratt & Whitney R-2800-71 engines. Span, 70 ft; length, 50 ft 9 in; gross weight, 35,000 lb; max speed, 330 mph; range, over 1,500 miles.

Douglas R6D-1 in the markings of Fleet Tactical Support Squadron VR-1. (*US Navy*)

DOUGLAS R6D LIFTMASTER

As part of the Navy contribution to MATS, orders were placed in 1950 for a version of the Douglas DC-6A similar to the C-118 purchased by the Air Force. Orders totalled 61, delivered by 1952 as R6D-1s, plus four R6D-1Zs in 1953 with executive interiors (128423–128433, 131567–131620). The R6Ds, redesignated C-118B and VC-118B in 1962, served with Fleet Tactical Support Squadrons VR-1 and VR-21. In 1958, 38 R6D-1s had been transferred to USAF as C-118As but continued to operate in MATS with Navy crews. Engines were 2,500 hp Pratt & Whitney R-2800-52Ws. Span, 117 ft 6 in; length, 107 ft; gross weight, 112,000 lb; max speed, 300 mph.

432

The first Elias EM-1 after modification of the lower wing. (*Elias & Brothers photo*)

ELIAS EM, EO

The firm of G. Elias & Brothers of Buffalo, NY, was founded shortly after World War I, and produced in 1922 the EM; a rugged two-seater developed for the short-lived Marine Expeditionary category intended to operate convertible wheel/floatplanes from advance-area fields with minimum surfacing and facilities in a multi-purpose observation–fighting–light bomber role.

In its original form the first of seven EM-1s (A5905–A5911) had unequal-span wings and a 300 hp Wright-Hispano H engine cooled by side radiators. The lower wing span was soon increased to match the upper and a nose radiator was fitted. The seaplane version was tested on a single float. The first EM-1 was delivered to the Marine Corps in 1922 The remaining EMs were completed as EM-2s with the 400 hp Liberty engine, the equal-span wings of the modified EM-1, but the original side radiator feature. EM-2 seaplanes were fitted with both single and twin floats. The first EM-2 was delivered to the Marines; the other five went to the Navy, one (A5908) as an observation type designated EO-1. EM-2 data: span, 39 ft 8 in; length, 28 ft 6 in; gross weight, 4,233 lb; max speed, 120 mph.

An Elias EM-2 seaplane with single main float. (*US Navy*)

433

The single Fairchild XJQ-1 was a commercial Model FC-2. (*US Navy*)

FAIRCHILD JQ (RQ), J2Q (R2Q)

These Fairchild monoplanes were typical of numerous established commercial designs that the Navy bought off-the-shelf for evaluation in the 1920s and 1930s. Some of these single acquisitions were followed by small-quantity purchases. Frequently given experimental designations, the single models were actually approved commercial types and were put to work as light transport and utility aircraft. The XJQ-1 of 1928 (A7978) was the five-seat Fairchild FC-2, a design already well-proven in Arctic and Canadian 'Bush' flying. Powered by a 220 hp Wright J-5 engine, the XJQ-1 retained the FC-2's folding wings, a novelty generally considered a monopoly of Navy carrier-based types. Re-powered with a 450 hp Pratt & Whitney engine, the XJQ-1 became XJQ-2 and later XRQ-2. The XJ2Q-1 (A8486) was an improved six-seat 1929 Fairchild known as the 71, also powered with the Wasp engine and retaining folding wing feature. This model was similar to the Army F-1/C-8 photographic and light transport model of 1930 and was later redesignated R2Q-1. R2Q-1 data: span, 50 ft; length, 32 ft 10 in; gross weight, 5,325 lb; max speed, 134 mph.

The sole Fairchild XJ2Q-1 was a commercial Model 71. *US Navy*)

Fairchild 24-Rs were used by the Coast Guard as J2K-1s. (*William T. Larkins*)

FAIRCHILD R2K, J2K, GK

The XR2K-1 of 1936 (9998) was a two-seat commercial Fairchild 22 parasol monoplane powered by a 145 hp Warner radial engine. It was procured by the Navy as a research vehicle for the National Advisory Committee for Aeronautics (NACA) and never operated as a naval aircraft. The K in the designation identified Kreider-Reisner, a company which had been absorbed by Fairchild in 1929 but managed to maintain the identity of its product line. The XRK-1 designation had already been assigned to Kinner (see page 421). Other Fairchild designs were identified in the Navy by the letter Q (see page 434). Later K-R designs purchased by the Navy were the JK-1 (see page 436) and the J2K-1 and J2K-2, which were commercial Fairchild 24-Rs. Two J2K-1s (V160, V161) and two J2Ks (V162, V163) were purchased for the Coast Guard in 1936. These were three-seat models powered by inverted air-cooled Ranger engines of 145 hp. A total of 13 improved four-seat Fairchild 24W-40s, similar to the USAAF's UC-61As with 165 hp Warner R-500 Super Scarab radial engines, were procured as GK-1 in 1940 and 1942 (7032–7034, 09787–09788, 09790–09797). GK-1 data: span, 36 ft 4 in; length, 23 ft 9 in; gross weight, 2,550 lb; cruising speed, 122 mph.

A Fairchild GK-1 with red stripes indicating its use as an instrument trainer. (*US Navy*)

435

The first Fairchild JK-1 in partial blue finish as a Command transport.
(*William L. Swisher*)

FAIRCHILD JK

The Navy acquired a single example of the civil Fairchild 45 (0800) in 1936, using it as a command aircraft for staff transport duties. At NAS Anacostia in 1937, it transferred to NAS San Diego in 1938 and to NAS Alameda when the latter opened in November 1939. Two more JK-1s (34112–34113) joined the Navy during World War II when civil machines were pressed into service. Power plant was a 320 hp Wright XR-760-6 radial. Span, 39 ft 6 in; length, 30 ft; gross weight, 19,500 lb; max speed, 167·5 mph.

One of eight Fairchild C-123Bs acquired by the US Coast Guard.

FAIRCHILD C-123 PROVIDER

During 1962 the Coast Guard obtained eight Fairchild C-123B Providers on loan from the USAF for use as cargo transports. They retained the original designations. Powered by two 2,300 hp Pratt & Whitney R-2800-99W radials, the C-123B originated as a design by Chase Aircraft, based on a cargo glider; a total of 300 were built for the USAF by Fairchild Aircraft Division. Span, 110 ft; length, 75 ft 9 in; gross weight, 71,000 lb; max speed, 245 mph.

One of four General Aviation PJ-1s showing the pusher engines. (*Howard Levy*)

GENERAL AVIATION FLB, PJ

The five flying lifeboats for the US Coast Guard marked the swan song of the American Fokker organization. The sleek twin-pusher flying-boats were designed by the Fokker Aircraft Corporation of America as Model AF-15 in a US Coast Guard design competition for a patrol and open-sea rescue boat. By the time that Fokker was declared the winner, the American company had become a division of the General Aviation Corporation and its manufacturing facilities were transferred to Dundalk, Maryland. No military designation was assigned at first, the Coast Guard simply calling them Flying Life Boats and using the letters FLB as a prefix to the serial numbers 51–55. The first was erroneously marked FLB-8 when it appeared in January 1932, but this was soon changed to the proper FLB-51.

The first FLB, named *Antares*, was sent to the Naval Aircraft Factory in 1933 and converted to a tractor design. This later became known as PJ-2 when standard designations were adopted after General Aviation had become North American Aviation. The FLBs were right in the middle of all the USCG serial number changes. After being FLB-51–55, they were re-serialled as 251–255 in 1935 and in 1936 became V112–115 for the PJ-1s and V116 for the PJ-2. PJ-1 specifications: power plant, two 420 hp Pratt & Whitney R-1340 Wasp. Span, 74 ft 2 in; length, 55 ft; gross weight, 11,200 lb; crew, 4.

The single General Aviation PJ-2 with its final serial number, V116, adopted in October 1936. (*US Navy*)

The Gallaudet D-1, serial A59, at Pensacola in mid-1916. (*US Navy*)

GALLAUDET D-1, D-4

One of the most unusual designs in the Navy inventory was the Gallaudet D-series, produced by the Gallaudet Engineering Company, Norwich, Conn. This featured the power plant amidships driving a geared-down propeller mounted on a ring encircling the fuselage. The D-1 (A59) was powered with two 150 hp Dusenberg engines mounted side by side. Following delivery of the D-1 in January 1917, the firm reorganized as Gallaudet Aircraft Corporation, and moved to Greenwich, Rhode Island. Two D-4s, powered with single Liberty engines, were then produced for the Navy (A2653, A2654). D-1 specifications: span, 48 ft; length, 33 ft; gross weight, 4,604 lb; max speed, 90 mph.

A Coast Guard Grumman J4F-1, one of a batch of 25 bought in 1941–2. (*Peter M. Bowers*)

GRUMMAN J4F WIDGEON

The Coast Guard obtained 25 G-44 light amphibians in 1941–2 as J4F-1s (V197–V221), later adding light bomb racks under the wings. Procurement of similar J4F-2s by the Navy began soon after, a total of 131 being acquired (09789, 09805–09816, 30151, 32937–32986, 33952–33957, 34585, 37711–37770). With a crew of two, the J4F-2s carried three passengers and were powered by two 200 hp Ranger L-440 engines. Span, 40 ft; length, 31 ft 1 in; gross weight, 4,500 lb; max speed, 165 mph.

A Grumman TC-4C with civil registration for early test flying. (*Grumman photo*)

GRUMMAN C-4 ACADEME

The Navy planned procurement of a version of the Grumman Gulf-stream in 1962 for use as a navigation trainer and transport, but procurement of this type, successively allocated the designations T-41A and TC-4B, was deferred. Meanwhile, a single VC-4A (1380) was delivered to the Coast Guard in 1963 as an executive transport. Late in 1966 the Navy ordered nine TC-4C flying classrooms, equipped to train bombardier/navigators for the A-6A Intruder. This version was first flown on June 14, 1967. The C-4s had two 2,185 hp Rolls-Royce Dart 529-8X engines. Span, 78 ft; length, 67 ft 11 in; gross weight, 36,000 lb; max speed, 350 mph.

The sole Grumman VC-11A Gulfstream II executive transport. (*Stephen P. Peltz*)

GRUMMAN C-11

The US Coast Guard has maintained two VIP transports in its fleet for most of the period since World War II. Originally, two Martin VC-3As were used (see page 454) but the fleet was modernized in 1963 with the addition of a single Grumman Gulfstream (as noted above). The Coast Guard's first, and up to 1975 only, pure jet aircraft is a single Gulfstream II acquired subsequently and designated VC-11A. Originally N862GA it was assigned the serial 01 in a new CG series for VIP aircraft, when the VC-4A became 02. The Gulfstream II is powered by two Rolls-Royce Spey Mk 511-8 turbofans. Span, 68 ft 10 in; length, 79 ft 11 in; gross weight, 62,000 lb; max cruising speed, 588 mph.

A Gyrodyne DSN-3 drone helicopter. (*Howard Levy*)

GYRODYNE H-50

Originally developed as the DSN-1, the QH-50A was the first Drone Anti-Submarine Helicopter (DASH) for the US Navy and was derived from the same manufacturer's YRON-1 Rotorcycles. The production model QH-50C with a 270 hp Boeing T50-BO-4 entered service in 1963 and carried two homing torpedoes. More than 300 were built, followed by QH-50Ds with 330 hp T50-BO-10 engines. Rotor diameter, 20 ft; overall length, 13 ft 1 in; gross weight, 2,296 lb; max speed, over 90 mph.

A Hiller HTE-1 in Navy blue finish. (*Hiller photo*)

HILLER HTE

Small quantities of Hiller UH-12 helicopters were acquired by the Navy following evaluation in 1950 of a single example of the UH-12A designated HTE-1 (125532). Powered by a Franklin O-335 engine, the HTE-1 had a tricycle undercarriage, and 16 more were bought by the Navy (128637–128652) followed in 1951 by 35 HTE-2s (129757–129791). Rotor diameter, 35 ft; length, 27 ft 6 in; gross weight, 2,500 lb, max speed, 84 mph.

The second Huff-Daland HN-1 taking off from Reflecting Pool, Washington, DC. (*US Navy*)

HUFF-DALAND HN, HO SERIES

Huff, Daland & Company of Ogdensburg, NY, was formed after World War I and was among the first American manufacturers to adopt welded steel-tube fuselage construction combined with thick wooden cantilever wings. In this choice it was heavily influenced by the war-time German Fokkers.

Its first Navy order was for three HN-1 trainers (A6349–A6351) powered with the war-surplus 180 hp Wright-Hispano E-2 engine. After the 200 hp Lawrance J-1 air-cooled radial engine appeared, an additional Navy order was placed with Huff-Daland for three HN-2s (A6701–A6703) with this power plant. In between the trainer orders was one for three near-duplicates of the HN-1 equipped as an unarmed observation type with the designation of HO-1 (A6560–A6562). Both the trainer and observation models could be fitted with twin floats. Delivery was in 1923.

In March 1927 Huff-Daland was reorganized as the Keystone Aircraft Corporation and the factory moved to Bristol, Pennsylvania. Keystone later merged with Loening and then was absorbed by Curtiss-Wright in 1929. HN-1 data: span, 33 ft; length, 28 ft 6 in.; gross weight, 2,545 lb; max speed, 96 mph.

A Huff-Daland HN-2 with Lawrance J-1 engine. (*Wayland Maxey*)

441

Green stripe around the rear fuselage of this Howard NH-1 indicates its
use as an instrument trainer. (*Peter M. Bowers*)

HOWARD GH, NH NIGHTINGALE

Based on Howard's pre-war DGA ('Damn Good Airplane'), the GH
and NH series were produced in quantity for Navy use from 1941 onwards.
Minor differences distinguished the three utility models, each powered by
a Pratt & Whitney R-985 engine; they comprised 34 GH-1s (04390–04395,
08006–08028, 09769–09770, 09775, 09779, 09781); 131 GH-2s (08029,
32336–32385, 32787–32866) and 115 GH-3s (44921–44922, 44935–44937,
44939, 44941–45049). Navy also used 205 NH-1 instrument trainers
(29376–29550, 44905–44920, 44923–44934, 44938, 44940). Span, 38 ft;
length, 25 ft 8 in; gross weight, 4,350 lb; max speed, 201 mph.

A Kaman HTK-1 showing intermeshing rotors. (*Gordon S. Williams*)

KAMAN HTK

First Kaman helicopters evaluated by the Navy were two K-225 prototypes (125446–125447), purchased in March and June 1950; one was later
re-engined with a 175 hp Boeing 502-2 to become the world's first turbine-
engined helicopter. A three-seat development, the K-240 with a 235 hp
Lycoming O-435-4 engine, was ordered on September 5, 1950, and 29
were delivered to the Navy as HTK-1 trainers (128653–128660, 129300–
129307, 137833–137835). They became TH-43Es in 1962. One other was
acquired for static test of the HTK-1K drone (138062). Rotor diameter,
41 ft each; length, 20 ft 6$\frac{1}{2}$ in; gross weight, 3,100 lb; max speed, 81 mph.

The Kaman HOK-1 was used exclusively by the Marines. (*Kaman photo*)

KAMAN HOK, HUK

The Kaman K-600 won a Navy design competition in 1950 leading to orders for two similar versions, the HOK-1 for Marine use and the ship-based HUK-1 for the Navy. They were powered by the 600 hp Pratt & Whitney R-1340-48 and R-1340-52 respectively. After two XHOK-1 prototypes (125477–125478) the Marines received 81 production aircraft (125528–125531, 129800–129840, 138098–138102, 139971–140001). The Navy took delivery of 24 HUK-1s (146304–146327) in 1958. The two types became UH-43C and OH-43D in 1962 and one was converted as the OH-43G drone. Rotor diameter, 47 ft each (intermeshing); overall length, 25 ft; gross weight, 6,800 lb; max speed, 110 mph.

A Keystone NK-1, modified with dihedral on the tailplane. (*US Navy*)

KEYSTONE NK

One of several entries in a 1928 competition for training aircraft, the XNK-1 had a Wright R-790 engine and could operate on wheels or floats. Three prototypes (A7941–A7943) were built, plus 16 production model NK-1s (A8053–A8068) with minor changes, delivered in 1930. Span, 37 ft; length, 27 ft (28 ft 7 in floatplane); gross weight, 2,658 lb (2,950 lb floatplane); max speed, 110 mph (105 mph floatplane).

443

A Keystone PK-1, based on the NAF PN-12, photographed in August 1931. (*US Navy*)

KEYSTONE PK

In its programme for production versions of the NAF PN-12, the Navy awarded a contract for 18 PK-1s (A8507–A8524) to the Keystone Aircraft Corporation, Bristol, Pa. Distinguishing features were twin rudders and fully-cowled 575 hp Wright R-1820-64 Cyclone engines in a lower location. Shortly before starting work on the PK-1, Keystone had merged with Loening to become Keystone-Loening and the new firm later became a division of Curtiss-Wright. The boats were delivered in 1931. Span, 72 ft; length, 48 ft 11 in; gross weight, 16,413 lb; max speed, 120 mph.

The Kinner XRK-1 re-engined with a Pratt & Whitney Wasp engine, in command colours. (*William L. Swisher*)

KINNER RK

Three examples of the Kinner Envoy, a four-passenger light transport, were acquired by the Navy in 1936 and assigned as staff transports. All were designated XRK-1 (9747–9749) although they were not experimental aircraft. When delivered, each XRK-1 had a 340 hp Kinner R-1044-2 engine but one aircraft (9747) was later re-engined with a 400 hp Pratt & Whitney R-985-38 radial. Span, 39 ft 8 in; length, 29 ft; gross weight, 4,000 lb; max speed, 171 mph.

Blue fuselage finish indicates use of this Lockheed XRO-1 as an executive transport. (*US Navy*)

LOCKHEED XRO

The first aircraft operated by the US Navy with a fully retractable undercarriage was not, as might have been expected, a high performance fighter but was a commercial model Lockheed Altair. Based at NAS Anacostia after delivery in October 1931 it was designated XRO-1 (9054) and was allocated as the personal transport of the Assistant Secretary for the Navy (at that time David Ingalls). In this capacity, the XRO-1, with a 645 hp Wright R-1820E Cyclone, replaced the Curtiss XO2C-2 (8845). Span, 42 ft 10 in; length, 27 ft 6 in; gross weight, 5,193 lb; max speed, 209 mph.

This Lockheed XR3O-1 was assigned to the Coast Guard as personal transport for the Secretary of the Treasury. (*Gordon S. Williams*)

LOCKHEED R2O, R3O

Two examples of the Lockheed Electra went to the Navy and Coast Guard in 1936, respectively XR2O-1 and XR3O-1. The XR2O-1 (0267) was a Model 10-A with Pratt & Whitney R-985-48 engines and was delivered on February 19, 1936, to serve as personal transport for the Secretary of the Navy. The XR3O-1 (383 and later V151) delivered to the Coast Guard on April 9, 1936, as personal transport for the Secretary of the Treasury, was a commercial Model 10-B with Wright R-975E-3s. Span, 55 ft; length, 38 ft 7 in; gross weight, 10,100 lb; max speed, 205 mph.

The Lockheed XJO-3 was used for tricycle undercarriage experiments in 1938. (*Lockheed photo*)

LOCKHEED JO

The Navy acquired a single Lockheed Model 12A in August 1937 as JO-1 (1053) and used it as a seven-seat transport. In the same year, two six-seat JO-2s (1048, 2541) went to the Navy and three others (1049–1051) to the Marine Corps, one each to Headquarters, VMJ-1 and VMJ-2. The single XJO-3 (1267) was acquired in October 1938 to test carrier-deck performance of a twin-engined aircraft with tricycle gear, and had a fixed undercarriage. Two 400 hp Pratt & Whitney R-985-48 engines. Span, 49 ft 6 in; length, 36 ft 4 in; gross weight, 8,400 lb; max speed, 217 mph.

A Lockheed R5O-1 with blue nacelles and rudders to indicate its executive transport status. (*Peter M. Bowers*)

LOCKHEED R5O

Nearly 100 examples of the Lockheed Model 18 Lodestar reached the Navy and Marines between 1940 and 1943. The series began with a single XR5O-1 (2101) and two R5O-1s (4249–4250) as command transports for the Navy and one (V188) for the USCG, a single R5O-2 (7303) and three R5O-3s (01006–01007, 27959). These, and 12 R5O-4s (05046–05050, 12447–12453) were primarily executive transports with four to seven passenger seats. The R5O-5 (12454–12491, 30148–30150) was the standard 12–14 seat transport and 35 R5O-6s (39612–39646) had benches for 18 paratroops. The R5O-1, R5O-4, R5O-5 and R5O-6 had Wright R-1820 engines; the R5O-2 and -3 had Pratt & Whitney R-1830s. Span, 65 ft 6 in; length, 49 ft 10 in; gross weight, 18,500 lb; max speed, 250 mph.

446

A Lockheed PBO-1 in original British camouflage in which it was taken over by Army Air Corps and assigned to Navy. (*US Navy*)

LOCKHEED PBO HUDSON

On September 25, 1941, the Navy initiated action to requisition 20 Lockheed Hudsons then in production for Britain's RAF. Powered by two Wright R-1820-40 Cyclone engines, the Hudsons were designated PBO-1 (03842–03861) and were required for operation from new land bases acquired by the Navy in Newfoundland. Delivered in October 1941, the PBO-1 equipped VP-82 at Argentia, Newfoundland, from which base they attacked and sank the first two U-boats destroyed by the Navy in World War II, on March 1 and March 15, 1942. Span, 65 ft 6 in; length, 44 ft 4 in; gross weight, 20,203 lb; max speed, 262 mph.

One of the two double-deck Lockheed XR6O-1 transports. (*E. M. Sommerich*)

LOCKHEED R6O CONSTITUTION

Two prototypes of the Lockheed Model 89 were ordered by the Navy after development of this large transport was started in 1942 at the instigation of Pan American Airways. An obvious relative of the Constellation, the XR6O-1 had a double-deck fuselage to carry 168 passengers and was powered by four 3,500 hp R-4360-22W engines. First flight was made on November 9, 1946, and both XR6O-1s (85163, 85164) served with VR-44 as transports until 1955. Span, 189 ft; length, 156 ft; gross weight, 184,000 lb; max speed, 300 mph.

447

Lockheed X-26B in use at the NTPS, Patuxent. (*Howard Levy*)

LOCKHEED X-26

The unique X-26B entered service with the US Naval Test Pilot School at Patuxent River, Md, in 1969, being one of two Lockheed QT-2PCs that had previously been tested by the Army in Vietnam. The latter were specially developed 'quiet' aeroplanes intended to reconnoitre enemy positions without giving away their own presence. Based on the Schweizer 2-32 sailplane, the QT-2PC was powered by a 100 hp Continental engine and when acquired by NTPS was designated X-26B as that unit already used three Schweizer 2-32s as X-26As (157932–157933 and 158818). Span, 57 ft 1 in; length, 30 ft 10 in; gross weight, 2,450 lb; best speed, 115 mph.

A McDonnell Douglas C-9B Skytrain II in US Navy markings. (*US Navy*)

McDONNELL DOUGLAS C-9 SKYTRAIN II

To increase its airlift capability, the US Navy selected the McDonnell Douglas DC-9 Srs 30 early in 1972, placing an order for five (later increased to eight, serials 159113–159120). Already purchased by the USAF as the C-9A aeromedical transport, the DC-9 was designated C-9B Skytrain II by the Navy, and the first example flew on February 7, 1973. Powered by 14,500 lb st Pratt & Whitney JT8D-9 turbofans, the C-9B was similar to the commercial DC-9, with a forward freight-loading door, and could accommodate up to 107 passengers or over 30,000 lb of freight. Deliveries began 8 May, 1973, when one C-98 each went to VR-1 at Norfolk and VR-30 at Alameda.

The third J. V. Martin K-IV, also known as the KF-1. (*National Archives 80-CF-423 14*)

J. V. MARTIN K-IV

The J. V. Martin K-IV was a seaplane variant of the diminutive 32 hp, 15 ft span K-III that had been evaluated by the Army in 1917. When submitted to the Navy in 1921 as a seaplane scout, it was necessary to increase the wing span and use a 60 hp Lawrance engine to carry the weight of the floats. The Navy bought three (A5840-A5842) as KF-1s. Span, 24 ft 2 in; length, 17 ft; gross weight, 940 lb; max speed, 82 mph.

The Martin S seaplane A69 (originally AH-19). (*US Navy*)

MARTIN S SEAPLANE

Glenn L. Martin was one of the first successful American aeroplane designers and manufacturers. In 1915 the Army bought six Model S seaplanes fitted with 125 hp Hall-Scott A-5 engines and the Navy bought one as AH-19 to replace the rejected Wright Aeroboat AH-19. When the Navy Martin was re-serialled A69, a second was bought as A70. Span, 46 ft 5 in; length, 29 ft 7 in; gross weight, 2,600 lb; max speed, 82 mph.

The Wright-Martin R with Hall-Scott A-5A engine. (*Glenn L. Martin photo*)

WRIGHT-MARTIN R

In September 1916 the Glenn L. Martin Company and the Wright interests combined with the Simplex Automobile Company and the General Aeronautical Corporation to form the Wright-Martin Company. Wright was not building aeroplanes at the time, but current Martin designs were delivered as Wright-Martin. Three Model R seaplanes (A288–A290), powered with 150 hp Hall-Scott A-5A engines, were delivered to the Navy in 1917. Span, 50 ft 7 in; length, 27 ft 2 in; gross weight, 2,888 lb; max speed, 86 mph.

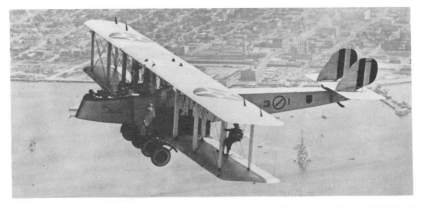

A Martin MT biplane about to drop paratroops over North Island, San Diego. (*US Navy*)

MARTIN MBT/MT

The Martin MB-1 was a revolutionary bomber built for the Army in 1918. The Navy bought two duplicates in 1920 as MBT (A5711–A5712) for Martin Bomber-Torpedo. Eight improved versions (A5713–A5720) were designated MT for Martin Torpedo. Used by the US Marines, the MTs were MB-1s with the longer wings of the Army MB-2. MT data: power plant, two 400 hp Liberty. Span, 71 ft 5 in; length, 45 ft 8 in; gross weight, 12,098 lb; max speed, 109 mph.

The Martin MO-1 observation monoplane was built to NAF design. (*National Archives 80-G-1192*)

MARTIN MO

The Navy had been impressed by the thick cantilever wings of the Dutch Fokkers of the early 1920s. In consequence, BuAer developed a single-engined, three-seat observation design powered with the new 435 hp Curtiss D-12, and awarded a contract to Martin for 36 as MO-1 (A6455–A6460, A6633–A6662). Structure was all-metal with fabric covering. A twin-float seaplane version tested in 1924 was unsuccessful. Span, 53 ft 1 in; length, 38 ft 1 in; gross weight, 4,642 lb; max speed, 104·5 mph.

A Martin PM-1 in initial form with open cockpits and uncowled engines. (*National Archives 80G-180228*)

MARTIN PM

The 30 Martin PM-1s (A8289–A8313, A8477–A8481) were 1930 production versions of the NAF PN-12 with 525 hp Wright R-1750D engines. Throughout their service life they picked up such minor refinements as ring cowlings and enclosed pilots' cockpits. Twenty-five improved PM-2s (A8662–A8686) with larger 575 hp R-1820-64 Cyclones were distinguished by twin vertical tail surfaces similar to those of the XP4N-2/PN-11. PM-2 data: span, 72 ft 10 in; length, 48 ft 5 in; gross weight, 16,964 lb; max speed, 116 mph.

451

One of the Martin M-130s used by the Navy in 1942. (*Pan American photo*)

MARTIN M-130

In 1942 the Navy met some of its requirements for overseas air transport by taking over existing airliners. Among these were the two surviving examples of Pan American Airways' famed Martin 130 flying-boats, collectively known as the *China Clippers*. While the Navy serial numbers 48230 and 48231 were assigned, the Martins were not given naval designations. Power plant, four 830 hp Pratt & Whitney R-1830s. Span, 130 ft; length, 90 ft 7 in; gross weight, 52,000 lb; 4 crew; max speed, 180 mph.

Martin JRM-2 *Caroline Mars* in service with VR-2. (*Peter M. Bowers*)

MARTIN PB2M/JRM MARS

The Navy's largest flying-boat was ordered on August 23, 1938, as the XPB2M-1 (1520). First flown on July 3, 1942, it was modified to a transport configuration as XPB2M-1R in December 1943. Twenty JRM-1 transports, with single in place of twin tail and other changes, were ordered in Janaury 1945 (76819–76838), but only five were built, plus one JRM-2 (76824) with higher (165,000 lb) weight. The JRM-1s were modified to the latter standard as JRM-3s and they were operated by VR-2 based at Alameda. The Mars had four 2,300 hp Wright R-3350-8 engines. Span, 200 ft; length, 120 ft 3 in; gross weight, 145,000 lb; max speed, 225 mph.

Martin JM-1 serving with Marine Corps in all-yellow target-tug markings
(*William T. Larkins*)

MARTIN JM MARAUDER

A total of 272 Marauders went to the Navy in World War II to serve primarily as target tugs. These included 225 JM-1s (66595–66794, 75183–75207) which were ex-USAAF AT-23B conversions from the operational B-26C, while 47 were JM-2s (90507–90521, 91962–91993) similar to the Air Force TB-26G. A few JM-1s were converted for reconnaissance duties as JM-1Ps. Both versions had 2,000 hp Pratt & Whitney R-2800-43 engines. Span, 71 ft; length, 58 ft 3 in; gross weight, 38,200 lb; max speed, 282 mph.

A Martin P4M-1 showing piston-and-jet nacelles. (*Douglas Olson*)

MARTIN P4M MERCATOR

One of several Navy attempts to combine piston and jet engines for maximum range with high-speed capability, this patrol bomber was ordered on July 6, 1944, when Martin received a contract for two XP4M-1s (02789–02790). First flight was made on September 20, 1946, the engines comprising two Pratt & Whitney R-4360-4 (-20A for production) radials and two 3,825 lb thrust Allison J33-A-17 (-10A for production) jets. Deliveries began on June 28, 1950, to VP-21, production totalling 19 (121451–121454, 122207–122209, 124362–124373). A few P4M-1Q, with ECM gear, were produced later by modifying P4M-1s and were used by VQ-1 in Japan. Span, 114 ft; length, 84 ft; gross weight, 83,378 lb; max speed, 410 mph at 20,100 ft.

One of two Martin 4-0-4 transports used by the Coast Guard as RM-1Z.
(*E. M. Sommerich Collection*)

MARTIN C-3

The Coast Guard acquired two commercial model Martin 4-0-4 transports in 1951 to be used in logistics support duties. Powered by two Pratt & Whitney R-2800-34 engines, these aircraft were ordered as RM-1G but went into service in 1952 designated RM-1Z with executive interiors. Still serving in 1962, they were then redesignated VC-3A in the new unified system and were eventually transferred to the Navy with the serials 158202–158203. Span, 93 ft 3 in; length, 74 ft 7 in; gross weight, 44,900 lb; max speed, 312 mph.

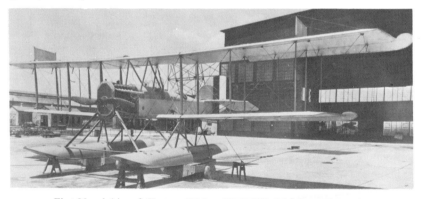

First Naval Aircraft Factory PT-1 at NAF Philadelphia. (*US Navy*)

NAVAL AIRCRAFT FACTORY PT

The PT torpedo seaplanes were unique examples of using war-surplus parts to create new designs. Developed during 1922, the 15 Liberty-powered PT-1s (A6034–A6048) and 18 PT-2s (A6326–A6343) used Curtiss R-6L fuselages and tails fitted to Curtiss HS-1L and HS-2L wings. The letters in the designation stood for Patrol-Torpedo. The PT-1 used the 62-ft wings of the HS-1L while the PT-2 used the longer HS-2L wing. PT-2: span, 74 ft; length, 34 ft 5 in; gross weight, 7,075 lb; max speed, 92 mph.

454

Naval Aircraft Factory TG-2 torpedo seaplane. (*US Navy*)

NAVAL AIRCRAFT FACTORY TG

In 1922 the NAF produced five variants of a basic seaplane gunnery trainer biplane designated TG-1 through TG-5 (A6344–A6348). TG-1 and TG-2 were powered with the 200 hp Liberty engine, TG-3 and TG-4 with the 200 hp Aeromarine T-6, and TG-5 was powered with the 180 hp Wright-Hispano E-4. Fuel for TG-2 and TG-5 was carried in the float; others carried fuel in the fuselage. Photo and data for TG-2: span, 36 ft; length, 30 ft; gross weight, 2,996 lb; max speed, 97 mph.

One of three Martin M2O-1s built to NAF design. (*US Navy*)

NAVAL AIRCRAFT FACTORY NO/MARTIN M2O

The NO-1 was a three-seat Bureau of Aeronautics observation seaplane, two of which were built in 1924 by the NAF (A6431–A6432), A third became NO-2 (A6433). Three were contracted to Martin as M2O-1 (A6452–A6454). Power plant, 435 hp Curtiss D-12. Span, 43 ft 6 in; length, 31 ft 10 in; gross weight, 4,173 lb; max speed, 104 mph.

A New Standard NT-1 trainer in all-yellow finish. (*Gordon S. Williams*)

NEW STANDARD NT

In 1930 the Navy purchased six New Standard D-29s that had been altered to Navy trainer specifications and designated NT-1 (A8583–A8588). This was a conventional design powered with the 100 hp commercial Kinner K-5 five-cylinder engine which was not given a Navy designation at the time. In its original form, the D-29/NT-1 featured a unique 'bathtub' cockpit in which student and instructor sat in tandem in a single cockpit. The drag and discomfort that this arrangement caused soon resulted in conversion to the traditional 'two hole' configuration.

Two NT-2s acquired by the Coast Guard were not derivatives of the NT-1, but were much earlier New Standard D-25A models, powered with the 245 hp Wright J-6-7 engine, which had been captured from smugglers. These aircraft were impressed into service in 1935, designated NT-2 for convenience, and assigned US Coast Guard serial numbers 311 and 312, which were soon changed to V123 and V124. Data for NT-1: span 30 ft; length, 24 ft 7½ in; gross weight, 1,800 lb; max speed, 99 mph.

The designation NT-2 was applied to two commercial New Standard D-25As captured from smugglers. (*David C. Cooke*)

A North American PBJ-1J, equivalent to the AAF's B-25J. (*North American photo*)

NORTH AMERICAN PBJ MITCHELL

The Navy's first contact with the Mitchell came in April 1942, when Lt Col Doolittle operated 16 Army Air Force B-25Bs off the USS *Hornet* to attack Tokyo. Acquisition of Mitchells by the Navy, starting in January 1943, followed the July 1942 agreement to allow the Navy to share the production of land-based bombers. The first squadron to receive these twin-engined bombers was VMB-413; they were designated PBJ-1, with a suffix corresponding to that of the B-25 series. The Navy received 50 PBJ-1C; 152 PBJ-1D; one PBJ-1G; 248 PBJ-1H and 255 PBJ-1J. All had 1,700 hp Wright R-2600 radial engines. Span, 67 ft 7 in; length, 52 ft 11 in; gross weight, 35,000 lb; max speed, 272 mph.

North American T-39D in December 1966. (*Lawrence S. Smalley*)

NORTH AMERICAN T-39

A version of the North American T-39 (Sabreliner) twin-jet was selected by the Navy in 1962 as a trainer for maritime radar operators. Initially designated T3J-1 and later T-39D, they carried Magnavox radar systems and were powered by two 3,000 lb s.t. Pratt & Whitney J60-P-3s. Deliveries to Naval Air Training Command HQ at NAS Pensacola began in August 1963, and a total of 42 was built for the Navy. Span, 44 ft 5 in; length, 43 ft 9 in; gross weight, 17,760 lb; max speed, 550 mph.

A North American AJ-2P reconnaissance aircraft in original configuration. (*US Navy*)

NORTH AMERICAN AJ SAVAGE

Ordered on June 24, 1946, the AJ-1 was a high-performance carrier-based nuclear strike aircraft powered by two 2,400 hp Pratt & Whitney R-2800-44W engines plus a 4,600 lb thrust Allison J33-A-19 turbojet in the tail. The first of three XAJ-1s (121460–121462) flew on July 3, 1948, and a production order for 55 AJ-1s (122590–122601, 124157–124184, 124850–124864) followed; deliveries to VC-5 began in September 1949. In the 55 AJ-2s (130405–130421, 134035–134072) engines were R-2800-48 and J33-A-10, fuel capacity was increased and other changes made; 30 AJ-2Ps (128043–128054, 129185–129195, 130422–130425, 134073–134075) were similar with five cameras in a redesigned nose. Four squadrons still flying AJ-2s in 1958 were re-equipped in 1959, but many of these aircraft were modified as flight-refuelling tankers, a nose-and-reel unit being fitted in the bomb-bay. Surviving AJ-1s and AJ-2s became A-2As and A-2Bs in 1962. Data for AJ-1: span, 75 ft 2 in; length, 63 ft; gross weight, 52,862 lb; max speed, 471 mph.

A North American AJ-2 modified as a flight-refuelling tanker. (*Chance Vought photo*)

458

North American Rockwell CT-39G staff transport. (*Rockwell photo*)

NORTH AMERICAN ROCKWELL CT-39

Starting in 1969, the Navy began to acquire a fleet of newer Sabreliners (see page 457) to provide rapid response airlift of high priority passengers and cargo. The initial order was for three commercial model Sabre Series 40s, with 3,300 lb Pratt & Whitney JT12A-8 engines and accommodation for up to ten passengers. A total of seven was acquired as CT-39Es (157352–157354, 158380–158383) after which 12 CT-39Gs were ordered, these being commercial Sabre 60s with longer fuselages and increased accommodation. Data for CT-39E: span, 44 ft 5¼ in; length, 43 ft 9 in; gross weight, 18,650 lb; max speed, 563 mph.

The Northrop RT-1 in Coast Guard service. (*Douglas photo*)

NORTHROP RT

Among the wide variety of aircraft used by the Coast Guard between the two World Wars was a single example of the Northrop Delta, a commercial transport. Designated RT-1 in service and bearing the serial 382, it was assigned as the personal aircraft for the Secretary of the Treasury in 1935, and was replaced in April 1936 by the Lockheed XR3O-1 (see page 422). Power plant, one 575 hp Wright R-1820 Cyclone engine. Span, 47 ft 4 in; length, 33 ft 1 in; gross weight, 7,350 lb; max speed, 219 mph.

A Northrop F2T-1 used by the Marine Corps for training. (*Clay Jansson*)

NORTHROP F2T

After the Northrop Corporation became a division of Douglas (see page 357), John Northrop formed a new Northrop Aircraft Corporation at Hawthorne, California, in 1939. The best-known World War II product of this company was the P-61 Black Widow twin-engined night fighter with four fixed 20 mm cannon and four 0·50-in guns in a power-driven turret. The Marines acquired five P-61As in 1946 and used them as F2T-1s, without armament, for training. Power plant, two 2,000 hp Pratt & Whitney R-2800-10s. Span, 66 ft; length, 48 ft 11 in; gross weight, 27,000 lb; max speed, 375 mph.

Northrop F-5E of the Naval Fighter Weapons School at Miramar.
(*Northrop photo*)

NORTHROP F-5E TIGER II (AND T-38)

During 1974, the US Navy took delivery of the first two of five F-5E Tiger IIs ordered for use at the Naval Fighter Weapons School, NAS Miramar, to simulate the flight characteristics of enemy fighters in air-to-air combat training of USN fighter pilots. Three more were delivered in 1975. Previously, the USN Test Pilot School at Patuxent had acquired five T-38A Talon trainers (from which the F-5 design was derived) from USAF stocks (158197–158201). Data for F-5E: power plant, two 5,000 lb st General Electric J85-GE-21 turbojets; span, 26 ft 8 in; length, 48 ft 3¾ in; gross weight, 24,080 lb; max speed, Mach=1·6 at 36,000 ft.

A Piasecki HRP-1. These 'Flying Bananas' often operated with fuselage covering removed.
(*US Navy*)

PIASECKI HRP

The well-known 'Flying Banana' made use of the tandem-rotor layout favoured by designer Frank Piasecki. Early model PV-3 was ordered by the Navy on February 1, 1944; only one of two prototype XHRP-1s (37968-37969) was accepted, followed by 20 HRP-1s (111809-111828) used by helicopter experimental units VX-3 and HMX-1 (Marines). Powered by a 600 hp Pratt & Whitney R-1340-AN-1 engine, the HRP-1 had a fabric-covered fuselage seating 10; the similar HRP-2 (Piasecki PV-17) had metal skin. Rotor diameter, 41 ft each; length, 54 ft; gross weight, 7,225 lb; max speed, 104 mph.

A Piasecki HUP-2 with forward hatch open for winching survivors from the sea.
(*US Navy*)

PIASECKI (VERTOL) HUP RETRIEVER

After trials with two prototype Model PV-14 XHJP-1 (37976-37977) helicopters for plane-guard and utility duties, the Navy ordered 32 similar PV-18 HUP-1s (124588-124594, 124915-124929, 126706-126715); with the 525 hp Continental R-975-34 engine and outrigger fins. Deliveries began in February 1949, and Squadron HU-2 received its first aircraft in January 1951. The 165 HUP-2s had an autopilot, no outer fins and 550 hp Continental R-975-42 engine. The HUP-2S had radar for anti-submarine duties. From the Army, the Navy acquired 50 HUP-3s (147582-147630; 149088) with R-975-46A engines. The HUP-2 and HUP-3 models became UH-25B and UH-25C respectively in 1962. Rotor diameter, 35 ft each; length, 32 ft; gross weight, 5,440 lb; max speed, 105 mph.

461

A camouflaged Piper NE-1 attached to Airship Squadron 32. (*Russell Ulrich*)

PIPER AE, HE, NE GRASSHOPPER

Primarily to serve at elimination training bases in World War II the Navy acquired 230 Piper NE-1s (26196–26425), basically similar to the Army L-4s with Continental O-170 engines. Twenty NE-2s (29669–29688) were similar. The Navy also acquired, in 1942, 100 HE-1 (30197–30296) ambulance versions of the Piper J-5C with Lycoming O-235-2 engines and capable of carrying one stretcher plus the pilot. These aircraft were redesignated AE-1 when the H designation was assigned to helicopters in 1943. Data for NE-1: span, 35 ft 3 in; length, 22 ft 3 in; gross weight, 1,100 lb; max speed, 95 mph.

Piper UO-1 for the Navy was an off-the-shelf purchase of the Aztec. (*Piper photo*)

PIPER UO

Twenty Piper Aztecs were purchased off-the-shelf by the Navy in February 1960 for use as utility transports. Powered by 250 hp Lycoming O-540-A1A engines, they differed from civilian models in having propeller anti-icing, an oxygen supply and additional radio. Delivered as UO-1s (149050–149069), they were later redesignated U-11A. Span, 37 ft; length, 27 ft 7 in; gross weight, 4,800 lb; max speed, 215 mph.

A Ryan NR-1 in non-standard colours, with polished metal fuselage and chrome yellow wings and tail. (*Ryan photo*)

RYAN NR RECRUIT

Following adoption of the Ryan ST-3 by the USAAC as a primary trainer the Navy ordered a batch of 100 on August 19, 1940. Designated NR-1 (4099–4198) they served until 1944; power was supplied by a 125 hp Kinner R-440-3 radial. The XNR-1 designation had previously been used for two Maxson twin-engined trainers and did not apply to prototypes of the Ryan design. Span, 30 ft 1 in; length, 22 ft 5 in; gross weight, 1,825 lb; max speed, 115 mph.

One of five US-built Viking OO-1s purchased by the Coast Guard. (*Howard Levy*)

SCHRECK/VIKING OO

In 1931 the Coast Guard bought a French-built Schreck FBA 17HT-4 flying-boat (No. 8, later V107), a wooden-hulled pusher derived from the FBA designs of World War I. When an Americanized version powered by a 240 hp Wright J-6-7 radial engine was built by the Viking Flying Boat Company, New Haven, Conn, the Coast Guard bought five under the designation OO-1 in 1936 (V152 to V156). Span, 43 ft 4 in; length, 29 ft 4 in; gross weight, 3,450 lb; max speed, 105 mph.

463

The Sikorsky XPBS-1, showing the nose and tail turrets. (*US Navy*)

SIKORSKY PBS

In the search for flying-boats of better performance and greater operational effectiveness, the Navy ordered a series of big, four-engined types during the thirties, starting with the Sikorsky XPBS-1 ordered on June 29, 1935. First flown on August 13, 1937, the XPBS-1 (9995) was powered by four 1,050 hp Pratt & Whitney XR-1830-68s and was the first American military aircraft to have a tail turret, as well as a nose turret and two waist guns. Span, 124 ft; length, 76 ft 2 in; gross weight, 48,540 lb; max speed, 227 mph.

A Sikorsky JRS-1 in service as a patrol bomber early in 1942. (*National Archives 80-G-12656*)

SIKORSKY JRS

Seventeen of these utility transport amphibians were purchased by the Navy between 1937 and 1939, including two for the Marine Corps. Similar to the commercial Sikorsky S-43, the all-metal JRS-1s had two 750 hp Pratt & Whitney R-1690-52 engines and seated up to 19. The Marine JRS-1s (1060, 1061) served with squadrons VMJ-1 and VMJ-2; the Navy had 15 (0504–0506, 1054–1059, 1062–1063, 1191–1194) including eight assigned to Utility Squadron One (VJ-1) at San Diego. Only one remained by the end of 1941. Span, 86 ft; length, 51 ft 1 in; gross weight, 19,096 lb; max speed, 190 mph.

An ex-Army Sikorsky R-4B serving the Navy in July 1947 as HNS-1. (*US Navy*)

SIKORSKY HNS

Following a Navy decision to procure helicopters for evaluation, made in July 1942, action was taken later the same year to obtain a single YR-4B from the Army, with the designation HNS-1 (46445); two more were requisitioned in March 1943 (46699–46700). The Navy accepted the first HNS-1 on October 30, 1943, and it reached the CGAS at Floyd Bennett Field a month later. To train helicopter pilots 22 more were obtained later (39033–39052, 75727–75728), the first 20 being ex-Army R-4Bs. The engine was a 200 hp Warner R-550-3. Rotor diameter, 38 ft; length, 35 ft 5 in; gross weight, 2,600 lb; max speed, 82 mph.

A Sikorsky HOS-1 at Floyd Bennett Field in 1946. (*Edgar Deigan*)

SIKORSKY HOS

Procurement of three Sikorsky XR-6As from the Army was initiated by the Navy in the second half of 1942, the designation XHOS-1 being assigned (46446–46448). First acceptance was in September 1944, the helicopter going to CGAS at Floyd Bennett Field for evaluation. A batch of 36 HOS-1s (75589–75624) was also acquired from Army R-6A production and used by Squadron VX-3. Power plant, one 240 hp Franklin O-405-9. Rotor diameter, 38 ft; length, 38 ft 3 in; gross weight, 2,590 lb; max speed, 96 mph.

Sikorsky HO3S-1 in all-blue finish. (*Harold G. Martin*)

SIKORSKY HO3S

Two three-seat Army R-5As were accepted by the Navy at Floyd Bennett Field in December 1945 for trials as HO2S-1s (75689–75690). Production orders were placed for developed four-seat versions of the same basic Sikorsky S-51 design, designated HO3S-1 and equivalent to the USAF's H-5Fs, with Pratt & Whitney R-985-AN-5 engines. Despite observation designation, the HO3S-1s served in a wide variety of roles, and gave outstanding service with Squadron HU-1 in Korea. Deliveries began in November 1946, and VX-3 received its first aircraft of this type in May 1947; at least 88 were procured (122508–122529, 122709–122728, 123118–123143, 124334–124353). One XHO3S-3 had a new rotor head and blades, in 1950. Rotor diameter, 49 ft; length, 44 ft 11½ in; gross weight, 5,500 lb; max speed, 103 mph.

An HO3S-1G, the first Coast Guard helicopter. (*Edgar Deigan*)

A Sikorsky HO5S-1G as used by the Coast Guard. (*J. Goodyear*)

SIKORSKY HO5S

First US helicopter to have metal rotor blades, the Sikorsky S-52 was developed as a two-seater with a 178 hp Franklin engine, flying for the first time on February 12, 1947. Developed into the four-seat S-52-2 with a 245 hp Franklin O-425-1 engine, it was ordered to replace the HO3S in Marine units, and deliveries to HMX-1 started in March 1952. Procurement of the HO5S-1 totalled 79 (125516–125527, 126696–126705, 128601 128620, 130101–130137); a few others served with the Coast Guard as HO5S-1Gs. Rotor diameter, 33 ft; length, 27 ft 5 in; gross weight, 2,700 lb; max speed, 110 mph.

A Coast Guard Sikorsky HH-52A demonstrating rescue technique. (*Sikorsky photo*)

SIKORSKY HH-52

The US Coast Guard selected a version of the commercial model S-62 in 1962 as a replacement for its HH-34 helicopters in the search and rescue role. Deliveries began in January 1963, and orders for the HH-52A version totalled 58 by early 1967. Powered by a 1,250 shp General Electric T58-GE-8 turboshaft engine, the HH-52A is fully amphibious and has several special features for the rescue role, including a folding platform along the side of the fuselage and a winch above the door. Rotor diameter, 53 ft; length, 44 ft 7 in; gross weight, 8,300 lb; max speed, 109 mph.

An all-yellow Spartan NP-1 as used at Reserve schools in 1940. (*Charles Schuler*)

SPARTAN NP

Spartan NP-1 biplane trainers were added to the Navy inventory by an order dated July 10, 1940, for 201 aircraft (3645–3845). They were modernized versions of the Spartan C-3 open three-seater produced by the Mid-Continent Aircraft Company in 1927 and were assigned to new Naval Reserve flight training schools opened in 1940 at NRAB Atlanta, NRAB Dallas and NRAB New Orleans. The engine was a 220 hp Lycoming R-680-8 radial. Span, 33 ft 9 in; length, 24 ft 3 in; gross weight, 3,006 lb; max speed, 108 mph.

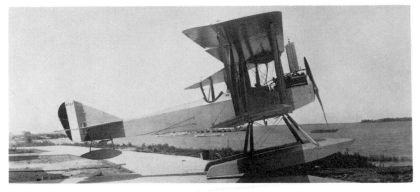

A Standard H-4-H at Langley Field, Hampton, Va. (*National Archives*)

STANDARD H-4-H

The Standard Aero Corporation was formed in 1916 in anticipation of America's eventual participation in World War I. First products at the company's Plainfield, NJ, plant were Model H-3 reconnaissance-trainers derived from earlier Sloan Aircraft models designed by Standard's chief engineer, Charles H. Day. First sales by Standard to the Navy were three improved H-4-H seaplane trainers (A137–A139) delivered in 1917. Power plant, 125 hp Hall-Scott A-5.

The Navy used two Stearman-Hammond JH-1s for radio-control experiments.
(*William T. Larkins*)

STEARMAN-HAMMOND JH

The Stearman-Hammond Y-1S was built as the result of a 1935 government design competition for a low-cost, safe and easy-to-fly aeroplane. Target cost was $700, but the production article cost approximately $7000. The Navy bought two of the all-metal pushers in 1937 as JH-1 utility aircraft (0908, 0909). The first tricycle landing-gear designs since the pre-World War I Curtiss pushers, the yellow-painted JH-1s were used for radio control experiments. Power plant, 150 hp Menasco XL-395-2. Span, 40 ft; length, 26 ft 11 in; gross weight, 2,150 lb; max speed, 130 mph.

The Coast Guard's Stinson RQ-1 with incorrect designation 'QR-1' on tail.
(*Fred E. Bamberger, Jr.*)

STINSON RQ, XR3Q

Two four-seat Stinson SR-5 Reliant cabin monoplanes were acquired in 1935, at a time when the Navy was acquiring commercial aircraft designs for light transport duties. One went to the Coast Guard as RQ-1 (381, later V149 and redesignated XR3Q-1). The other went to the Navy as XR3Q-1 (9718). The R2Q was a Fairchild design (see page 411). Power plant, 225 hp Lycoming R-680-6. Span, 41 ft; length, 27 ft 3 in; gross weight, 3,550 lb; max speed, 133 mph.

The Sturtevant S seaplane, serial number A81. (*US Navy*)

STURTEVANT S

The Sturtevant S, built by the Sturtevant Aeroplane Company, Boston, Mass., was a two-seat seaplane with all-steel frame, powered by a 150 hp Sturtevant 5-A engine. One was bought in 1916 as AH-24. When this was redesignated A76, five others were ordered as A77–A81. These were followed by another order for six (A128–A133). Because of the standardization of designs brought about by World War I, Sturtevant, along with other small manufacturers, ceased to develop its own designs and concentrated on building sub-assemblies for the major manufacturers. Span, 48 ft 7½ in; other data unavailable.

A Temco TT-1 used by the Navy to investigate jet trainers. (*US Navy*)

TEMCO TT PINTO

Developed as a private venture, the Temco Pinto primary jet trainer made its first flight on March 26, 1956, powered by a 920 lb s.t. Continental J69 turbojet. The Navy evaluated the prototype and ordered 14 as TT-1s (144223–144236) to conduct a full-scale study on the feasibility of using a jet aircraft for primary training. With J69-T-9 engines, they were delivered in 1957. Span, 29 ft 10 in; length, 30 ft; gross weight, 4,400 lb; max speed, 325 mph.

A Thomas-Morse SH-4 observation seaplane. (*Thomas-Morse photo*)

THOMAS-MORSE SH-4

The SH-4 was one of the first aircraft obtained by the Navy that had practical military experience behind its design, several predecessor models having been sold to Britain in 1915–16 by the Thomas Aeroplane Co before its merger with Morse Chain Company to form Thomas-Morse. Fourteen SH-4s (A134–A136, A396–A406), powered with 100 hp Thomas engines, were bought by the Navy as observation and trainer types in 1917. Span, 44 ft; length, 29 ft 9 in. Performance not available.

A Thomas-Morse S-4C serving with the Navy in September 1920, carrying original Army serial number but in Navy grey finish. (*US Navy*)

THOMAS-MORSE S-4B, S-4C

The S-4s were the Army's standard single-seat advanced trainers in 1917–18. The S-4B used the 100 hp American-built Gnome rotary engine while most of the later S-4Cs used the more reliable 80 hp Le Rhône. Principal recognition feature was the use of sweptback ailerons on the S-4B. The Navy's use of ten S-4Bs (A3235–A3244) and four S-4Cs (A5855–A5858) was for fighter-pilot training, and armament was principally a camera gun. S-4C: span, 26 ft 6 in; length, 19 ft 10 in; gross weight, 1,330 lb; top speed, 97 mph.

One of the six Thomas-Morse S-5 seaplanes, serial A762, in June 1918. (*US Navy*)

THOMAS-MORSE S-5

The six S-5s (A757–A762) were seaplane versions of the S-4B trainers built for the army and were delivered in 1917 to the Navy in Army olive drab colouring. The arrangement of the three wooden floats was copied from the British Sopwith Baby, which the S-5 closely resembled. Single-seat scout seaplanes did not work well for the Navy and further development was abandoned. Power plant, 100 hp American-built Gnome rotary engine. Span, 26 ft 6 in; length, 22 ft 9 in; gross weight, 1,500 lb; max speed, 95 mph.

One of eleven Thomas-Morse MB-3As used by Marine Corps. (*US Navy*)

THOMAS-MORSE MB-3

The MB-3, designed in 1918, was to be America's fighter contribution to World War I. The Army ordered 50 examples from the designing company Thomas-Morse in 1919 and 200 improved MB-3As from Boeing in 1920. Eleven of the MB-3s (A6060–A6070) were transferred to the Navy in 1921 for use by the US Marines as advanced trainers. Power plant, 300 hp Wright-Hispano. Span, 26 ft; length, 20 ft; gross weight, 1,818 lb; max speed, 152 mph.

472

A Timm N2T-1 trainer with plastic-bonded plywood fuselage. (*E. M. Sommerich*)

TIMM N2T TUTOR

Timm Aircraft Corp of Van Nuys, Calif., developed a form of construction using plastic-bonded plywood, known as Aeromold, and applied it to a two-seat primary training monoplane, the PT-160K. After official trials, a similar type was adopted by the US Navy as the N2T-1, similar to the commercial PT-175K. Orders were placed for 262 N2T-1s (05875–05876, 32387–32636, 39182–39191), this version having a 220 hp Continental R-670-4 engine. They were delivered in 1943. Span, 36 ft; length, 24 ft 10 in; gross weight, 2,725 lb; max speed, 144 mph.

Chance Vought XF6U-1 third prototype in production configuration. (*Howard Levy*)

VOUGHT F6U PIRATE

First jet fighter by Chance Vought, the Pirate made its first flight on October 2, 1946, powered by a 3,000 lb s.t. Westinghouse J34-WE-22. The Navy ordered three XF6U-1s (33532–33534) on December 29, 1944, and accepted the first of 30 production model F6U-1s (122478–122507) in August 1949. The F6U-1 had a 4,225 lb s.t. (with afterburner) J34-WE-30 engine; one was modified to F6U-1P (122483) with cameras, and 35 more on order were cancelled. Span, 32 ft 10 in; length, 37 ft 7 in; gross weight, 12,570 lb; max speed, 564 mph at 20,000 ft.

473

The Waco XJW-1 with trapeze hook for operation from the USS *Macon*.

WACO XJW, J2W

The Waco Aircraft Company, Troy, Ohio, was one of the leading American manufacturers of open-cockpit training and sports biplanes since 1926. The name Waco resulted from use of the initials of a previous company, the Weaver Aircraft Company, as an acronym. In 1934 two standard Waco UBF three-seat open-cockpit biplanes were modified and delivered to the Navy as XJW-1 (9521, 9522) utility aircraft for the airship USS *Macon*. These were fitted with hooks for engaging the trapeze on the airship, and also served as trainers for checking out new pilots in hook-on procedures.

Three 1936 cabin model EQC-6s were procured for the Coast Guard as J2W-1 (V157–V159). Early in World War II, three additional cabin Wacos were acquired from private owners, but as in the case of other Navy procurements of this type, they were not given standard Navy designations. J2W data: power plant, one 320 hp Wright R-760-E2. Span, 35 ft; length 26 ft 2 in; max speed, 176 mph.

One of the three Waco J2W-1s used by the Coast Guard. (*Joseph Nieto*)

A Vultee SNV-2, with reversed designation/serial number on tail. (*A. U. Schmidt*)

VULTEE SNV VALIANT

Procurement of 1,350 SNV-1s (02983–03182, 05675–05874, 12492–12991, 34135–34584) by the Navy began with an order dated August 28, 1940. These Vultee Model 74s had Pratt & Whitney R-985-AN-1 engines and were equivalent to the Army Air Corps BT-13As. The Navy also obtained 650 SNV-2s (44038–44187, 52050–52549), which differed only in having 24-volt electric systems and were equivalent to the BT-13B. Span, 42 ft 2 in; length, 28 ft 8 in; gross weight, 4,360 lb; max speed, 166 mph.

The Wright F3W-1 in 1926 with original engine and wings. (*US Navy*)

WRIGHT F3W

The Wright Aeronautical Company, mainly a builder of engines, produced the experimental F3W-1 (A7223) Apache fighter in 1926. Originally powered with the unsuccessful 450 hp Wright P-1 radial engine, it was used by the Navy, ironically, as a test-best for the competing 420 hp Pratt & Whitney Wasp engine. Span, 27 ft 4 in; length, 22 ft; gross weight, 2,128 lb; max speed, 162 mph.

A Wright Model C Seaplane operated by the Navy as its B series. (*Wright photo*)

WRIGHT SEAPLANES

While the Wright Brothers had invented the first practical aeroplane and were able to maintain a lead over other fliers for nearly seven years, they clung stubbornly to their original aerodynamic and structural concepts and soon found themselves passed by competitors who had adopted more advanced ideas. Orville Wright carried on the business after the death of Wilbur in 1912 and was able to sell three two-seat Model C-H seaplanes to the Navy, which identified them as B-1 to B-3 (later AH-4 to AH-6). These were similar to the open Model B of 1911, with chain-drive to two pusher propellers from a single 60 hp Wright engine, and were sometimes flown as landplanes to improve their marginal performance.

One 1914 Wright Model G Aeroboat was tested by the Navy as AH-19, but was rejected for poor performance. The last Wright purchased from the original company was the full-fuselage Model K tractor seaplane (AH-23, later A51), still with chain-drive and warping wings. C-H data: span, 38 ft; length, 29 ft 9 in; gross weight, 1,610 lb; max speed approx 50 mph.

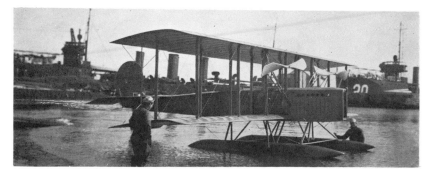

The Wright Model K, serial A51, with chain-drive to propellers. (*US Navy*)

FOREIGN AIRCRAFT

Included in this section are details and illustrations of aircraft of foreign design and manufacture purchased by or used on loan to the US Navy. The majority are French and British designs used by US Naval forces in Europe during World War I. These aircraft were not assigned regular US Navy serial numbers, but operated with the serials of the countries that supplied them. Only those examples shipped to the US during or after the war for test and evaluation were given Navy serial numbers. This practice applied to war-booty aircraft acquired for test and the numerous European types purchased from the manufacturers in the early 1920s.

In the between-war years, the Navy bought several foreign aircraft for the use of its Naval Attachés in Europe. These were given serial numbers in their proper sequence and were operated under the original manufacturer's designations. Unfortunately, illustrations of some of these interesting types in their Naval markings are not available.

During World War II a large number of Canadian-built aircraft were added to the Navy inventory, but these were American designs built in Canada, mostly with lend–lease funds, and carried standard American designations. These are not regarded as foreign aircraft.

The US Navy's first Blackburn Swift. (*US Navy*)

BLACKBURN SWIFT

The Blackburn Swift was developed as a British carrier-based torpedo-plane shortly after World War I, with a three-bank 450 hp Napier Lion engine. The Navy bought two (A6056, A6057) for evaluation in 1921. Span, 46 ft; length, 35 ft 6 in; gross weight, 6,000 lb; max speed, 115 mph.

The second Bristol Bulldog supplied to the US Navy. (*US Navy*)

BRISTOL BULLDOG

The Bristol Bulldog was a standard British fighter of the early 1930s; the Navy bought one Bulldog IIA (A8485) for evaluation in November 1929, but it crashed on the first flight. A second Bulldog IIA (A8607) was purchased in April 1930. Power plant was a 500 hp Bristol Jupiter VIIF. Span, 33 ft 10 in; length, 24 ft 10 in; wing area, 306·5 sq ft; gross weight, 2,183 lb; max speed, 140 mph.

A US Navy Caproni Ca-44 serving with the Northern Bombing Group Night Wing at Orly. (*USAAF photo*)

CAPRONI Ca-44

In 1918 the Navy procured 19 Italian-built Caproni Ca-44 bombers for use by the Northern Bombing Group. This organization had been trained by the British, and its members had obtained operational experience in British squadrons. The first independent action of the group was with Capronis on August 15, 1918, when they were used against German installations at Ostend. The unique twin-fuselage Caproni was typical of several similar models then in production, but the Ca-44 was powered with six-cylinder Fiat engines that proved to be especially troublesome and seriously handicapped operations. Span, 76 ft 10 in; length, 41 ft 2 in; gross weight, 12,350 lb; max speed, 103 mph.

One of two Caspar U-1 seaplanes tested by the Navy. (*US Navy*)

CASPAR U-1

In 1922, when the Navy was interested in extremely small scouting types for operation from submarines and destroyers, it ordered two examples of the all-wood German Caspar U-1 (A6434, A6435). Powered with a 60 hp Siemens air-cooled radial engine, the U-1 was of cantilever construction and could be disassembled quickly for stowage aboard a submarine. Span, 23 ft 9 in; length, 20 ft 10 in; gross weight, 1,125 lb; max speed, 90 mph.

479

A de Havilland 9A as used by the Navy in France. (*Imperial War Museum*)

DE HAVILLAND D.H.9A

Among the British aircraft supplied to the US Navy for use by the Northern Bombing Group in France in 1918 were 54 D.H.9A two-seat bombers powered by 400 hp American Liberty engines. These were not assigned regular US Navy serial numbers, but flew with American roundels being painted over the British on the wings and the fuselage roundel being painted out and replaced by station markings. Span, 42 ft 5 in; length, 30 ft; gross weight, 4,645 lb; max speed, 114 mph at 10,000 ft.

A de Havilland 60 Cirrus Moth of the type used by the Naval Air Attaché in London in 1927. (*Gordon Swanborough*)

DE HAVILLAND XDH-60

The famous British de Havilland D.H.60 Moth was introduced in February 1925, and soon became the most widely used club and training aeroplane in Britain. Powered with a 60 hp Cirrus air-cooled engine, the early Moths were of wooden construction. Needing an economical aeroplane for use by the Naval Air Attaché in London, the US Navy bought one Moth in 1927 and assigned the designation XDH-60 (A7564). Span, 30 ft; length, 23 ft 8 in; gross weight, 1,550 lb; max speed, 102 mph.

The de Havilland XDH-80 attached to the US Embassy in London. (*Flight photo*)

DE HAVILLAND XDH-80

The de Havilland D.H.80 Puss Moth was a notable monoplane follow-on to the ubiquitous D.H.60 Moth. A compact three-seater with folding wings, it was powered with a 130 hp de Havilland Gipsy Major inverted air-cooled engine. The Navy bought one example in 1934 as a replacement for the Moth used by the Naval Attaché in London, and designated it XDH-80 (A8877). This remained in Navy service until impressed into the RAF in 1939. Span, 35 ft 9 in; length, 25 ft; gross weight, 1,900 lb; max speed, 128 mph.

A de Havilland U-1B from NAS Quonset Point in June 1966. (*Denis Hughes*)

DE HAVILLAND CANADA UC OTTER

Examples of the de Havilland Otter, purchased in large numbers by the US Army as U-1s, were first ordered by the Navy in 1955, primarily to serve in the Antarctic. The first four UC-1s (142424–142427) were assigned to VX-6 there in 1956. Procurement of a further 14, later designated U-1B, included some for other countries (144259–144261, 144669–144674, 147574–147576, 148322–148323). Power plant, one 600 hp Pratt & Whitney R-1340-AN-3. Span, 58 ft; length, 41 ft 10 in; gross weight, 7,600 lb; max speed, 160 mph.

A de Havilland L-20 Beaver in Navy markings.

DE HAVILLAND CANADA L-20 BEAVER

At least one de Havilland Beaver was operated by the Navy in the early 1960s, having been acquired from the US Army, which operated the type in large numbers. Powered by a single 450 hp Pratt & Whitney R-985-AN-1 engine, the eight-seat Beaver was a product of the Canadian de Havilland company which flew the first civil prototype on August 16, 1947. Span, 48 ft; length, 30 ft 3 in; gross weight, 5,100 lb; max speed, 163 mph.

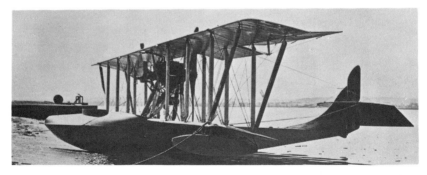

A Donnet-Denhaut flying-boat in the US. (*US. Navy*)

DONNET-DENHAUT FLYING-BOAT

US Navy patrol operations along the coast of France in World War I, initiated with Tellier flying-boats (page 469) were supplemented by larger quantities of French Donnet-Denhaut flying-boats, which were directly comparable designs with 200 hp Hispano-Suiza pusher engines, wooden hulls, and accommodation for two or three crew members. Out of a total of 58 Donnets procured by 1918, two were sent to the US, where they acquired Navy serial numbers A5652 and A5653. Span, 53 ft 7 in; length, 35 ft 5 in; gross weight, 3,860 lb; max speed 72 mph.

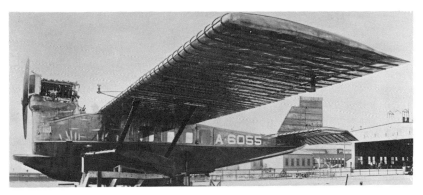

The single Dornier CsII tested by the Navy. (*US Navy*

DORNIER CsII

The Dornier CsII was an early post-war attempt by the German company to develop an all-metal commercial flying-boat. In spite of its rather odd configuration, with the pilot in an open cockpit behind the engine and the passengers in a cabin under the wing, it developed into the successful Delphin series. The US Navy bought one example (A6055) to study its metal construction in 1920. Power plant was a 185 hp BMW IIIa. Span, 55 ft 10 in; length, 33 ft 7 in; wing area, 505 sq ft; gross weight, 4,400 lb; max speed 93 mph.

An Italian built FBA Model H. (*FBA photo*)

FBA TYPE H

The Navy operated 11 FBA (Franco-British Aviation) flying-boats, of French manufacture, from its bases in France in 1918 and at least six others of Italian manufacture from its base at Porto Corsini, Italy. These were three-seat reconnaissance and light bomber aircraft with 180 hp Hispano-Suiza engines. Span, 47 ft 7 in; length, 33 ft 2 in; gross weight, 3,218 lb; max speed, 90 mph.

A Fokker D.VII used by the Marine Corps. (*US Navy*)

FOKKER D.VII

After the Armistice, the Army brought 142 Fokker D.VIIs into the United States. These were the finest German fighters in service at war's end, and were used in numbers throughout the Air Service as trainers for several years. Twelve were to have been transferred to the Navy for use by the Marine Corps, but only six (A5843–A5848) were used. These remained in the training role at Quantico, Virginia, until 1924. The first influence of the D.VII on subsequent US Naval aircraft design was to be noted in the Boeing FB-1 of 1925 (page 52). Power plant was a 180 hp Mercedes or 185 hp BMW. Span, 29 ft 3 in; length, 23 ft; gross weight, 1,993 lb; max speed, 124 mph.

One of three Fokker C.Is at Quantico in 1922. (*US Navy*)

FOKKER C.I

In 1921 the Navy purchased three Fokker C.Is from the Dutch Fokker factory. Actually, these had been built in Germany at the close of World War I and had been taken into Holland when Fokker returned to his native land and established a new company. The C.I was essentially a D.VII with slightly lengthened fuselage and wings, and a 185 hp BMW. The C.Is (A5887–A5889) were used by the US Marines at Quantico. Span, 34 ft 10 in; length, 23 ft 8 in; gross weight, 2,576 lb; max speed, 112 mph.

A Fokker FT-1 at NAS Hampton Roads in February 1925. (*US Navy*)

FOKKER FT-1

During the early post-war years, when it was experimenting with European aircraft, the Navy bought three Dutch Fokker T.III torpedoplanes and assigned the designation of FT-1 for Fokker Torpedo (A6008–6010). A twin-float seaplane, the FT-1 provided interesting comparison with such contemporary American Navy monoplanes with cantilever wings as the Curtiss CT (page 425) and the Martin MO-1 (page 450). Although Dutch-built, the FT-1 used the American Liberty engine. Span, 65 ft; length, 51 ft; gross weight, 7,300 lb; max speed, 104 mph.

One of three Handley Page HPS-1s before delivery to the US Navy. (*Crown Copyright*)

HANDLEY PAGE HPS

The British Handley Page HP 21 was a radical innovation for a shipboard aircraft in 1921. It was far ahead of its time aerodynamically and featured full-span leading edge slots and trailing edge flaps. The US Navy ordered three (A6402–A6404) and assigned the designation of HPS-1 for Handley Page Scout at the time when it was buying numbers of European aircraft for evaluation. Power plant, 230 hp Bentley BR.2 rotary. Span, 29 ft 3 in; length, 21 ft 5 in; gross weight, 1,995 lb; max speed 120 mph.

Hawker Siddeley AV-8A of the first USMC Harrier squadron, VMA-513.
(*Hawker Siddeley photo*)

HAWKER SIDDELEY AV-8 HARRIER

Selection of the Harrier by the US Marine Corps marked the first export sale achieved by the British VTOL strike fighter that emerged from a six-year development programme starting with the original P.1127 prototype that was first hovered on October 21, 1960. The US participated, with Britain and Germany, in a tri-partite evaluation of a batch of improved P.1127s known as Kestrels, and six of these aircraft were subsequently transferred to the USA for continued testing under the designation XV-6A (after the earlier allocation of the designation VZ-12 to two Kestrels, not taken up).

Primary mission for which the Harrier was selected by the US Marine Corps was close air support, with some interdiction and reconnaissance capability. The Harrier's ability to operate from the decks of ships (not necessarily aircraft carriers) or from rudimentary landing sites made it especially attractive for the support of ground forces in beach-head assaults. The first Marine Corps order was placed in 1969 and contracts eventually totalled 102 single-seat AV-8As and eight two-seat TAV-8As; in addition, the US government contracted during 1973 to purchase six AV-8A and two TAV-8A on behalf of the Spanish Navy. The AV-8 designation, incidentally, did not conform strictly to the Defense Department procedure, it being 'A-8' as an attack aircraft with 'V' added as a type symbol, whereas correct application of the system would have placed it in the 'V' series numerically, with 'A' added as a missions symbol.

Deliveries of the AV-8A to the USA began in January 1971, following a formal hand-over in the UK on November 20, 1970, and the first of three designated Harrier squadrons, VMA-513, began working up on the type at Beaufort in April 1971. The first 10 aircraft were delivered with engines of Pegasus 102 standard rated at 20,000 lb st and designated F402-RR-400; subsequently they were up-rated to Pegasus 103 standard rated at 21,500 lb st and designated F402-RR-401, and all later aircraft had this engine also. The second and third USMC squadrons are VMA-542 and VMA-231.

TECHNICAL DATA (AV-8A)

Manufacturer: Hawker Siddeley Aviation Ltd, Kingston-upon-Thames, Surrey.

Type: V/STOL close support and reconnaissance aircraft.

Accommodation: Pilot only or (TAV-8A) two in tandem.

Power plant: One 21,500 lb s.t. Rolls-Royce F402-RR-401 Pegasus turbofan.

Dimensions: Span, 25 ft 3 in; span (with ferry tips), 29 ft 8 in; overall length, 45 ft 6 in; overall length (TAV-8A), 55 ft 9½ in; height, 11 ft 3 in; height (TAV-8A), 13 ft 8 in; wing area, 201·1 sq ft; wing area (ferry tips), 216 sq ft.

Weights: Operating weight empty, with crew, 12,200 lb; max take-off weight (VTO), 17,930 lb; max take-off weight (STO), 21,150 lb.

Performance: Max speed, over 730 mph EAS at low level; speed with typical external loads, 650 mph at 1,000 ft; typical cruising speed, 560 mph at 20,000 ft; tactical radius (hi-lo-hi), 260 miles; operational ceiling, over 50,000 ft.

Armament: Provision for two 30 mm gun pods under fuselage; four wing hard points for bombs. rocket pods, Sidewinder AAMs, Bullpup ASMs, up to a total of 5,300 lb.

Serial numbers: AV-8A: 158384–158395; 158694–158711; 158948–158977; 159230–159259; 159366–159377. TAV-8A: 159378–159385.

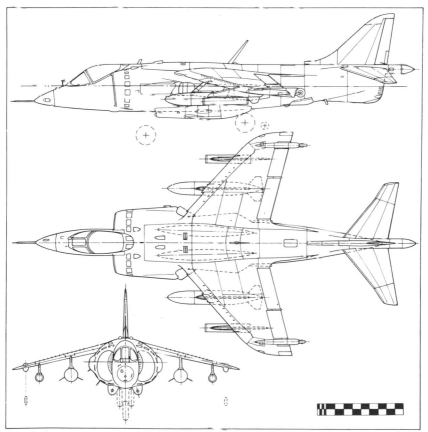

487

A Hanriot HD-1 taking off from platform over gun turret on USS *Mississippi.*
(*Bowers Collection*)

HANRIOT HD-1, 2

The Navy acquired 26 French Hanriot HD-2 seaplane fighters for use in Europe during World War I. Except for the 130 hp Clerget engine and a large rudder, these were seaplane versions of the well-known Hanriot HD-1 (for Hanriot-Dupont) landplane. After the Armistice 10 HD-2s were shipped to the US Naval Aircraft Factory for reconditioning and conversion to HD-1 landplanes, and were assigned serials A5620–A5629. Fitted with flotation bags and hydrovanes ahead of the wheel undercarriage, they were flown from battleships. Span, 28 ft 6 in; length, 19 ft 8 in; gross weight, 1,521 lb; max speed, 115 mph.

A Junkers-Larsen JL-6 floatplane during evaluation at NAS Anacostia, in July 1920.
(*US Navy*)

JUNKERS-LARSEN JL-6

The Junkers-Larsen JL-6 was an Americanized version of the German Junkers-F 13, a revolutionary all-metal single-engine airliner developed in 1919. Powered by 243 hp BMW IIIa engines, three were acquired by the Navy (A5867–A5869) and evaluated in the utility role on both wheels and floats. Span, 48 ft 6 in; length, 31 ft 6 in; gross weight, 3,644 lb; max speed, 101 mph.

A Levy-Lepen HB-2 flying-boat as used by the Navy. (*S.A.F.A.R.A.*)

LEVY-LEPEN HB-2

The Navy operated 12 French two/three seat Lévy-Lepen HB-2 reconnaissance flying-boats from its Le Croisac base in France in 1918. These were in operation from June 7, 1918, until the Armistice. Three were brought to the US and were assigned Navy serial numbers A5650–A5651 and A5657. Power plant, 300 hp Renault. Span, 60 ft 8 in; length, 40 ft 8 in; gross weight, 5,181 lb; max speed, 93 mph.

A Macchi M.5 during tests at Hampton Roads, Va, in 1917.
(*National Archives SC1278*)

MACCHI M.5

During its operations from the Italian base of Porto Corsini from July 1918 until the Armistice, the US Navy operated at least eight single-seat Italian Macchi M.5 fighter flying-boats. These were used to escort bombers and with the French Hanriots (see page 486) were the only seaplane fighters flown in combat by American pilots in either of the World Wars. One M.5 had been sent to the US for evaluation in 1917. Power plant, 160 hp Isotta-Fraschini V-4B. Span, 39 ft; length, 26 ft 2 in; gross weight, 2,266 lb; max speed, 118 mph.

489

A Macchi M.8 flying-boat as used by the Navy in Europe. (*Macchi photo*)

MACCHI M.8

At least eight Italian Macchi M.8 two-seat light bombing flying-boats were used by US Naval forces operating from Porto Corsini on the Adriatic sea, which became an active station in July 1918. These two-seaters were used several times for bombing raids against the Austrian bases at Pola and were highly successful in this little-known phase of Naval aviation history. Power plant, 160 hp Isotta-Fraschini V-4B. Span, 45 ft 5 in; length, 29 ft 7 in; gross weight, 3,100 lb; max speed, 103 mph.

A Macchi M.16 floatplane in the US. (*US Navy*)

MACCHI M.16

The Italian Macchi M.16 was an ultra-light messenger type aeroplane developed during 1918. Powered with a 30 hp Anzani air-cooled radial engine, it could be used on wheels or twin floats. The US Navy acquired three (A6005–A6007) during 1922 for evaluation with other American and foreign ultra-lights in the Ships' Scout role. The wire bracing of the M.16 wings and associated rigging problems ruled it out of consideration for stowage aboard submarines. Span, 19 ft 8 in; length, 13 ft 10 in; gross weight, 572 lb; max speed, 82·5 mph.

490

A Morane-Saulnier AR-1 trainer in Navy service. (*US Navy*)

MORANE-SAULNIER AR-1

The Morane-Saulnier was a simple low-cost training plane developed from the famous Morane Parasols of 1914–15, and was powered with the reliable war-surplus 80 hp Le Rhône rotary engine. The Navy bought six (A5976–A5981) in 1921. Although no longer used by the military, the AR-1 remained in civil club and school use in France into the 1930s. Span, 34 ft 7 in; length, 22 ft 2 in; gross weight, 1,380 lb; max speed, 82 mph.

A 1920 photo of a Nieuport 28 with British Grain flotation gear installed. (*US Navy*)

NIEUPORT 28

Although a fighter of French manufacture and widely used by the AEF, the Nieuport 28 was not used by US Naval forces in France during World War I. Twelve were obtained in August 1919, however, from the supply that the Army had brought back to the US, and assigned serials A5794–A5805. These were assigned to the fleet and flew from platforms built over the forward turrets of battleships, eight of which were so equipped. Flotation bags were fitted, along with hydrovanes to prevent nosing over in case of a landing at sea. Power plant was a 160 hp Gnome. Span, 26 ft 3 in; gross weight, 1,625 lb; max speed, 122 mph.

A ski-equipped Noorduyn JA-1 *en route* to the Antarctic in 1946. (*US Navy*)

NOORDUYN JA-1 NORSEMAN

The Navy acquired three of these eight-seat Canadian transports (57992–57994), each powered by a 600 hp Pratt & Whitney R-1340-AN-1 engine. Span, 51 ft 6 in; length 32 ft 4 in; gross weight, 7,440 lb; max speed, 162 mph.

A US Navy Parnall Panther with Grain flotation gear. (*Fred C. Dickey, Jr.*)

PARNALL PANTHER

Two of these British carrier-based fighters were purchased by the Navy in 1919 (A5751–A5752). Powered by the 230 hp Bentley BR.2 rotary, they had hydrovanes and flotation gear. Span, 29 ft 6 in; length, 25 ft; gross weight, 2,550 lb; max speed, 114 mph.

492

An S.E.5A with British serial number but US Navy markings. (*US Navy*)

R.A.E. S.E.5A

British-built S.E.5A fighters were among the foreign types acquired by the US Navy in 1918 in France. They served in original British colours, with the RAF roundels overpainted in American colour sequence and at least one S.E.5A in these markings operated off a turret on the USS *Mississippi* after the end of the War. Power plant, one 200 hp Hispano-Suiza. Span, 26 ft 7½ in; length, 20 ft 11 in; gross weight, 2,058 lb; max speed, 123 mph.

An Italian-built Fokker C-VE as used by the US Naval Attaché in Rome

ROMEO Ro 1

The Dutch Fokker C.VE of 1926 was an extremely versatile military two-seater and was licensed to several other manufacturers throughout Europe. Among these was the Officine Ferroviarie Meridionali, better known as Romeo, of Naples, Italy. It produced the C.VE under the designation of Ro 1 and powered it with an Italian-built version of the 450 hp Bristol Jupiter engine. The US Navy purchased an Ro 1 (A7565) for the use of the US Naval Attaché in Rome in 1928. Span, 50 ft 2 in; length, 31 ft 2 in; gross weight, 5,004 lb; max speed, 141 mph.

A Sopwith Baby seaplane, with original British serial N3709 and Navy markings.
(*US Navy*)

SOPWITH BABY SEAPLANE

An unspecified number of British Sopwith Baby seaplanes, including Gnome- and Clerget-powered versions, were obtained by the US Navy in Europe in 1917–18 for training. Four (A869–A872) were sent to the States for evaluation. Those remaining in Europe continued to operate under their British serial numbers. The single-seat seaplane scout design, with a 110 hp Clerget engine, was investigated by a few American manufacturers, but the type was not accepted for service. Span, 25 ft 8 in; length, 22 ft 10 in; gross weight, 1,580 lb; max speed, 92 mph.

A Navy Sopwith Camel in France. (*US Air Force*)

SOPWITH CAMEL

The Sopwith Camel was one of the most successful British fighters of 1917–18, and numbers were supplied to both the US Army and Navy for use in France. Mounting two 0·303-in machine guns, the Camel was powered by the 130 hp Clerget rotary engine. After the Armistice, the Navy operated six (A5658–A5659, A5721–A5722, A5729–A5730), sometimes from platforms built over the forward turret guns of battleships. Span, 28 ft; length, 18 ft 9 in; gross weight, 1,453 lb; max speed, 113 mph at 6,500 ft.

494

A Sopwith 1½-Strutter at Guantanamo, with hydrovanes for emergency water alighting. (*US Navy*)

SOPWITH 1½-STRUTTER

Among the Navy's British purchases during World War 1 were two-seat Sopwith 1½-Strutters, widely used as observation and training types by both France and Britain. The one example shipped to the States received serial A5660. An additional 21 obtained from the US Army after the war became A5725–A5728 and A5734–A5750. These were used as light observation types. Power plant was a 130 hp Clerget. Span, 33 ft 6 in; length, 25 ft 3 in; gross weight, 2,150 lb; max speed, 100 mph.

A Tellier flying-boat in France. (*US Navy*)

TELLIER FLYING-BOAT

Using some of the 34 French-built Tellier flying-boats, the Navy began patrol operations from Le Croisac, France, on November 18, 1917. With the 200 hp Hispano-Suiza engine, these aircraft were operated with French serial numbers and only the one example sent to the US for evaluation received a Navy serial number, A5648. Span, 51 ft 3 in; length, 38 ft 10 in; gross weight, 3,745 lb; max speed, 75 mph.

The US Navy's sole Vickers Viking amphibian. (*US Navy*)

VICKERS VIKING

One British Vickers Viking IV flying-boat amphibian (A6073) was procured in 1921 for evaluation. The Navy had had little previous experience with the type other than adapting retractable wheels to established flying-boat designs (see page 112) or fitting wheels into the floats of seaplanes. The Viking, powered by a 450 hp Napier Lion, was procured at a time when the Navy was buying a variety of European designs. Span, 50 ft; length, 33 ft 6 in; gross weight, 5,600 lb; max speed, 111·2 mph.

The Wright WP-1, built by Dornier in Switzerland. (*US Navy*)

WRIGHT WP-1

Designed in Germany as a result of experience with the all-metal D I biplane fighter of 1918, the Dornier Falke was built by the Swiss branch of the company because of the restrictions then in effect on the construction of military aircraft in Germany. Fitted with a 300 hp Wright H-3 (licence-built Hispano-Suiza) engine, the Falke was imported into the US by the Wright Aeronautical Company and entered in a 1923 Navy fighter competition as the WP-1 (A6748) for Wright Pursuit. Although possessed of remarkable performance, the WP-1 was not adopted because the Navy was not yet ready for either monoplanes or all-metal construction. Span, 32 ft 10 in; length, 24 ft 6 in; gross weight, 2,765 lb; max speed, 162 mph.

APPENDIX C

GLIDERS

From the late 1920s, when experiments were conducted in releasing man-carrying gliders from the airship *Los Angeles*, to the end of World War II, the Navy carried on a small-scale glider programme. This received impetus in 1941 after the early German successes with military gliders during the war. The Navy followed the Army's lead in this line and adopted standard Army models for training. These were given regular Navy type designations such as N for trainer or R for cargo, but the designations were prefixed by the letter L, one of the few letters not already in use as a type designation, to identify the designs as gliders.

At first, some were merely established two-seat commercial sailplane designs produced with no change other than markings for use by the military. The great performance difference between the sailplane-type trainers and the low-performance cargo-type gliders that the pilots would fly in service soon led both services to abandon the sailplanes in favour of liaison-type aeroplanes from which the engine had been removed and replaced with an extra forward seat. These had gliding and landing characteristics similar to those of the cargo models.

Although widely used in Europe as the first stage in teaching potential military pilots to fly, the gliders were little used for primary instruction by the US services. Flight fundamentals were taught much more quickly in powered aircraft and the training gliders were used mainly for transition training.

While the Navy used mostly Army models of both training and cargo gliders, it undertook the development of specialized seaplane gliders on its own. Basic designs were worked out by the Bureau of Aeronautics and then turned over to industry for building of test and production models.

A Franklin PS-2 in flight over NAS Pensacola. (*Capt Ralph S. Barnaby*)

FRANKLIN PS-2

American industry soon improved on the German secondary gliders imported in the early 1930s and developed models specifically designed to American conditions. The Franklin PS-2 (for Primary–Secondary) featured a steel-tube fuselage and a built-in single landing wheel. The Navy ordered six (9401–9402, 9614–9617) for evaluation and training without assigning standard designations. Span, 36 ft; wing area, 180 sq ft; gross weight, 400 lb; glide ratio, 15:1.

The first production Pratt-Read glider bearing the designation XLNE-1.
(*Capt Ralph S. Barnaby*)

PRATT-READ LNE-1

The LNE-1 was an all-wood training sailplane with side-by-side seating. After test of a civil-registered prototype in 1942, the Navy bought it (31505) and ordered 100 production versions as LNE-1 (31506–31585, 34115–34134). The letter E, already in use by Piper, was assigned to this new manufacturer before Piper became a Navy glider supplier. When withdrawn from service, the LNE-1s were turned over to the Army as TG-32s prior to surplus sales. Span, 54 ft 6 in; wing area, 230 sq ft; gross weight, 1,150 lb; glide ratio, 26:1.

The first Navy glider, a German Prufling. (*National Archives 80-CF-421-4*)

PRUFLING

The Navy's first glider was a German-built Prufling secondary trainer (A8546) bought from the American Motorless Aviation Corporation glider school at the instigation of Lt Ralph Barnaby for experiments in launching gliders from airships. This was an open-cockpit wooden single-seater, used without a regular naval designation. The first release from the airship *Los Angeles* was made by Barnaby on January 31, 1930. Span, 32 ft 8 in; length, 17 ft 9 in; gross weight, 400 lb; glide ratio, 15:1.

A Schweizer LNS-1 during take-off. (*Schweizer photo*)

SCHWEIZER LNS-1

In 1941 the Army tested the 1938 model Schweizer 2-8 all-metal sail-plane, manufactured in Elmira, NY, as the XTG-2 for training glider pilots. Production orders for 39 TG-2s and TG-2As, each with its own transportation trailer, followed. In 1942 the Navy ordered 13 as LNS-1 (02979–02980, 04380–04389, 26426) for the new Marine Corps glider training programme at Cherry Point, NC. Colour was trainer yellow with polished metal wing leading edges. Span, 52 ft; length, 23 ft 3 in; wing area, 214 sq ft; gross weight, 860 lb; glide ratio, 23:1.

A Taylorcraft TG-6 of the type acquired by the Navy. (*USAF photo*)

TAYLORCRAFT LNT-1

In 1943 the Navy followed the Army's lead in abandoning sailplane-type gliders for primary and familiarization training and adopted modified lightplanes from which the 65 hp Continental engines had been removed and replaced by an extra seat in the nose to maintain balance. The Navy bought 35 Army Taylorcraft TG-6s, 10 as XLNT-1 (36428–36430, 67800 - 67806) and 25 as LNT-1 (87763–87787). These had sinking speeds and glide ratios more compatible with the cargo-type gliders the pilots were eventually to fly. Span, 35 ft 5 in; length, 25 ft 2 in; gross weight, 1,260 lb; aeroplane towing speed, 140 mph.

A Waco LRW-1 used at the NAF to develop automatic pilots for towed gliders.
(*Capt Ralph S. Barnaby*)

WACO LRW-1

In 1943 before abandoning its glider programme, the Navy acquired a total of 13 Army Waco CG-4A troop-transport gliders under the designation of LRW-1 (37639–37648, 44319, 69990–69991). These could accommodate 15 fully-armed troops or carry one jeep or a 75-mm field gun, which could be loaded and unloaded through the hinged-up nose section. Troops normally loaded through side doors at the rear of the cabin. Span, 83 ft 8 in; length, 48 ft 4 in; gross weight, 7,500 lb; aeroplane towing speed, 125 mph.

One of the two Allied XLRA-1 prototypes built at the instigation of the
Navy for operation off the water. (*Edgar Deigan*)

EXPERIMENTAL TRANSPORT GLIDERS

Before abandoning the idea of transporting personnel and cargo in gliders, the Navy developed a unique wooden 12-seat flying-boat transport glider in which the low wing, resting in the water, provided the necessary lateral support when afloat. Two XLRA-1 prototypes (11647–11648) were built by Allied Aviation Corporation, and of four XLRQ-1s ordered from Bristol Aeronautical Corporation, two were built (11651–11652). Production orders for 100 from each firm were cancelled, along with two further prototypes which were to have been built by the Navy. Even larger amphibious gliders, with twin hulls and accommodating 22 troops, were ordered in 1942 from AGA Aviation (XLRG-1) and Snead & Co (XLRH-1) but were not built. Data for XLRA-1: span, 72 ft; length, 40 ft.

A Bristol XLRQ-1 flying-boat glider to the same specification as the XLRA-1 at the top of the page. The wing roots provided lateral stability on the water.
(*Capt Ralph S. Barnaby*)

APPENDIX D

BALLOONS AND AIRSHIPS

COMPARED to other major powers and the US Army, the Navy made a small and late start in airship activity when a contract for its first Blimp, the DN-1, was signed in 1915. This notably unsuccessful craft did not make its first flight until April 1917. By the end of World War I the Navy had a lighter-than-air arm nearly as good as those of England and France. In 1921 an agreement was worked out between the services whereby the Army Air Service took over responsibility for coastal patrol with non-rigid airships while the Navy concentrated on rigid scouting designs. The Navy was allowed only such non-rigids as were necessary to train the rigid airship crews or conduct experiments. Since the other nations allowed their airship fleets to phase out during the 1920s, this left the US Army supreme in the field with the US Navy in second place. In 1937 the Army transferred its existing airships to the Navy, which then became the world's exclusive operator of military airships. This fleet grew to 167 during World War II, the largest ever assembled, but was largely de-activated after V-J Day, with only two squadrons remaining in service by the 1950s.

Increasing transfer of the off-shore patrol function to long-range four-engined aeroplanes and other anti-submarine activity and utility work to helicopters eliminated the principal functions of the surviving airships in the post-war years. By 1960 only a few remained. Two of the latest fleet models were granted a brief reprieve from retirement when they were used as flying aeronautical research laboratories entirely unconnected with the development or deployment of airships.

While unique in their long endurance and ability to remain motionless over one spot, the airships suffered from some major disadvantages. Attack from enemy aircraft never became a problem, so the slow speed was not a handicap in this respect. However, airships were highly susceptible to the vagaries of weather in flight and on the ground. Their altitude range was quite limited and their reaction to strong thermal currents such as found over heated land masses made it logical to operate them in the more stable air found over the relatively constant-temperature sea. They also had the inherent 'gasbag' problems of lift variations due to sudden temperature changes; loss of gas through valving, leakage, or altitude change; the necessity to valve gas to compensate for weight lost through fuel consumption; and reduced military load resulting from the require-ment to carry enough disposable ballast to maintain the required lift-to-weight ratio through all foreseeable variations.

They also required large ground-handling crews during take-off and landing, which were a waste of manpower in that this specialized team work was required for relatively few hours a month. Finally, there was the shelter handicap, extremely large buildings with open approach areas being required for most service and maintenance work.

LIGHTER-THAN-AIR DESIGNATIONS

Disregarding spherical free balloons used for training, captive observation balloons and the unmanned barrage balloons of World War II, the Navy operated only two basic lighter-than-air (LTA) types—the non-rigid pressure airships popularly called blimps and large rigid-frame designs patterned on the German Zeppelins. As a class, lighter-than-air craft were designated by the letter Z ahead of the designation. In the case of rigid airships, the Z was actually used as part of the designation painted on the craft, such as ZR-1 for the USS *Shenandoah*. Except for the undelivered ZR-2, the rigids were given individual names in the manner of surface ships. As with the surface ships, the rigid airship designation also served as a serial number.

The two types were distinguished by further letters, R for the rigids and N for the non-rigids. At first, the rigids were merely numbered consecutively. By the late 1920s the primary mission was included in the designation, for example, ZRS-5 (S-for-Scout) for the USS *Macon*, the Navy's fifth (and last) rigid.

The letter N identified the non-rigids but never became part of the working designation although it was used in general references to the type. The Navy's first airship, delivered before adopting this system, was the DN-1, the letters standing for Dirigible, Non-rigid. As soon as the Navy began ordering non-rigids in quantity early in World War I, a model-letter designating system was adopted, starting with B (the Navy's second airship design). These model letters were followed by sequential numbers. Subsequent models became C and D, etc. There was considerable difference of detail within these model designations, but size and power remained relatively the same for each. When different manufacturers were involved in building the same basic model, the same letter was used for all, individual identification being by the number. With minor deviations, the letter designations for non-rigids reached N before the programme ended in the early 1960s.

Several changes in non-rigid designations were made during and after World War II. As a security measure, the model letters and sequence numbers, which had been painted boldly on the sides of the envelopes since the beginning, were deleted, being replaced by the words 'US NAVY.' The actual designation of each non-rigid was retained, however, and was supplemented by a letter designating the mission, such as ZPK—for a K-model used for patrol; ZSG—for a G-model used as a scout; and ZNL—for an L-model used in training. The mission was also reflected in the administrative paperwork, general class references being ZNP for

503

patrol and ZNN for trainers. The letter N-for-Trainer reverted to T in 1948 The letters U-for-Utility, S-for-Scout and H-for-Search-Rescue were added to the list after World War II. Where specific aircraft were concerned, the model designation was included, as ZNP-K-9. As new models developed from the K series (ZPK) after World War II, their progression was reflected in the designation in the manner of different aeroplane models, ZP2K-, ZP3K-, etc.

In 1954 the basic model-letter system was abandoned in favour of aeroplane-type designations reflecting the mission and model of the airship and its manufacturer, as ZS2G-1 for a second scout model from Goodyear (the Navy's only regular supplier of complete non-rigids since 1919; other companies provided components such as spare envelopes). Existing designations were actually changed, the XZPN-1 (prototype of the N-model) becoming the ZPG-1, the ZP4K becoming the ZSG-4 and the ZP5K becoming ZS2G-1. The airships lasted just long enough to be included in the aircraft redesignation system of 1962, the ZPG-2 then becoming the SZ-1B in a new Z-for-Airship series (see page 491).

The notable exceptions to the airship designation system were the two Army non-rigids TC-13 and TC-14, taken into the Navy in 1937 and operated under their original designations, and the ZMC-2, a metalclad built in 1929. The letters MC stood for Metal Clad and the -2 was not a sequential number but stood for the unique ship's 200,000 cubic foot capacity.

BALLOONS

From 1917 the Navy operated kite balloons for observation purposes from ships and shore stations, but these were largely replaced by observation aeroplanes in the 1920s. Eventually designated as class ZK, a total of 117 was procured in a variety of volumes from 16,000 to 50,000 cubic feet. Ninety-nine were direct purchases from American manufacturers as follows:

Goodyear Tire & Rubber Company	65
Goodrich Tire and Rubber Company	29
Connecticut Aircraft Company	3
Air Cruisers, Inc	2

An additional 17 were obtained from World War I allies: 13 from England and two each from France and Italy, plus one from the US Army. Forty-eight Goodyear and Goodrich models were shipped overseas in 1918.

Although kite balloons vanished from the inventory in the early 1930s, they reappeared in World War II in an entirely different role, again known as ZKs. Extensive use was made of unmanned captive barrage balloons. These were flown over logical targets or towed by ships as a deterrent to

Left, an observation kite balloon at Lakehurst Naval Air Station (*US Navy*) and, right, a 90,000 cu ft free balloon carrying a crew of six. (*National Archives 80-CF-41132-3*)

attack by dive-bombers or low-level attack aircraft, the gasbags themselves and their steel cables creating a hazard for the attacking aeroplanes. This system was developed in England early in the war and hundreds were supplied to the US services by both British and US manufacturers.

Spherical free balloons were used for training to the end of the airship programme. These balloons were known as ZF, but did not have model designations or sequential numbers. They were listed merely by their volumes in cubic feet and their serial numbers. Serials were assigned to them in sequence of procurement along with the aeroplane acquisitions. From 1917 through 1944 the Navy procured a total of 66 free balloons with displacements ranging from 9,000 to 90,000 cubic feet. These were used for training and occasional racing and were of conventional rubberized cloth construction with open baskets suspended below the envelopes. They should not be confused with the post-World War II high-altitude research balloons. Procurement was from six sources as follows:

Goodyear Tire & Rubber Company . , . . 34	
Connecticut Aircraft Company 18	
Meadowcraft Company 5	
Air Cruisers, Inc 4	
Lakehurst Naval Air Station 3	
Naval Aircraft Factory 2	

NON-RIGID AIRSHIPS

The Navy acquired a total of 241 non-rigid airships, including 21 from World War I allies, between 1917 and 1958. These airships were exactly what the name implied, having no supporting structure to maintain their shape. The nose, however, was stiffened by a ribbed structure rather resembling that of an umbrella.

505

The overall gasbag or envelope shape of the non-rigid was maintained by a combination of the lifting gas with which it was inflated and air pressure in one or more ballonets inside the main envelope. In forward flight, the ballonet was pressurized by ram air coming in through one or more airscoops; in hovering flight or on the ground, the ballonet was pressurized by a motor-driven blower. The ballonet also helped to conserve costly lifting gas. As the gas expanded with altitude, it acted on the ballonet, compressing it and forcing the air from the air scoops to maintain a constant gas volume within the envelope. When two ballonets were used, longitudinal trim of the non-rigid could be maintained by pumping air from one located forward to one located aft, or vice versa, to alter the actual distribution of lifting gas within the envelope.

The earliest non-rigids had passenger or control cars suspended from the envelopes by cables. This enhanced observation, but was essentially a fire precaution since the power plants were attached to the cars and the airships were inflated with highly inflammable hydrogen. It was several years after the adoption of helium (a US Government monopoly) that the cars were attached directly to the envelopes.

An inaccurate legend has long existed concerning the origin of the word Blimp, which is certainly one of the most appropriate nicknames ever applied to an object. The term was believed to have originated from a class of British airship used early in World War I, the B, limp (i.e. non-rigid). The development of this designation into a word is entirely logical. However, the British non-rigids were not designated as limp. The currently accepted origin is traced to an event on December 5, 1915, when Lt A. D. Cunningham, Royal Navy, snapped his fingers against the taut envelope of a non-rigid during an inspection. The resulting sound was 'Blimp,' and Cunningham's subsequent comments produced the spoken word that became a famous name.

Although the envelope was the most conspicuous feature of the blimp, it, along with the tail surfaces, was merely an accessory. The identity of the airship was determined by the number of the car, or gondola. Envelopes frequently wore out or were damaged, and were replaced, sometimes with one of a different volume produced by a different manufacturer.

The blimps, which were given serial numbers in sequence with aeroplane procurement, did their most valuable work on coastal patrols and in guarding shipping convoys against submarine attack in both World Wars. Forty-eight were authorized by Congress in the naval aircraft programme of June 15, 1940. In spite of this, the Navy had only 10 non-rigids operating at the time of Pearl Harbor. In June 1942 the authorized total was increased to 200 and 167 were actually delivered by the end of the war.

Other than training, the principal use of the blimps in World War II was coastal patrol and convoy duty, and it is significant that during the war no convoys ever lost a ship to enemy submarine action when escorted by blimps. Blimp stations were set up along both coasts of the United States, in the Caribbean and South Atlantic and in the Mediterranean.

506

The ships used in the Mediterranean were delivered by air across either the North or South Atlantic. Fitted with magnetic airborne detectors, they were able to guard the Straits of Gibraltar in instrument weather that grounded the aeroplanes used on anti-submarine duty. With the blimps on duty, not a single German U-boat is known to have passed through the Straits.

The blimp's unique flight characteristics suited them to a variety of other duties, such as picking up downed airmen from the jungle, spotting and then destroying mines by gunfire, and retrieving practice torpedoes for re-use. Experiments were also conducted in using radio-controlled gliders loaded with a depth-bomb as a glide-bomb against submarines.

Only one blimp was lost to enemy action—on July 18, 1943, K-74 engaged the German submarine U-134 in a machine-gun duel off the Florida coast. The U-boat's superior firepower soon riddled the blimp and it slowly fell into the sea. All of the crew but one were rescued.

DN-1

The Navy's first airship was the DN-1 (Dirigible, non-rigid; or Dirigible, Navy, No. 1) built by the Connecticut Aircraft Company. This was built before the Navy standardized airship designations, but became the A-type by implication since the next model procured was designated B.

Connecticut was low bidder among several, and a contract was signed for the DN-1 on June 1, 1915, for $45,646. Delivery was to have been later the same year, but inexperience on the part of both the builder and the Navy delayed delivery until December 1916. Fitted with two 140 hp Sturtevant engines driving swivelling propellers intended to assist vertical movement (an idea readopted years later for the ZRS-4 and ZRS-5) the DN-1 was considerably overweight. In order for it to fly, it was necessary to remove one engine.

The Connecticut DN-1 afloat. (*US Navy*)
507

The car or gondola was made in the form of a boat, and it was intended that the DN-1 operate from water in the manner of a seaplane. A special floating hangar was built for it at Pensacola, Florida. The first of three flights was made on April 20, 1917, and the last on April 29, after which it was scrapped.

Length, 175 ft; volume, 150,000 cu ft; no performance figures available.

B-SERIES

The Navy's first production airship was the B. This was designed by the Navy and initial orders for 16 were placed with Goodyear, Goodrich and Connecticut Aircraft Company.

With no design or operating experience of its own (the DN-1 had not yet flown), the Navy B was heavily influenced by what little information was available from war-time England before the United States became an ally and a legitimate recipient of classified technical information. The car was a slightly modified two-place aeroplane fuselage with separate cockpits and a Curtiss or Hall-Scott water-cooled engine in the nose driving a single tractor airscrew. The wheeled undercarriage was replaced by pneumatic flotation bags. Structural details and dimensions differed throughout the production of the Bs. Principal recognition feature was the combination of an aeroplane-fuselage car, relatively small size, and the absence of a top vertical fin although some Bs did carry a very small top fin. Others had double fins and rudders on the underside.

The initial B order went to Goodyear in March 1917, for B-1 through B-9 (A235–A243). These were 160 ft long, 31·5 ft in diameter, and had a volume of 77,000 cu ft. A 100 hp Curtiss OXX-2 motor drove them at 45 mph. Goodyear B-ships used the finger-patch method of attaching the car suspension lines to the envelope. B-10 through B-14 (A244–A248) were ordered from Goodrich. These were longer and fatter, having a

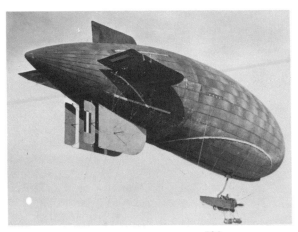

A Connecticut B-series blimp with five fins and rope suspension lines. (*National Archives 80-CF-4119-4*)

The Goodyear B-20 with three fins and cable suspension. (*US Navy*)

length of 167 ft, diameter of 33 ft, and a volume of 80,000 cu ft. Speed
was 50 mph with the OXX-2. The cars for B-11 through B-13 were provided by Curtiss. B-15 and B-16 (A249–A250) were built by Connecticut,
with a length of 156 ft, diameter of 35 ft, and a volume of 75,000 cu ft.
A 100 hp Hall-Scott A-7a drove them at 47 mph. Goodrich and Connecticut used the belly-band method of attaching the suspension lines.

Delivery of the first 16 Bs was from June 1917 to July 1918. Later,
Goodyear rebuilt three Bs which were redesignated B-17 to B-19 and re-
serialled A5464–A5465, A5467. Goodyear also built an entirely new B-ship
as B-20 (A5257) with a volume of 84,000 cu ft and an OXX-3 engine.

C-SERIES

The C-series was a highly successful design intended for coastal patrol
and convoy work and exerted great influence on subsequent US non-rigids
in both the Army and the Navy. Ten (A4118–A4127) were delivered;
C-1, C-3, C-4, C-5, C-7 and C-8 by Goodyear and C-2, C-6, C-9 and C-10
by Goodrich. The C-1 first flew on September 30, 1918.

Normal crew was four in a boat-like car with one pusher engine mounted
on each side. Published specifications state these to be 150 hp Wright-
Hispanos but all available photographs reveal them to be 200 hp Hall-
Scott L-6s. The cars for the entire C-series were built by the Burgess
division of Curtiss.

With a worthwhile useful load, the C-ships were involved in a number of
interesting experiments. On December 12, 1918, C-1 made the first successful release of an aeroplane from a non-rigid airship. In February 1919,
C-3 was refuelled while aloft from a surface ship at sea and on May 14–15,
1919, C-5 made a 1,400-mile nonstop flight from New York to Newfoundland as the first stage of a projected transatlantic flight intended to beat
the NC flying-boats across by several days. However, a gale at St John's
ripped the unmanned C-5 from its moorings and it was lost.

On December 1, 1921, C-7 became the first airship in the world to fly
with helium gas in place of hydrogen. After this successful test, helium

The Goodyear-built C-7, first airship to be flown with helium gas. (*US Navy*)

was decreed for all US military airships, a great fire protection measure that justified the slightly decreased lift and was made possible by the US monopoly of the world's natural helium resources. Because of its value in airships and the Navy's impending rigid airship programme, the Navy was given control of the natural gas wells. C-2 and C-4 were turned over to the US Army in 1921, and C-2 subsequently became the first airship to cross the American continent from coast to coast.

Length, 192 ft; diameter, 42 ft; volume, 181,000 cu ft; max speed, 60 mph; full speed endurance, 10 hr; static ceiling, 8,000 ft.

D-SERIES

The D-series was a slight refinement of the C-series. The initial 1919 order was for five (A4450–A4454), D-1, D-3 and D-4 from Goodyear and D-2 and D-5 from Goodrich. Notable differences were the increased length, a six-foot parallel section having been spliced into the centre of an otherwise standard C-envelope, and a redesigned car with the engines moved to the aft end to improve crew comfort. To make more room for

The Goodrich-built D-2 after its transfer to US Army; fuel tanks are suspended from the envelope. (*US Army Air Service*)

510

the crew, the fuel tanks of D-1 to D-5 were suspended from the sides of the envelope, five tanks to a side.

D-6 (A5972) was a different ship, with a C-type envelope built by Goodyear in 1920 and a special enclosed car built by the Naval Aircraft Factory. Fuel tanks on D-6 were carried in the car. D-1 was destroyed by fire in June 1920 and D-6 in August 1921. The others were turned over to the Army in 1921.

Length, 198 ft; diameter, 42 ft; volume, 189,000 cu ft; power plant, two 120 hp Union; max speed, 56 mph; full speed endurance, 12 hr; static ceiling, 8,000 ft.

E- AND F-SERIES

Only one airship was built in each of these classes, which were identical except for power plant and minor details. Both were built by Goodyear as commercial ships in 1919 and were subsequently purchased by the

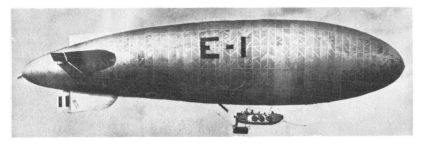

The Goodyear E-1 at Pensacola in September 1920. (*US Navy*)

Navy. The two-seat E-1 (A4109) was fitted with a single 150 hp Thomas engine mounted in the rear of the open car while the three-seat F-1 (A6348) was powered with a 120 hp Union engine. Following delivery of these ships, Goodyear became the sole supplier of complete non-rigid airships in production quantities to the US Navy.

Length, 162 ft; diameter, 33·5 ft; volume, 95,000 cu ft; max speed, 56 mph (E-1), 52 mph (F-1).

G-SERIES

There were three different G-designs, and aviation historians have been understandably confused by them. The first G-ship was designed by the Navy in 1919 as a 322,000 cubic-footer, but was never built. The second G-design was the G-1, (9999 in the first serial number series) formerly the *Defender* of Goodyear's famous advertising fleet. This had been built in 1929 and was delivered to the Navy in 1935. With a length of 184 ft, diameter of 45 ft, and volume of 178,000 cu ft, it had accommodation for eight passengers and crew. Original power plants were two pusher Wright

511

Goodyear G-6 carrying a recruiting message, 1953. (*Douglas D. Olson*)

J-6-5s of 165 hp, later replaced by 225 hp J-6-7s. Cruising speed 62 mph and range 1,175 miles.

Seven additional G-types, G-2 to G-8, were ordered in the expansion programme of June 1942. As with some other World War II blimps, these appear to have been delivered without serial numbers. There was no direct developmental link between the wartime G-trainers and the G-1, general similarity of dimensions and arrangement making the similar designation logical. The later G-ships were six-seaters with 220 hp Continental R-670-4 engines. Some of these remained in post-war service as late as 1959. Length, 192 ft; diameter, 45 ft; volume, 196,000 cu ft; max speed, 60 mph.

H-SERIES

The H-1 (A5973) was built by Goodyear and delivered to the Navy in April 1921. Essentially a powered two-seat observation balloon built along the lines of the commercial Goodyear 'Pony Blimp', it had a length of 95 ft, diameter of 28 ft, and a volume of only 35,000 cu ft. Power plant, one 50 hp Lawrance radial engine mounted as a pusher. The H-1 was destroyed by fire on August 31, 1921.

The two-man H-1 could be flown as a captive balloon or as a dirigible. (*US Navy*)

512

J-SERIES

Although the J-series ran to four numbers, only three blimps were operated as such. Principal difference from earlier C- and D-types was the use of tractor, rather than the traditional pusher engine arrangements. The Navy purchased two from Goodyear in 1922, but only the J-1 (A6111) was accepted although the car of J-2 (A6112) was delivered. When inflated with hydrogen, the J-1 could carry a crew of 10. With helium, the capacity of J-1 and subsequent Js was six or seven. In 1926 the Army TC-2 airship, essentially similar to the Navy J-type, was obtained and its car was fitted to a J-envelope and redesignated J-3 (A7382). In November 1927, the J-2 car was fitted to another J-envelope and redesignated J-4. For a while, J-3 and J-4 were the only non-rigids operating in the Navy. J-3 was lost at sea while searching for survivors of the USS *Akron* in April 1933.

The J-4 after alighting on water on its boat-like car, which had a removable enclosure fitted in 1933. (*National Archives*)

J-4 was turned over to the US Army after the Navy acquired more modern blimps but was returned to the Navy in 1936 and was scrapped in 1940.

Length, 196 ft; diameter, 45 ft; volume, 210,000 cu ft; max speed, 60 mph; range at full speed, 1,070 miles. Power plants, 150 hp water-cooled Wright-Hispanos, 220 hp air-cooled Wright J-5 and 225 hp Wright J-6-7.

K-SERIES

Like the G-ships, the K-series is confusing to historians because of the long period of procurement and the fact that several different designs were included in the same letter-series.

K-1 (9992) was designed by the Navy as an entirely new experimental blimp. The envelope, essentially an enlarged J-type, was ordered from

513

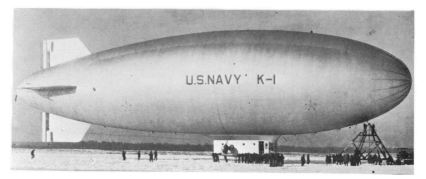

K-1 was the Navy's first blimp with the car attached directly to the envelope.
(*US Navy*)

Goodyear. The car, designed to attach directly to the bottom of the envelope, was built by the Naval Aircraft Factory in 1931. The outstanding feature of K-1 was the use of gaseous instead of liquid fuel, contained in a special 51,700 cu ft ballonet. Two Wright J-6-9 tractor engines of 330 hp were used. As a result of these advanced features, the Navy found itself with a fine patrol-type airship, far superior to the contemporary Army TC-series, that it couldn't use as such because of the Army's monopoly of the coastal patrol mission. K-1 carried on experimental and training work including the development of mobile mooring masts, and was scrapped in 1941. Length, 218 ft; diameter, 54 ft; volume, 319,900 cu ft; max speed, 65 mph.

The development of patrol blimps was undertaken seriously after the Navy took over all LTA operations in 1937, and K-2 (1211 in the second serial number series) was the result. This was a combined design effort of Goodyear and the Navy and was a 404,000 cu ft ship, the largest ever built at the time, and only superficially resembled the K-1. Power plants were 550 hp Pratt & Whitney R-1340-AN-2 Wasp engines driving tractor airscrews. First flight was in December 1938.

Subsequent orders are rather confused. Navy records show larger 416,000 cu ft K-5 to K-8 ordered with serial numbers 7025–7028 in the second serial series, while K-3 and K-4 were ordered with serials 01729 and 01730 in the third series. K-3, K-4, K-7 and K-8 are listed as training types while all other Ks through K-135 are patrol types. K-3 to K-8 used two 420 hp Wright R-975-28 engines, while K-9 and on reverted to the Pratt & Whitney R-1340-AN-2s of K-2. K-14 to K-35 had larger envelopes of 425,000 cu ft volume. Some later war-time modifications included envelopes of 456,000 cu ft. K-9 to K-29 had serial numbers 04359–04379 in the third series, K-30 to K-74 had serials 30152–30196. Other Ks were procured without serials.

Post-war modifications and rebuilding fitted still larger 527,000 cu ft envelopes to some war-time K-ships. These were type-designated ZP2K

A Goodyear ZP3K (ZSG-3) in 1956, with recruiting slogan. (*Gordon S. Williams*)

and ZP3K, and later ZSG-2 and ZSG-3, while the unmodified types became ZPK. The original K-numbers remained in use on the individual ships.

New models evolving from the K-series after the war were designated ZP4K and ZP5K, but these were redesignated ZSG-4 and ZS2G-1 (which see). Specifications for K-14 to K-135: length, 251·7 ft; diameter 62·5 ft; volume, 425,000 cu ft; crew, three officers and nine men; max speed, 75 mph; range, 2,000 miles at 40 kt.

L-SERIES

The L-1 trainer (1210) was ordered in 1937. L-2 and L-3 (7029, 7030) were ordered in the expansion programme of 1940. L-4 to L-8 were acquired from the Goodyear advertising fleet, the crews enlisting in the Navy along with their ships. L-4, the former *Volunteer*, had serial 09801 in the third series and L-5 was 09802. The other three advertising ships and 14 additional production models, L-9 to L-22, were procured without serial numbers. Length, 149 ft; diameter, 39 ft; volume, 123,000 cu ft; max speed, 60 mph; crew, four; power plant, 145 hp Warner R-500-2/6.

The L-8 was part of the Goodyear advertising fleet before passing into Navy service. (*National Archives 80-G-16180*)

515

One of the four M-type airships delivered in 1944.

M-SERIES

Of 22 M-ships ordered, only four were delivered because of the improved anti-submarine situation in 1944. M-1 (48239) first flew in October 1943. It was fitted with an unusually long car, 117 ft, which had to be built in three articulated sections to match the flexibility of the envelope. The volume of M-1 was 625,000 cu ft with a length of 310 ft. Volume of M-2 to M-4 (48240–48242) was 647,000 cu ft, later increased to 725,000 cu ft. Last flight of M-4 was in September 1956. Two 600 hp tractor Pratt & Whitney R-1340-AN-2 engines gave the 10-man-plus airships a speed of 75 mph.

N-SERIES

The single ZPN-1, which became ZPG-1 in 1954, was the prototype of a greatly improved long-range patrol series. With a volume of 875,000 cu ft and a length of 324 ft, the 14-man N-1 carried its two 800 hp Wright R-1300-2 Cyclone 7 engines inside the double-decked car. The tractor propellers were driven through gears and extension shafts and were cross-connected so that one engine could drive both propellers for low-speed cruising. The most noticeable feature of the N-1, last Navy blimp to be

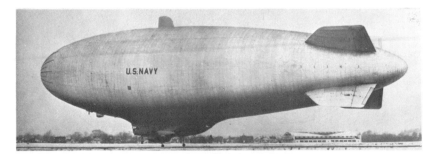

A ZPG-2W, showing the radome on top of the envelope. (*Goodyear photo*)

516

identified by a class letter and a sequential number, was the angled position of the tail surfaces, a departure from the traditional vertical-horizontal form. Maximum speed of the N-1, which was delivered in 1951, was 90 mph.

Sufficient changes were made in the following N-class blimps to justify a new designation, ZP2N. This was changed in the redesignations of 1954 to ZPG-2. Twelve 1,011,000 cu-ft ships, each 343 ft long, were ordered as ZP2N-1 (126716–126719, 135445–135448, 141559–141562), the first one taking to the air in March 1953. These had great endurance capability. In 1954, one set a world's non-refuelling endurance record of 200 hours. Three years later another stayed aloft for 264 hours (11 days) on an 8,216-mile cruise that took it from its base at Weymouth, Mass., to Key West, Florida, by way of Africa.

First and second Goodyear ZPG-3Ws in November 1958. (*Goodyear photo*)

Five additional ZP2Ns, adapted to Airborne Early Warning duties, were ordered as ZP2N-1W (13782, 139918, 141334, 141335, 141563) and later became ZPG-2W. These were dimensionally similar to the ZP2N-1s, but carried a crew of 21 and search radar antennae inside the envelope. They could be distinguished from the ZP2N-1s by an additional radome on top of the envelope. Maximum speed of the ZP2N-1W/ZPG-2W was 80 mph.

Final examples of the N-series, the last two of which were the last Navy airships delivered, were four ZPG-3Ws (144242–144243, 146296–146297). These were also the largest non-rigid airships ever built, having a length of 403 ft and a volume of 1,516,000 cu ft. The car, radar and crew accommodation were similar to the ZP2N-1W/ZPG-2W except for larger engines, 1,525 hp Wright R-1820-88 Cyclones which produced a maximum speed of 90 mph. The first ZPG-3W flew in July 1958. A fatal accident in June 1960, in which a ZPG-3W collapsed in the air, hastened the already imminent end of the Navy airship programme, and on June 28, 1961, announcement was made of the end of Naval airship operations.

517

One of the post-war ZP4K (ZSG-4) series. (*Peter M. Bowers*)

ZP4K/ZSG-4

The 15 ZP4Ks (131919–131926, 133639, 134019–134024) were post-war developments of the K-series, with length increased to 266 ft and a volume of 527,000 cu ft. 133639 was redesignated XZSG-4 and used for development work. Car and power plants were essentially the same, two 550 hp tractor Pratt & Whitney R-1340-46s providing a maximum speed of 80 mph. Fundamental changes in operating technique were introduced on this model. Previously, the pilot had handled the elevators and the co-pilot the rudders. The ZP4K/ZSG-4s gave both of these functions to one man. They were also fitted with power winches that could transfer supplies and crew members to and from the ground without making it necessary to land the ship. First flight was made in December 1953.

ZP5K/ZS2G-1

This was an entirely new scouting class bearing little resemblance to the war-time K-ships even though ordered under the K designation. Nineteen were built (137486–137491, 137811, 141564–141570, 144239–144241, 146294–146295). 137811, 146294 and 146295 were redesignated XZS2G-1 and assigned to developmental work, the others becoming

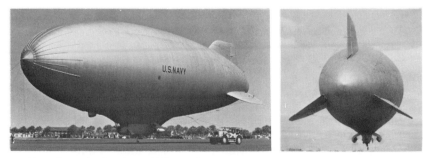

Two views of a Goodyear ZP5K (ZS2G-1). (*US Navy*, left, *Goodyear photo*, right)

ZS2G-1s. First flight was on July 22, 1954, so the new class did not see service under its original K-series designation.

Larger than the Ks, with a length of 285 ft and a volume of 650,000 cu ft, this model carried a crew of eight and was fitted with more powerful Wright R-1300-4 Cyclone 7 engines of 800 hp in outboard nacelles. Tractor propellers permitted a maximum speed of more than 85 mph. The distinctive recognition feature of this class alone was the use of an inverted-Y arrangement for the tail surfaces. This was intended to increase the all-weather capabilities of the type by preventing the accumulation of snow on the stabilizers and elevators.

ZMC-2

The ZMC-2 (A8282), built by the Aircraft Development Corporation of Detroit, Michigan, was one of the world's most unusual airships. Although built for the US Navy, its designation did not fit into the normal sequence of airship type letters. The MC in the designation stood for Metal Clad and the digit for the capacity of 200,000 cu ft.

Although the thin metal skin, which was supported by a series of transverse metal frames and longitudinal stiffeners, gave the appearance of being a rigid structure, the ZMC-2 was a pressure-type airship like other

The unusual metalclad ZMC-2. (*National Archives 80-G-21724*)

blimps. The helium gas was in direct contact with the 0·080-in-thick metal skin, the seams of which were made gas-tight by 3,000,000 $\frac{1}{20}$-in rivets with extremely close spacing. Gas pressure was maintained by air in two ballonets. Design and development work began in 1922, after the company was formed for the specific purpose of developing a metal airship. By 1926 the Navy was convinced of the practicability of the design and ordered one ship, which was first flown in August 1929, and delivered a month later. Principal recognition feature, other than the extremely low ratio of length to diameter, was the use of eight tail surfaces instead of the conventional four. Original power plants were 220 hp

The former Army TC-13 after transfer to the US Navy in 1937. (*National Archives 80-CF-41255-1*)

Wright J-5 Whirlwinds. Length, 149 ft 5 in; diameter, 52 ft 8 in; volume, 22,600 cu ft; max speed, 62 mph.

The ZMC-2 gave trouble-free service for 10 years. By arrangement, its last flight was made on August 19, 1939, the 10th anniversary of its first flight. After being used for ground tests, it was scrapped in 1941.

TC-13, TC-14

When the Army abandoned its airship programme at the end of the Fiscal Year 1937 (June 30, 1937), it turned its four remaining airships over to the Navy. These were the TC-10 and TC-11, generally similar to the obsolete J-series, and the relatively new TC-13 and TC-14. Only the last two saw Navy service, under their original army designations and without Navy serial numbers.

The TC-13 was built by Goodyear in 1933. Powered by two 325 hp Pratt & Whitney R-985 Wasp Jr engines, it had a length of 233 ft, diameter of 54 ft, and volume of 360,000 cu ft. A crew of eight was carried, and an observer could be lowered 1,000 ft below the ship in a Cloud Car

AT-13 was one of two Astra-Torres airships operated by the Navy in France in 1918. (*National Archives 80-HAG-5K-1*)

that permitted observation and telephone communication while the airship remained concealed in cloud.

The near-duplicate TC-14 was fitted with an envelope made by Air Cruisers, Inc, and a car made by Mercury Corporation. Both ships had been deflated and stored at the end of their Army service. TC-14 was shipped to Lakehurst and inflated in 1938, but TC-13 was not put back into service until 1940. Both then served until scrapped in 1943.

IMPORTED NON-RIGID AIRSHIPS

The US Navy used some French and British blimps in Europe during 1918 and purchased other non-American types after the Armistice. Those operated in Europe did so under their original designations and did not acquire US Navy serial numbers. Serials were assigned to the foreign-built ships imported into the United States after the war.

The British SST (Submarine Scout Twin) serving with the US Navy in British markings; note the US flag at stern, and the external fuel tanks. (*Boeing photo*)

French Airships—The Navy began coastal patrols in March 1918, starting with the French-built Astra-Torres AT-1 and Zodiac-Vedette VZ-3. These were soon supplemented by the VZ-7, VZ-13, AT-13, and the Chalais-Meudon *Capitaine Caussin*. AT-1, VZ-7 and VZ-13 were taken to the States, where they were assigned serial numbers A5472, A5592 and A5593, respectively. One additional Zodiac, the largest non-rigid built at the time, was imported in 1919. This was the only one delivered on a contract for three placed in June 1918, and was immediately turned over to the Army, which rebuilt it and operated it as ZD-US-1, DZ-1 and, finally, RN-1 over a period of four years.

British Airships—Six British blimps were transferred to the US Navy in 1918: the NS-14 North Sea, two SS Submarine Scouts, one SST Submarine Scout Twin and two SSZ Submarine Scout Zero types, the SSZ-23 and SSZ-24. The SSZ-23 and the NS-14 reached the States and acquired serial numbers A5472 and A5580, respectively.

Italian Airship—In March 1919 the Navy approved the test flights of the small non-rigid O-1 built by the Stabilimento Construzioni Dirigibili & Aerostati of Rome and had it shipped to the US, where it was commissioned with serial number A5587.

521

RIGID AIRSHIPS

The general success of the German Naval Zeppelins as auxiliaries of the fleet during World War I awakened US Navy interest in the same type of craft. Britain had embarked on a rigid airship programme even before the war, but lagged far behind the Germans in rigid airship technology although remaining ahead in the non-rigids. The capture of several Zeppelins relatively intact enabled the British to catch up and gave the US a starting-point for a rigid airship programme.

Of the Navy's five rigids, only three were American-built. The first one ready for service was the ZR-2, built in England as the R-38, but this was destroyed on a test flight prior to delivery. The ZR-3, named USS *Los Angeles*, was built at the Zeppelin plant in Germany.

Unlike the blimps, the rigids maintained their aerodynamic shape at all times because of their rigid aluminium framework. The lifting gas was carried in a series of separate ballonets inside the hull so that damage to one would not result in a total loss of lift. Because of the fire hazard with hydrogen, engines and control cars were all mounted outside of the envelope. Not until the US Navy ordered rigids designed specifically to use helium were the engines moved inside.

The Navy's rigid airships never had a chance to prove their worth in warfare, and their peace-time operations, handicapped by small numbers and budget limitations, did little to earn the rigid a secure place in naval operations since they never really got out of the research and development stage. The most successful rigid by far was the German-built ZR-3, which was used mainly for research and training from its delivery in 1924. The USS *Akron* and *Macon*, ZRS-4 and ZRS-5, were intended to work with the fleet as scouts, and even carried aeroplanes to assist in this mission, but the operational goals were never fully achieved. The aeroplanes carried by these two airships, since they were single-seaters with armament and fighter designations, have erroneously been considered as defensive fighters. Their primary function was to serve as scouts in the manner of the seaplanes catapulted from cruisers, operating beyond the horizons of the mother ship and extending its effective scouting area.

Although enthusiastic boosters within the Navy tried to get appropriations for additional rigid airships after the loss of the USS *Macon* in 1935 and the scrapping of the USS *Los Angeles* in 1939, the programme died with their passing. Ironically, the last two German Zeppelins, entirely unsuited to military purposes, were ordered to be scrapped in 1940.

ZR-1

The Navy's first rigid airship design, put on paper late in 1916, was to use a wooden structure and be inflated with helium gas. This design was abandoned in favour of a more conventional hydrogen-inflated aluminium structure based on the German high-altitude bomber L49, which had been forced down in France in October 1917 and captured intact. The

USS *Shenandoah*, the first of the US Navy's rigid airships. (*International Newsreel*)

new Navy design, ZR-1, eventually to be named USS *Shenandoah*, was designed by the Bureau of Aeronautics and built at the Naval Aircraft Factory in the Philadelphia Navy Yard. The parts were then transported to the Naval Air Station at Lakehurst, NJ, where the ZR-1 was assembled in the only suitable hangar in the country.

Powered with six 300 hp Packard engines and manned by a crew of 23, the ZR-1 made its first flight on September 4, 1923. The performance was not up to the original design specifications because the ZR-1 was inflated with helium instead of the hydrogen that it had originally been designed for. The change from hydrogen to helium may have been partly responsible for the eventual destruction of the ship in a storm near Marietta, Ohio, on September 3, 1925. To save weight and prevent the loss of the still rare and costly helium, seven of 15 manoeuvring valves and 10 of 18 automatic gas valves were removed. It is believed that the reduced gas-loss rate permitted cells to burst when the ship was carried above its pressure altitude in the storm. (For a period when the Navy had both the ZR-1 and the ZR-3, helium was in such short supply that both ships could not be operational at the same time; it was necessary to deflate one and transfer its helium to the other.)

During its two-year career, the ZR-1 made 57 flights, including a 9,317-mile double crossing of the United States. Although intended as a scout, it was used largely for test work, including mooring to a mast on the airship tender USS *Patoka*. During a weight reduction programme, the engine installed in the control car was removed. Original specifications: length, 680 ft; diameter, 79 ft; volume, 2,115,000 cu ft; max speed, 60 mph.

ZR-2, ZR-3

To speed up its rigid airship programme, the Navy decided to buy a new rigid, to be designated ZR-2, which was then under construction in England as the R.38. In addition, the Navy was to receive two war-time German Zeppelins as war reparations. None of these ships was received; the R.38 broke up on a trial flight before being handed over to the Navy and the Zeppelins were sabotaged in their sheds by their German crews.

As a result, the Inter-Allied reparations commission agreed that the Zeppelin company should build a single modern replacement airship for

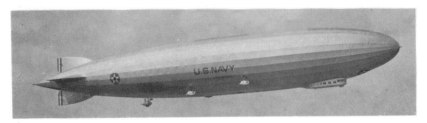

The USS *Los Angeles* in flight. (*Courtesy H. B. Cranshaw*)

the Navy. This proved to be the salvation of the company, for even though it had built two purely commercial ships right after the war, they had been confiscated by the Allies and pressure was being applied to break up the company and scatter its technicians. The LZ.126, to cite its factory designation, was built essentially as a commercial design since, as war reparations, it was not to be offensive military or naval equipment. Although the US Navy intended to operate it with helium, the LZ.126/ZR-3, later to be named USS *Los Angeles*, was inflated with hydrogen for its makers' tests and the 5,060-mile delivery flight from Friedrichshafen to Lakehurst, which was made during October 12–14, 1924.

The ZR-3 made 331 flights and accumulated 5,368 flight hours prior to retirement in June 1932. Among the more interesting experiments conducted were those involving the hook-on operation of aeroplanes. Following the loss of the ZRS-4 USS *Akron*, the ZR-3 was recommissioned but saw limited use. It was finally scrapped in 1939. Power plant, five 400 hp Maybach; length, 658 ft; diameter, 90 ft 6 in; volume, 2,472,000 cu ft; max speed, 76 mph.

ZRS-4 *AKRON*, ZRS-5 *MACON*

In 1926 the Navy was granted authority to procure two large rigid airships for scouting purposes. A design competition was announced, and the Goodyear-Zeppelin Corporation of Akron, Ohio, was declared the winner and given a contract in 1928 for the construction of the two ships, designated ZRS-4 and ZRS-5 and eventually to be named the USS *Akron* and *Macon*. Goodyear-Zeppelin was a subsidiary of both the

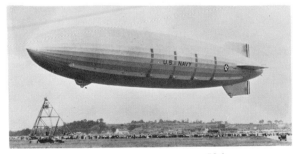

ZRS-4 *Akron*, designed to carry its own fighters. (*Goodyear photo*)

524

Goodyear Tire & Rubber Company, long experienced in the building of balloons and non-rigid airships, and Luftschiffbau Zeppelin GmbH, the original German Zeppelin firm.

Both ships were designed from the start to use helium gas, a fact that made possible several other design innovations that would have been hazardous on a hydrogen-filled ship. The eight power plants were located inside the hull and drove the propellers through shafts and gearing. These, in addition to providing the normal longitudinal thrust, could be rotated to provide thrust in upward, downward, or reverse directions. In addition, a 60-ft by 75-ft hangar was provided for the stowage of up to four small scout or fighter-type aeroplanes. These could be brought aboard in flight by means of a retractable trapeze assembly beneath the airship that was engaged by a hook on the aeroplane.

The ZRS-5 *Macon*, showing the aft control station in the lower fin. (*US Navy*)

A unique water-recovery system was incorporated to avoid the need to valve gas as fuel was consumed. Most of the moisture in the engine exhaust gases was recovered through a series of condensers running vertically on the side of the hull above each engine compartment.

Construction of ZRS-4 was begun on November 7, 1928, and she was commissioned on October 27, 1931. Stationed at Lakehurst, the *Akron* was lost at sea during a storm on the night of April 3–4, 1933. There were only three survivors. The ship's complement of four Curtiss F9C-2 Sparrowhawk aeroplanes (see page 132) was not aboard.

ZRS-5 was an exact duplicate of ZRS-4 except for minor refinements and weight economies that made the ship 8,000 lb lighter. She was commissioned on June 23, 1933, and assigned to the new Naval Airship Station at Sunnyvale, California, which had been named Moffett Field in honour of Rear Admiral William A. Moffett, Chief of the Bureau of Aeronautics, who had been lost on ZRS-4. ZRS-5, after operating with known structural damage for several months, was lost off the California coast on February 11, 1935, with the loss of only two lives. This loss ended the active rigid airship programme in the Navy. Although a smaller replacement rigid was authorized, it was not built. ZRS-5 specifications: power plant, eight 560 hp German-built Maybach engines; length, 785 ft; diameter, 132·9 ft; volume, 6,500,000 cu ft; normal crew, 60; max speed, 84 mph; range, 9,200 miles at 60 mph.

ENGINE NAMES

THROUGHOUT this book, engines are normally referred to only by their official designations, which comprised, in the case of piston engines, a letter indicating configuration and a number based on the cubic capacity of the cylinders; and in the case of gas turbines, a letter or letters indicating the class of engine and a sequential type number. The latter numbers start at 30 in each class and even numbers are used for engines developed on Navy contracts while Air Force engines have odd numbers.

Some, but not all, of these engines also have popular names assigned by the manufacturers. The following list shows the popular names for engines used in Navy aircraft:

Warner	R-500	Super Scarab
Wright	R-760	Whirlwind 7
Wright	R-975	Whirlwind 9
Pratt & Whitney	R-985	Wasp Junior
Pratt & Whitney	R-1340	Wasp
Wright	R-1510	Whirlwind 14
Pratt & Whitney	R-1535	Twin Wasp Junior
Pratt & Whitney	R-1690	Hornet
Wright	R-1750	Cyclone 9
Wright	R-1820	Cyclone 9
Pratt & Whitney	R-1830	Twin Wasp
Pratt & Whitney	R-2000	Twin Wasp
Pratt & Whitney	R-2180	Twin Wasp
Wright	R-2600	Cyclone 14
Pratt & Whitney	R-2800	Double Wasp
Wright	R-3350	Cyclone 18
Pratt & Whitney	R-4360	Wasp Major
Pratt & Whitney	J42	Nene
Pratt & Whitney	J48	Tay
Wright	J65	Sapphire
Allison	TF41	Spey

DESIGNATION INDEX

THE purpose of this section is to record the designations applied to Navy (including Marine Corps and USCG) aircraft since the introduction of a logical designation system in 1922 (see page 4). Each type or class category used since 1922 is included here in alphabetical order, and the aircraft designated in each of these categories are listed alphabetically by manufacturer's symbol. Since 1962, a unified designation system has been in use for all US armed services, and for the aircraft categories in use from that date on, only those designations applicable to USN, USMC and USCG aircraft are included in this Index.

For the purposes of this index, status prefix letters, including X for experimental types, have been omitted, as have special purpose suffix letters and type sequence numbers. Thus, for example, the XTB3F-1S is indexed under the TB heading as TB3F. Aircraft designated early in 1922 with the manufacturer's symbol *preceding* the type symbol are included in the same alphabetical sequence as if they conformed to the later practice, and are indicated by an asterisk. A careful study of the chapter on aircraft designations, commencing on page 4, will make the use of this index easier.

In those cases where an aircraft is described in the body of this book, an appropriate page reference is given in bold type. For those experimental types not described in this volume, brief details of the type are given, together with the Navy serial numbers wherever possible.

A—AMBULANCE, 1943

AE	PIPER	Cub variant. HE redesignated. **Page 462**

A—ATTACK, 1946–1962

AD	DOUGLAS	Skyraider. BT2D redesignated. To A-1, 1962. **Page 176**
A2D	DOUGLAS	Skyshark with T40-A-6 turbopropr. Seven built, 1947–1953.
A3D	DOUGLAS	Skywarrior. To A-3, 1962. **Page 186**
A4D	DOUGLAS	Skyhawk. To A-4, 1962. **Page 305**
AF	GRUMMAN	Guardian. **Page 226**
A2F	GRUMMAN	Intruder. To A-6, 1962. **Page 247**
AH	McDONNELL	Initial designation for F4H/F-4 Phantom II. **Page 301**
AJ	NORTH AMERICAN	Savage. To A-2, 1962. **Page 457**
A2J	NORTH AMERICAN	Developed AJ with two T40-A-6 turboprops. Two only (124439–40).
A3J	NORTH AMERICAN	Vigilante. To A-5, 1962. **Page 352**
AM	MARTIN	Mauler. BTM redesignated. **Page 321**
AU	CHANCE VOUGHT	USMC variant of F4U Corsair. **Page 406**
A2U	CHANCE VOUGHT	Projected attack development of F7U.

A—ATTACK, 1962 →

A-1	DOUGLAS	AD redesignated. **Page 176**
A-2	NORTH AMERICAN	AJ redesignated. **Page 458**
A-3	DOUGLAS	A3D redesignated. **Page 186**
A-4	DOUGLAS	A4D redesignated. **Page 305**
A-5	NORTH AMERICAN	A3J redesignated. **Page 352**
A-6	GRUMMAN	A2F redesignated. **Pages 247 and 250**
A-7	LTV	Corsair II; variant of F-8 Crusader. **Page 292**
A-8	HAWKER SIDDELEY	USMC Harrier, as AV-8A. **Page 486**

B—BOMBER, 1941–1943

BD	DOUGLAS	USAAF A-20 Boston. **Page 431**
BG	GREAT LAKES	BuAer Design No. 110. **Page 193**
B2G	GREAT LAKES	Improved BG with retractable u/c, 700 hp R-1535-82 engine. One (9722) only.
BM	MARTIN	Production of T5M. **Page 314**
BN	NAF	Prototype (A8643) cancelled before completion.
BT	NORTHROP	**Page 357**
B2T	NORTHROP	N-9M flying-wing for Navy training (XB2T-1 only).
BY	CONSOLIDATED	1932 prototype (A8921) monoplane, 600 hp R-1820-78.
B2Y	CONSOLIDATED	BuAer Design No. 110 prototype (9221) biplane, 700 hp R-1535-64.

B—BOMBER, 1962 →

B-26	DOUGLAS	JD redesignated. **Page 432**

BF—BOMBER-FIGHTER, 1934–1937

BFB	BOEING	F6B redesignated, 1935. Serial 8975 only.
BFC	CURTISS	F11C-2 redesignated, 1934. **Page 140**
BF2C	CURTISS	F11C-3 redesignated, 1934. **Page 141**

BT—BOMBER-TORPEDO, 1942–1945

BTC	CURTISS	1946 prototypes (31401–2) R-4360-14 engine.
BT2C	CURTISS	1946 prototypes, R-3350-24W engine. Nine built (50879–87).
BTD	DOUGLAS	Destroyer, single-seat variant of SB2D. **Page 431**
BT2D	DOUGLAS	Prototypes of Skyraider. To AD for production. **Page 176**
BTK	KAISER-FLEETWINGS	1945 prototypes, R-2800-34W engine. Five only (44313–14, 90484–6).
BTM	MARTIN	Prototypes of Mauler. To AM for production. **Page 321**

C—CARGO/TRANSPORT, 1962 →

C-1	GRUMMAN	TF redesignated. **Page 243**
C-2	GRUMMAN	Greyhound. COD version of E-2. **Page 246**
C-3	MARTIN	RM redesignated. **Page 454**
C-4	GRUMMAN	Gulfstream as trainer and for USCG. **Page 439**
C-9	McDONNELL DOUGLAS	DC-9 Skytrain II for USN. **Page 448**
C-11	GRUMMAN	Single commercial Gulfstream II as VC-11A USCG command transport. **Page 439**
C-45	BEECH	JRB/SNB redesignated. **Page 41**
C-47	DOUGLAS	R4D redesignated. **Page 170**
C-54	DOUGLAS	R5D redesignated. **Page 174**
C-117	DOUGLAS	R4D-8 redesignated. **Page 170**
C-118	DOUGLAS	R6D redesignated. **Page 432**
C-119	FAIRCHILD	R4Q redesignated. **Page 189**
C-121	LOCKHEED	R7V/WV redesignated. **Page 272**
C-123	FAIRCHILD	Provider for USCG. **Page 446**
C-130	LOCKHEED	UV/GV/R84 redesignated. **Page 275**
C-131	CONVAIR	R4Y redesignated. **Page 422**
C-140	LOCKHEED	JetStar UV-1 redesignated. None procured by USN.

DS—DRONE ANTI-SUBMARINE, 1959–1962
DSN   GYRODYNE   Drone anti-submarine. To H-50, 1962. **Page 440**

E—SPECIAL ELECTRONICS, 1962 →
E-1	GRUMMAN	WF redesignated. **Page 242**
E-2	GRUMMAN	W2F redesignated. **Page 244**

F—FIGHTER, 1922–1962
FA	GENERAL AVIATION	BuAer 96 design. Prototype (A8732) in 1932 with 450 hp R-1340C.
F2A	BREWSTER	Model 39 Buffalo. **Page 71**
F3A	BREWSTER	Alternative source of F4U. **Page 404**
FB	BOEING	Model 15, as USAAC PW-9. **Page 55**
F2B	BOEING	Model 69, FB development. **Page 58**
F3B	BOEING	Model 74, improved F2B. **Page 59**
F4B	BOEING	Model 99, as USAAC P-12. **Page 62**
F5B	BOEING	Model 205 monoplane (A8640) tested in 1930. 500 hp R-1340D.
F6B	BOEING	1933 prototype (8975) based on F4B with R-1535-44. Became BFB.
F7B	BOEING	Low-wing monoplane prototype (9378) in 1933 with R-1340-30.
F8B	BOEING	Model 400 prototypes (57984–6) with 2,500 hp R-4360-10.
CF*	CURTISS	Paper designation for CR racer. **Page 120**
F2C	CURTISS	Paper designation for R2C racer. **Page 125**
F3C	CURTISS	Paper designation for R3C racer. **Page 125**
F4C	CURTISS	Final version of NAF-designed TS-1. **Page 332**
F5C	—	Not used for Curtiss fighter to avoid confusion with F-5 boats.
F6C	CURTISS	Hawk, as USAAC PW-8. **Page 128**
F7C	CURTISS	**Page 426**
F8C	CURTISS	Falcon. Also operated as OC and O2C. **Pages 131, 135**
F9C	CURTISS	Sparrowhawk airship-borne fighter. **Page 137**
F10C	CURTISS	Single O2C-2 re-engined with Cyclone. **Page 136**
F11C	CURTISS	Goshawk dive-bomber. To BFC, BF2C. **Page 140**
F12C	CURTISS	Initial designation for SBC Helldiver. **Page 146**
F13C	CURTISS	High-wing monoplane/biplane prototype (9343) in 1934 with 600 hp R-1510-94.
F14C	CURTISS	1944 prototype (03183) with 2,300 hp R-3350-16.
F15C	CURTISS	1946 prototypes (01213–5) with R-2800-34W and Goblin jet.
FD	DOUGLAS	BuAer Design No. 113. Prototype (9223) in 1933 with 700 hp R-1535-64.
FD	McDONNELL	Phantom, initial designation. Redesignated FH. **Page 295**
F2D	McDONNELL	Banshee, initial designation. Redesignated F2H. **Page 297**
F3D	DOUGLAS	Skynight. **Page 182**
F4D	DOUGLAS	Skyray. **Page 184**
F5D	DOUGLAS	Developed F4D; prototypes only. **Page 185**
F6D	DOUGLAS	Missileer project in 1961 with two TF30-P-2 turbofans.
FF	GRUMMAN	**Page 195**
F2F	GRUMMAN	**Page 197**
F3F	GRUMMAN	Improved F2F. **Page 199**
F4F	GRUMMAN	Wildcat. **Page 205**
F5F	GRUMMAN	Skyrocket prototype (1442) in 1940 with 40s.
F6F	GRUMMAN	Hellcat. **Page 217**
F7F	GRUMMAN	Tigercat. **Page 220**
F8F	GRUMMAN	Bearcat. **Page 223**

F9F	GRUMMAN	Panther and Cougar family. **Pages 231, 233**
F10F	GRUMMAN	Jaguar variable-sweep prototype (128311) with 11,600 lb J40-WE-8.
F11F	GRUMMAN	Tiger. F9F-9 redesignated. **Page 237**
F12F	GRUMMAN	Projected all-weather development of F11F. Two (143401–2) cancelled.
FG	EBERHART	1927 prototype (A7944) biplane with 425 hp R-1340.
FG	GENERAL MOTORS	Alternative production source for F4U. **Page 405**
F2G	EBERHART	FG (A7944) tested in 1928 as single-float seaplane.
F2G	GENERAL MOTORS	Improved FG. **Page 406**
FH	HALL	1929 prototype (A8009) with 450 hp R-1340B.
FH	McDONNELL	Phantom. FD redesignated. **Page 295**
F2H	McDONNELL	Banshee. F2D redesignated. To F-2, 1962. **Page 297**
F3H	McDONNELL	Demon. To F-3, 1962. **Page 299**
F4H	McDONNELL	Phantom II. To F-4, 1962. **Page 301**
FJ	BERLINER JOYCE	BuAer 96 design. Prototype (8288) in 1930 with R-1340.
FJ	NORTH AMERICAN	Fury family. To F-1, 1962. **Pages 344 and 346**
F2J	BERLINER JOYCE	1934 prototype (8973) based on FJ with 625 hp R-1510-92.
F3J	BERLINER JOYCE	1934 prototype (9224) with 625 hp R-1510-26.
FL	LOENING	Cancelled project (9346).
FL	BELL	Airabonita version of USAAF P-39 (1588) with tail-down u/c.
F2L	BELL	Seven P-39 Airacobras (91102–3, 122447–51) as target drones.
FM	GENERAL MOTORS	Alternative production source for F4F. **Page 208**
F2M	GENERAL MOTORS	Project only with R-1820-70W engine.
F3M	GENERAL MOTORS	Alternative production source for F8F. **Page 223**
FN	NAF	Cancelled project (A8978).
FN	SEVERSKY	Version of USAAC P-35 for USN test in 1937.
FO	LOCKHEED	Four USAAF F-5Bs (01209–01212) used by USN in North Africa.
FO	LOCKHEED	Initial designation for FV tail-sitters.
FR	RYAN	Fireball. **Page 362**
F2R	RYAN	FR development with XT31-GE-2 turboprop and J31-GE-3 jet. One only (39661).
FS	SUPERMARINE	Paper designation for Spitfire.
FT	NORTHROP	Low-wing prototype (9400) in 1934 with 625 hp R-1510-26.
F2T	NORTHROP	USAAF P-61 for Marine Corps. **Page 459**
UF*	VOUGHT	Initial designation for UO-1. **Page 386**
FU	VOUGHT	Variant of UO. **Page 358**
F2U	VOUGHT	1929 prototype (A7692) biplane with 450 hp R-1340.
F3U	VOUGHT	BuAer Design No. 113. Produced as SBU. **Page 397**
F4U	VOUGHT	Corsair. **Page 403**
F5U	VOUGHT	Circular-wing prototypes (33958–9) built but not flown. Two R-2000-2s.
F6U	VOUGHT	Pirate. **Page 473**
F7U	VOUGHT	Cutlass. **Page 408**
F8U	VOUGHT	Crusader. To F-8, 1962. **Page 289**
FV	VICKERS	Projected second production source of F6F, 1942.
FV	LOCKHEED	VTO tail-sitters (138657–8) with T40-A-14 turboprop, 1954.
FW		No information.
F2W	WRIGHT	Two (A6743–4) biplane racers in 1923 with 70 hp Wright T-3.
F3W	WRIGHT	Apache. **Page 475**
FY	CONVAIR	VTO tail-sitters (138648–50) with T40-A-14 turboprop, 1954.
F2Y	CONVAIR	Sea Dart of 1953, water based with hydro-skis, two J34-WE-42s. Four built.

F—FIGHTER, 1962 →

F-1	NORTH AMERICAN	FJ-3/4 redesignated. **Page 346**
F-2	McDONNELL	F2H redesignated. **Page 297**
F-3	McDONNELL	F3H redesignated. **Page 299**
F-4	McDONNELL	F4H redesignated. **Page 299**
F-5	NORTHROP	Tiger II USAF fighter; four used by Navy. **Page 460**
F-6	DOUGLAS	F4D redesignated. **Page 184**
F-7	CONVAIR	F2Y redesignated.
F-8	LTV	F8U redesignated. **Page 289**
F-9	GRUMMAN	F9F-5/8 redesignated. **Page 233**
F-10	DOUGLAS	F3D redesignated. **Page 182**
F-11	GRUMMAN	F11F redesignated. **Page 237**
F-14	GRUMMAN	Tomcat variable geometry fighter. **Page 252**
F-18	McDONNELL DOUGLAS	Developed Northrop F-17 for USN/USMC.
F-111	GRUMMAN	Navy version of General Dynamics TFX winner. Prototypes 151970–151974 only.

G—TRANSPORT (SINGLE-ENGINED), 1939–1941

GB	BEECH	Commercial Model 17. **Page 412**
GH	HOWARD	Commercial DGA-15 variant. **Page 442**
GK	FAIRCHILD	Commercial Model 24W. **Page 435**
GQ	STINSON	Commercial Model SR-10. Two only (09798–9).

G—TANKER, 1958-1962

GV	LOCKHEED	Hercules tanker for USMC. To C-130, 1962. **Page 448**

H—HOSPITAL, 1929-1942

HE	PIPER	Cub variant. To AE, 1943. **Page 462**
HL	LOENING	Ambulance version of OL series. **Page 287**

H—HELICOPTER, 1962 →

H-1	BELL	USMC Iroquois, redesignated from HU-1. **Page 48**
H-2	KAMAN	HU2K redesignated. **Page 256**
H-3	SIKORSKY	HSS-2 redesignated. **Page 373**
H-13	BELL	HTL/HUL redesignated. **Page 45**
H-19	SIKORSKY	HRS/HO4S redesignated. **Page 366**
H-25	BOEING-VERTOL	HUP redesignated. **Page 461**
H-34	SIKORSKY	HSS-1/HUS redesignated. **Page 370**
H-37	SIKORSKY	HR2S redesignated. **Page 368**
H-43	KAMAN	HOK redesignated. **Page 443**
H-46	BOEING-VERTOL	HRB redesignated. **Page 68**
H-50	GYRODYNE	DSN redesignated. **Page 440**
H-51	LOCKHEED	Tri-service evaluation of two L-186s with PT6 turbo-shaft engines.
H-52	SIKORSKY	HU2S redesignated. **Page 467**
H-53	SIKORSKY	Sikorsky S-65 assault transport for USMC. **Page 377**
H-57	BELL	JetRanger as light turbine trainer. **Page 414**

HC—CRANE HELICOPTER, 1952–1955

HCH	McDONNELL	1951 proposal for ship-to-shore crane. Three (138654–6) cancelled.

HJ—UTILITY HELICOPTER, 1944–1949

HJD	McDONNELL	Initial designation of HJH (see below).
HJH	McDONNELL	Prototype (44318) twin rotor with two R-985s, tested 1947.
HJP	PIASECKI	Initial designation of HUP; two prototypes.
HJS	SIKORSKY	Two prototypes (30368, 30370) in 1948 with R-975-34 engine.

HN—TRAINING HELICOPTER, 1944–1948
HNS   SIKORSKY   As Army R-4. **Page 465**

HO—OBSERVATION HELICOPTER, 1944–1962
HOE	HILLER	HJ-1 Hornet with rotor-tip ramjets. Three prototypes (138651–3) in 1951.
HOG	GYRODYNE	Initial designation for RON.
HOK	KAMAN	K-600 for USMC. To H-43, 1962. **Page 443**
HOS	SIKORSKY	As Army R-6. **Page 465**
HO2S	SIKORSKY	Evaluation of three Army R-5As. **Page 466**
HO3S	SIKORSKY	S-51 version developed from HO2S. **Page 466**
HO4S	SIKORSKY	S-55 version. **Page 366**
HO5S	SIKORSKY	S-52-2 for USMC. **Page 467**

HR—TRANSPORT HELICOPTER, 1944–1962
HRB	BOEING-VERTOL	Model 107-II Sea Knight. To H-46, 1962. **Page 68**
HRH	McDONNELL	Heavy assault project, two XT56-A-4s (133736–38). Cancelled.
HRP	PIASECKI	Model PV-3 'Flying Banana'. **Page 461**
HRS	SIKORSKY	S-55 version. To H-19, 1962. **Page 366**
HR2S	SIKORSKY	S-56 for USMC. To H-37, 1962. **Page 368**
HR3S	SIKORSKY	Projected transport variant of S-61/HSS-2. Not procured.

HS—ANTI-SUBMARINE HELICOPTER, 1951–1962
HSL	BELL	Model 61. **Page 413**
HSS	SIKORSKY	S-58 Seabat. To H-34, 1962. **Page 370**

HT—TRAINING HELICOPTER, 1948–1962
HTE	HILLER	UH-12A. **Page 440**
HTK	KAMAN	K-240. **Page 442**
HTL	BELL	Model 47 variant. **Page 45**

HU—UTILITY HELICOPTER, 1950–1962
HUK	KAMAN	K-600 for Navy. To H-43, 1962. **Page 443**
HU2K	KAMAN	Seasprite. To H-2, 1962. **Page 256**
HUL	BELL	Model 47 variant. **Page 45**
HUM	McCULLOCH	Evaluation of MC-4A (133817–8) with 200 hp Franklin.
HUP	PIASECKI	Retriever. **Page 461**
HUS	SIKORSKY	S-58 Seahorse. To H-34, 1962. **Page 370**
HU2S	SIKORSKY	Initial designation for H-52. **Page 467**

J—TRANSPORT, 1926–1931
JA	ATLANTIC	Fokker Super Universal (A8012) evaluated by Navy, 1928.
JR	FORD	Ford Tri-Motor variants. To RR, 1931. **Page 191**

J—GENERAL UTILITY, 1931–1955
JA	NOORDUYN	Norseman. **Page 490**
JB	BEECH	Model 17. **Page 412**
JD	DOUGLAS	Invader. **Page 432**
JE	BELLANCA	Senior Pacemaker. **Page 414**
JF	GRUMMAN	Duck amphibian. **Page 202**
J2F	GRUMMAN	Variant of JF. **Page 203**
J3F	GRUMMAN	Model G-21 for evaluation. Produced as JRF. **Page 211**
J4F	GRUMMAN	Widgeon. **Page 438**
JH	STEARMAN-HAMMOND	Civil model Y-1S. **Page 469**
JK	FAIRCHILD	Civil model 45. **Page 436**

J2K	FAIRCHILD	Civil model 24. **Page 435**
JL	COLUMBIA	Monoplane amphibian with 1,350 hp R-1820-56. Two prototypes (31399–31400).
JM	MARTIN	Marauder, as USAAF B-26. **Page 453**
JO	LOCKHEED	Model 12A. **Page 446**
JQ	FAIRCHILD	Civil model FC-2. **Page 434**
J2Q	FAIRCHILD	Civil model 71. **Page 434**
JW	WACO	Civil model UBF. **Page 474**
J2W	WACO	Civil model EQC-6 for USCG. **Page 474**

JR—UTILITY TRANSPORT, 1935–1955

JRB	BEECH	Model 18 variant. **Page 41**
JRC	CESSNA	Model T-50. **Page 418**
JRF	GRUMMAN	Model G-21. **Page 211**
JR2F	GRUMMAN	Model G-64 prototypes. Production as UF. **Page 229**
JRK	NASH KELVINATOR	Proposed four-engined 37-seat boat.
JRM	MARTIN	Production variant of PB2M Mars. **Page 452**
JRS	SIKORSKY	Model S-43. **Page 464**
JR2S	SIKORSKY	Model VS-44A variant. Three only (12390–2).

LB—BOMB-CARRYING GLIDER, 1941–1945

LBE	PRATT-READ (GOULD)	Three only (85290–2).
LBP	PIPER	Alternative production source of LBE Glomb.
LBT	TAYLORCRAFT	Twenty-five produced (85265–89), as LBE.

LN—TRAINING GLIDER, 1941–1945

LNE	PRATT-READ (GOULD)	**Page 472**
LNP	PIPER	As Army TG-8.
LNR	AERONCA	As Army TG-5.
LNS	SCHWEIZER	As Army TG-2 for USMC. **Page 499**
LNT	TAYLORCRAFT	As Army TG-6. **Page 500**

LR—TRANSPORT GLIDER, 1941–1945

LRA	ALLIED	Amphibious 12-seat prototypes. **Page 501**
LR2A	ALLIED	Proposed development of LRA-1 (31503–4).
LRG	AGA	Twin hull, 24-seat amphibious project, 1942. Not built.
LRH	SNEAD	Twin hull, 24-seat amphibious project, 1942. Not built.
LRN	NAF	Two only (36431–2).
LR2N	NAF	Single hull, 24-seat amphibious project, 1944. Not built.
LRQ	BRISTOL	Amphibious 12-seat prototypes. **Page 501**
LRW	WACO	As Army CG-4A. **Page 500**
LR2W	WACO	Two only (85094 -5). No information.

M—MARINE EXPEDITIONARY, 1922–1923

| EM* | ELIAS | **Page 433** |
| NM* | NAF | All-metal prototype (A6450) with 325 hp Packard 1A-1237, tested in 1925. |

N—TRAINER, 1922–1960

NB	BOEING	Model 21. **Page 53**
N2B	BOEING	Model 81, prototype (A8010) only tested in 1928–29.
NC	—	Not used in Trainer series. NC flying-boats. **Page 328**
N2C	CURTISS	Fledgling. **Page 133**
NE	PIPER	Grasshopper. **Page 462**
HN*	HUFF-DALAND	**Page 441**
NH	HOWARD	DGA-15 instrument trainers. **Page 442**
NJ	NORTH AMERICAN	NA-28, as USAAC BT-9. **Page 341**
NK	KEYSTONE	**Page 443**

NL	LANGLEY	Prototype (39056) twin-engined design with plastic structure.
N2M	MARTIN	Single example (A6800).
N2N	NAF	Three examples (A6693–5) with 200 hp Lawrence J-1 engine.
N3N	NAF	'Yellow Peril' biplane. **Page 338**
N4N	NAF	No information.
N5N	NAF	Prototype low-wing monoplane (1521) with R-760-6.
NP	SPARTAN	Based on commercial C-3. **Page 468**
NQ	FAIRCHILD	Two prototypes (75725–6) in 1946 with 320 hp R-680-10.
NR	MAXSON	Twin-engined trainer prototypes (1756–1757).
NR	RYAN	Recruit. **Page 463**
NS	STEARMAN	Model 73. **Page 380**
N2S	STEARMAN	Improved NS. **Page 380**
NT	NEW STANDARD	Model D-25 and D-29. **Page 456**
N2T	TIMM	Tutor. **Page 473**
NW	WRIGHT	Biplane (A6544) for 1922/23 races with 750 hp Wright Tornado.
NY	CONSOLIDATED	Navy version of Army PT-1. **Page 75**
N2Y	CONSOLIDATED	Fleet I biplanes. **Page 420**
N3Y	CONSOLIDATED	Developed NY in 1929. **Page 77**
N4Y	CONSOLIDATED	Model 21. **Page 420**

O—OBSERVATION, 1922–1962

OB	BOEING	Amphibian project to BuAer designs. Not built. **Page 156**
O2B	BOEING	Variant of de Havilland DH-4. **Page 156**
OC	CURTISS	Falcon. F8C-1 redesignated. **Page 131**
O2C	CURTISS	Falcon. F8C-5 redesignated. **Page 135**
O3C	CURTISS	Prototype of SOC Seagull series. **Page 143**
OD	DOUGLAS	USAAC O-2 for USMC. **Page 429**
O2D	DOUGLAS	1934 prototype (9412).
EO*	ELIAS	Variant of EM. **Page 433**
OE	CESSNA	Bird Dog. **Page 419**
OF	GRUMMAN	Projected Marine Corps purchase of OV-1 Mohawk.
HO*	HUFF-DALAND	Variant of HN. **Page 441**
OJ	BERLINER-JOYCE	BuAer Design No. 86. **Page 415**
OK	KEYSTONE	BuAer Design No. 86, prototype only (8357).
OL	LOENING	Biplane amphibian. **Page 286**
O2L	LOENING	Developed OL in 1932 (8525) with R-1340.
MO*	MARTIN	Production of BuAer design. **Page 451**
M2O*	MARTIN	Variant of NO. **Page 455**
NO*	NAF	**Page 455**
O2N	NAF	Initial designation of OSN (0385).
OO	VIKING	Variant of Schreck FBA. **Page 463**
OP	PITCAIRN	Three autogiros (8850, A8976–7) for evaluation, 1931.
UO*	VOUGHT	Improved VE-7/9. **Page 386**
O2U	VOUGHT	Corsair. **Page 390**
O3U	VOUGHT	Corsair. Improved O2U. **Page 393**
O4U	VOUGHT	Prototype (8641) in 1931; second dissimilar prototype in 1932 carried same serial.
O5U	VOUGHT	Prototype (9399) in 1934 with R-1340-12.
OY	CONVAIR	Sentinel. **Page 421**
OZ	PENN ACFT SYNDICATE	XN2Y-2 (8602) rebuilt as floatplane, winged autogiro.

O—OBSERVATION, 1962 →

O-1	CESSNA	OE redesignated. **Page 419**

OS—OBSERVATION SCOUT, 1935–1945

OSE	EDO	Single-seat shipboard seaplane with Ranger V-770. Twelve only (44316–17, 75210–15, 75625–28).
OSN	NAF	Prototype (0385) in 1938 with R-1340-36.
OS2N	NAF	Production of OS2U. **Page 401**
OSS	STEARMAN	Prototype (1052) in 1938 with R-1340-36.
OSU	VOUGHT	Final O3U-6 (0016) with modifications, tested in 1937.
OS2U	VOUGHT	Kingfisher. **Page 401**

P—PURSUIT, 1923

WP	WRIGHT	Dornier Falke for evaluation. **Page 496**

P—PATROL, 1923–1962

PB	BOEING	One only (A6881) with two 800 hp Packards.
PB	BOEING	B-17 for Navy and USCG. **Page 66**
P2B	BOEING	B-29 for Navy. **Page 416**
P3B	BOEING	Model 466 in 1947 with two Allison XT-40 turbo-props; not built.
PD	DOUGLAS	Production derivative of PN-12. **Page 428**
P2D	DOUGLAS	Final variant of T2D. **Page 161**
P3D	DOUGLAS	Prototype monoplane (9613) with R-1830 engines.
PF	GRUMMAN	Interim designation for UF. **Page 229**
PH	HALL	Variant of PN-11 design. **Page 254**
P2H	HALL	One only (A8729) biplane with four 600 hp V-1570-54s.
PJ	NORTH AMERICAN	Fokker/General Aviation F2B redesignated. **Page 437**
PK	KEYSTONE	Production derivative of PN-12. **Page 444**
PM	MARTIN	Production derivative of PN-12. **Page 451**
P2M	MARTIN	Development of PY. **Page 316**
P3M	MARTIN	Production version of P2M. **Page 317**
P4M	MARTIN	Mercator. **Page 453**
P5M	MARTIN	Marlin. To P-5, 1962. **Page 323**
P6M	MARTIN	Model 275 Seamaster with four J75-P-2 jets Eight prototypes and three P6M-2s.
PN	NAF	Developed from F-5L. **Page 334**
P2N	NAF	Paper designation for surviving NC boats. **Page 329**
P3N	NAF	No information.
P4N	NAF	Developed PN. **Page 336**
PO	LOCKHEED	Initial designation for WV Constellation. **Page 273**
PS	SIKORSKY	S-38 for Navy and Marine Corps. **Page 364**
P2S	SIKORSKY	One only (A8642) with two 450 hp R-1340-88s.
PV	LOCKHEED	Ventura and Harpoon. **Page 259**
P2V	LOCKHEED	Neptune. To P-2, 1962. **Page 263**
P3V	LOCKHEED	Orion. To P-3, 1962. **Page 278**
PY	CONSOLIDATED	Monoplane boat to NAF design with two/three 450 hp R-1340-38s (8011).
P2Y	CONSOLIDATED	Developed from PY. **Page 78**
P3Y	CONSOLIDATED	Prototype for PBY Catalina. **Page 80**
P4Y	CONSOLIDATED	Model 31 Corregidor boat (27852) with two 2,300 hp R-3350-8s. Production (44705–44904) cancelled.
P4Y	CONVAIR	PB4Y redesignated. To P-4 in 1962. **Page 90**
P5Y	CONVAIR	Two prototype boats (121455–6) with T40-A-4 turbo-props.
P6Y	CONVAIR	Three-engined boat project in 1956. Not built.

P—PATROL, 1962 →

P-2	LOCKHEED	P2V redesignated. **Page 263**
P-3	LOCKHEED	P3V redesignated. **Page 278**
P-4	CONVAIR	P4Y redesignated. **Page 90**
P-5	MARTIN	P5M redesignated. **Page 323**

PB—PATROL BOMBER, 1935–1962

PBB	BOEING	Sea Ranger prototype (3144) with two R-3350-8s. Production cancelled.
PB2B	BOEING	Canadian production of PBY. **Page 82**
PBJ	NORTH AMERICAN	Mitchell, as USAAF B-25. **Page 457**
PBM	MARTIN	Mariner. **Page 318**
PB2M	MARTIN	Mars. Initial designation. To JRM. **Page 452**
PBN	NAF	Variant of PBY Catalina. **Page 82**
PBO	LOCKHEED	Hudson. **Page 447**
PBS	SIKORSKY	**Page 464**
PBV	VICKERS	Canadian production of PBY. **Page 82**
PBY	CONSOLIDATED	Model 28 Catalina. **Page 80**
PB2Y	CONSOLIDATED	Model 29 Coronado. **Page 85**
PB3Y	CONSOLIDATED	Boat project with four R-2800-18s in 1942. Cancelled.
PB4Y	CONVAIR	Land-based Liberator and Privateer. To P4Y and P-4. **Page 87**

PTB—PATROL TORPEDO-BOMBER, 1937

PTBH	HALL	Seaplane prototype (9721) with two 800 hp XR-1830-60s.

R—RACER, 1922–1928

BR*	BEE LINE	Low-wing monoplanes (A6429–30) in 1922.
CR*	CURTISS	Biplanes for 1921 Pulitzer. Also CF. **Page 120**
R2C	CURTISS	Biplanes for 1923 Pulitzer. Also F2C. **Page 125**
R3C	CURTISS	Biplanes for 1925 Pulitzer. Also F3C. **Page 125**

R—TRANSPORT, 1931–1962

RA	ATLANTIC	TA redesignated. **Page 39**
RB	BUDD	Connestoga with stainless-steel structure, two 1,200 hp R-1830-92s. Seventeen built (39292–39308).
RC	CURTISS	Kingbird. **Page 426**
R4C	CURTISS	Condor. **Page 427**
R5C	CURTISS	Commando, as USAAF C-46. **Page 428**
RD	DOUGLAS	Dolphin for USCG and Navy. **Page 163**
R2D	DOUGLAS	DC-2. **Page 430**
R3D	DOUGLAS	DC-5. **Page 430**
R4D	DOUGLAS	Skytrain. To C-47, 1962. **Page 170**
R5D	DOUGLAS	Skymaster. To C-54, 1962. **Page 174**
R6D	DOUGLAS	DC-6A. To C-118, 1962. **Page 432**
RE	BELLANCA	Skyrocket. **Page 414**
RK	KINNER	Envoy. **Page 444**
R2K	FAIRCHILD	Fairchild 22. **Page 435**
RM	MARTIN	Model 4-O-4 for USCG. To C-3, 1962. **Page 454**
RO	LOCKHEED	Altair. **Page 445**
R2O	LOCKHEED	Electra for Navy. **Page 445**
R3O	LOCKHEED	Electra for USCG. **Page 445**
R4O	LOCKHEED	Single Model 14 (1441) as Navy staff transport.
R5O	LOCKHEED	Lodestar. **Page 446**
R6O	LOCKHEED	Constitution. To R6V. **Page 447**
R7O	LOCKHEED	Initial designation for R7V Constellation. **Page 272**
RQ	STINSON	Reliant. To R3Q. **Page 469**
RQ	FAIRCHILD	JQ redesignated. **Page 434**
R2Q	FAIRCHILD	J2Q redesignated. **Page 434**
R3Q	STINSON	Reliant. **Page 469**
R4Q	FAIRCHILD	Packet. To C-119 in 1962. **Page 189**
RR	FORD	Tri-Motor. JR redesignated. **Page 191**
RS	SIKORSKY	PS redesignated. **Page 364**
RT	NORTHROP	Delta for USCG. **Page 459**
R6V	LOCKHEED	Constitution. R60 redesignated. **Page 447**
R7V	LOCKHEED	Constellation. R70 redesignated. **Page 272**
R8V	LOCKHEED	Initial designation of C-130 for USCG. **Page 275**

RY	CONVAIR	Variants of PB4Y. **Page 89**
R2Y	CONVAIR	Model 39 (09803) based on PB4Y with new fuselage, tested in 1945.
R3Y	CONVAIR	Tradewind. **Page 422**
R4Y	CONVAIR	CV-340 variants. **Page 422**

RO—ROTORCYCLE, 1954–1959

| ROE | HILLER | One-man observation helicopter, 43 hp Nelson engine. Twelve built. |
| RON | GYRODYNE | One-man observation helicopter, Nelson or 62 hp Solar engine. |

S—SCOUT, 1922–1946

AS*	AEROMARINE	Seaplane scouts. **Page 411**
CS*	CURTISS	**Page 123**
SC	CURTISS	Martin production version of CS. **Page 123**
SC	CURTISS	Seahawk monoplane. **Page 153**
S2C	CURTISS	Naval version of YA-10 Shrike (9377) with R-1510-28.
S3C	CURTISS	XF10C-1 redesignated. **Page 136**
S4C	CURTISS	Interim designation for SBC Helldiver. **Page 146**
SDW	DAYTON-WRIGHT	Modification of DT-2. **Page 160**
SE	BELLANCA	1932 prototype (A9186) monoplane.
SF	GRUMMAN	Variant of FF. **Page 195**
SG	GREAT LAKES	1932 prototype (A8974) amphibian.
SL	LOENING	Sub-borne boat (A8696) with Scarab; later Menasco B 6 engine.
S2L	LOENING	1932 prototype (A8971) amphibian boat.
MS*	MARTIN	BuAer Design No. 20 submarine scout. **Page 423**
SS	SIKORSKY	1933 prototype (A8972) amphibian boat with one R-1340.
SU	VOUGHT	Corsair. O3U redesignated. **Page 394**
XS*	COX-KLEMIN	BuAer Design No. 20 submarine scout. **Page 423**

S—ANTI-SUBMARINE, 1946–1962

| S2F | GRUMMAN | Tracker. To S-2, 1962. **Page 239** |
| S2U | VOUGHT | Project, same class as S2F. |

S—ANTI-SUBMARINE, 1962 →

| S-2 | GRUMMAN | S2F redesignated. **Page 238** |
| S-3 | LOCKHEED | Viking. **Page 281** |

SB—SCOUT BOMBER, 1934–1946

SBA	BREWSTER	**Page 417**
SB2A	BREWSTER	Model 340 Buccaneer. **Page 73**
SBC	CURTISS	Helldiver biplane. **Page 146**
SB2C	CURTISS	Helldiver monoplane. **Page 150**
SB3C	CURTISS	Two prototypes (03743–4) ordered, not completed.
SBD	DOUGLAS	Dauntless. Production development of BT-1. **Page 167**
SB2D	DOUGLAS	Two-seat prototypes of BTD. **Page 431**
SBF	GRUMMAN	1936 prototype (9996) derived from SF with 700 hp R-1535-72.
SBF	CANADIAN FAIRCHILD	Alternative production source for SB2C. **Page 151**
SBN	NAF	Production of SBA. **Page 417**
SBU	VOUGHT	BuAer Design No. 113. **Page 397**
SB2U	VOUGHT	Vindicator. **Page 399**
SB3U	VOUGHT	Prototype (9834) as SBU with retractable u/c.
SBW	CCF	Alternative production source for SB2C. **Page 151**

SN—SCOUT TRAINER, 1939–1948

SNB	BEECH	Model 18 variant. **Page 41**
SNC	CURTISS	CW-21 variant. **Page 427**
SNJ	NORTH AMERICAN	Texan, as USAAC AT-G. **Page 341**
SN2J	NORTH AMERICAN	Prototypes (121449–50) based on SNJ; 1,100 hp R-1820-78.
SNV	VULTEE	Valiant. **Page 475**

SO—SCOUT–OBSERVATION, 1934–1946

SOC	CURTISS	Seagull. Production version of O3C. **Page 143**
SO2C	CURTISS	Improved SOC, prototype (0950) only.
SO3C	CURTISS	Seamew monoplane. **Page 148**
SOE	BELLANCA	Prototype (9728) cruiser-based scout with R-1820-84.
SOK	FAIRCHILD	Proposed version of PAA transport XA-942. Prototype (9724) cancelled.
SON	NAF	Production of SOC Seagull by second source. **Page 144**
SOR	RYAN	Planned production of SO3C-1 by Ryan. Cancelled. **Page 149**
SO2U	VOUGHT	Monoplane prototype (1440) with 550 hp Ranger XV-770-4.

T—TORPEDO, 1922–1935

TB	BOEING	**Page 415**
CT*	CURTISS	Model 24. **Page 425**
DT*	DOUGLAS	**Page 158**
T2D	DOUGLAS	Production of NAF XTN-1. **Page 161**
T3D	DOUGLAS	1931 prototype (A8730) biplane with R-1860/R-1830.
TE	DETROIT	Developed TG-1 ordered from Detroit but built by Great Lakes (subsidiary company) as TG-2. **Page 313**
FT*	FOKKER	T.III for evaluation. **Page 487**
TG	GREAT LAKES	Production of T4M variant. **Page 312**
MT*	MARTIN	Improved MBT. **Page 450**
T2M	MARTIN	Paper designation for SC-2.
T3M	MARTIN	Improved SC. **Page 310**
T4M	MARTIN	Improved T3M. **Page 311**
T5M	MARTIN	BuAer Design No. 77 prototypes. **Page 314**
T6M	MARTIN	1930 prototype (A8411) biplane with 575 hp R-1860.
TN	NAF	1926 prototype (A7027) for T2D production.
T2N	NAF	BuAer Design No. 77 prototype, as T5M. **Page 314**
ST*	STOUT	Prototype monoplanes (A5899–901) with two 300 hp V-1237s. Three more cancelled.

T—TRANSPORT, 1927–1930

TA	ATLANTIC	Fokker F.VII built in US. Redesignated RA. **Page 39**

T—TRAINER, 1948–1962

TE	EDO	Single-float seaplane, as OSE. Prototypes only (75216–7, 75629–32).
TF	GRUMMAN	Variant of S2F. To C-1, 1962. **Page 243**
TJ	NORTH AMERICAN	SNJ-8 redesignated TJ-8. Cancelled 1952.
T2J	NORTH AMERICAN	Buckeye. To T-2, 1962. **Page 359**
T3J	NORTH AMERICAN	Initial designation for T-39. **Page 457**
TO	LOCKHEED	Initial designation for TV, as USAF F-80C and T-33. **Page 269**
TT	TEMCO	Pinto. **Page 470**
TV	LOCKHEED	TO redesignated. **Page 269**
T2V	LOCKHEED	Seastar. **Page 269**

538

T—TRAINER, 1962 →

T-1	LOCKHEED	T2V redesignated. **Page 269**
T-2	NORTH AMERICAN	T2J redesignated. **Page 359**
T-28	NORTH AMERICAN	Trojan. **Page 350**
T-29	CONVAIR	CV-340 variants on loan from USAF.
T-33	LOCKHEED	TO/TV redesignated. **Page 269**
T-34	BEECH	Mentor. **Page 412**
T-38	NORTHROP	Talon USAF trainer; five acquired by USN. **Page 460**
T-39	NORTH AMERICAN	T3J redesignated. **Page 457**
T-41 ·	GRUMMAN	Preliminary designation for TC-4B, re-assigned to Cessna 172 for USAF and Army. **Page 439**
T-44	BEECH	King Air B90 as multi-engine trainer. 15 ordered 1976, more later.

TB—TORPEDO BOMBER, 1935–1946

TBD	DOUGLAS	Devastator monoplane. **Page 165**
TB2D	DOUGLAS	Two 1945 prototypes (36933–4) with XR-4360-8s; 23 cancelled.
TBF	GRUMMAN	Avenger. **Page 213**
TB2F	GRUMMAN	Variant of F7F design. Not built. **Page 226**
TB3F	GRUMMAN	Initial designation for AF. **Page 226**
TBG	GREAT LAKES	1935 biplane (9723) with 800 hp XR-1830-60.
TBM	GENERAL MOTORS	Alternative production source for TBF. **Page 213**
TBU	VOUGHT	Sea Wolf prototypes. **Page 422**
TBY	CONVAIR	Production of TBU. **Page 318**

TD—TARGET DRONE, 1942–1946

TDC	CULVER	As USAAF PQ-8. **Page 424**
TD2C	CULVER	As USAAF PQ-14. **Page 424**
TD3C	CULVER	As USAAF PQ-15.
TD4C	CULVER	To UC-1.
TDD	RADIOPLANE	Pilotless, as USAAF OQ-2A, -3, -7.
TD2D	McDONNELL	Pilotless, pulsejet-powered Katydid.
TD3D	FRANKFORT	Pilotless, as USAAF OQ-16.
TD4D	RADIOPLANE	Pilotless, as USAAF OQ-17.
TDL	BELL	USAAF P-39Q (42-20807) tested as XTDL-1 drone.
TDN	NAF	Radio-controlled, TV-directed assault drone.
TD2N	NAF	Gorgon variant by Martin. Nineteen built.
TD3N	NAF	Gorgon variant with pulsejet engine. Eight built.
TDR	INTERSTATE	Production of TDN. Two 220 hp Lycoming. Provision for check pilot. Operational in 1943.
TD2R	INTERSTATE	Improved TDR, 450 hp engines. Two only (33921–33922).
TD3R	INTERSTATE	As USAAF BQ-6. Twelve built.

TS—TORPEDO–BOMBER/SCOUT, 1943

TSF	GRUMMAN	Two prototypes (84055–6) cancelled.

U—UTILITY, 1955–1962

UC	DE HAVILLAND	Otter, To U-1, 1962. **Page 481**
UF	GRUMMAN	G-64 Albatross. JR2F redesignated. **Page 229**
UO	PIPER	Aztec. To U-11, 1962. **Page 462**
UV	LOCKHEED	Initial designation (UV-1L) for C-130BL Hercules. **Page 275**
UV	LOCKHEED	Proposed purchase of two Jet Stars (149820–1) in 1961.

U—UNPILOTED DRONE, 1946–1955

UC	CULVER	TD4C redesignated. Two only (120339–40).

U—UTILITY, 1962 →

U-1	DE HAVILLAND	UC redesignated. **Page 481**
U-6	DE HAVILLAND	L-20A Beaver redesignated. **Page 482**
U-8	BEECH	Projected purchase in 1964 of Twin Bonanza as instrument trainer.
U-11	PIPER	UO redesignated. **Page 462**
U-16	GRUMMAN	UF redesignated. **Page 229**

V—VTOL/STOL, 1962 →

V-8	HAWKER SIDDELEY	Harrier V/STOL attack fighter for USMC. **Page 486**
V-10	NORTH AMERICAN	Bronco observation twin for USN and USMC. **Page 355**
V-12	FAIRCHILD HILLER	Proposed purchase of 19 OV-12A Porters (157102–157120). Contract cancelled.
V-12	ROCKWELL	NR-356 thrust-augmented wing prototypes of V/STOL fighter for testing in 1976, as XFV-12A.
V-16	McDONNELL/HSA	Advanced Harrier AV-16A derivative of AV-8A. **Page 486**

W—EARLY WARNING, 1952–1962

WF	GRUMMAN	Variant of S2F. To E-1, 1962. **Page 242**
W2F	GRUMMAN	Hawkeye. To E-2, 1962. **Page 244**
WU	VOUGHT	Two prototypes (133780–1) cancelled before completion.
WV	LOCKHEED	Constellation as Warning Star AEW aircraft. Became EC-121, 1962. **Page 272**

X—SPECIAL RESEARCH, 1948

X-22	HILLER	Model D2127 with four tilting ducted airscrews; four YT58-GE-8D engines. Tri-service programme under USN direction.
X-26	SCHWEIZER	Two X-26A Schweizer 2-32 sailplanes used by Naval Test Pilot School and one X-26B Lockheed QT-2PC 'Quiet Thruster' for NTPS use. **Page 448**
X-28	PEREIRA	Homebuilt Osprey flying-boat (158786) tested at NADC, 1971.

GENERAL INDEX

THE arrangement of the material in this volume is such that it is largely self-indexing, with the products of one manufacturer grouped in chronological order in the body of the book and in the appendices. The designation index which starts on page 501 provides page reference numbers to all the aircraft types described in this volume which have been designated in the standard Navy systems used since 1922. This general index covers the earlier non-standard designations, aircraft names, individuals, organizations and events.

A

Acadame, Grumman, 439
Acosta, Bert, 121
Aeromarine
  Model 39, 37
  Model 40, 411
  Model 700, 410
Akron, USS, 137, 138, 420, 524
Albatross, Grumman, 229
America, Curtiss, 106
Antarctic expedition, 427
Anvil, Project, 88
Apache, Wright, 475
AR-1, Morane Saulnier, 491
Arctic Expedition, 1925, 286
  1926, 39
Avenger, Grumman, 218

B

Baby, Sopwith, 391
Badoeng Strait, USS, 363
Bairoko, USS, 363
Banshee, McDonnell, 297
Barnaby, Ralph, Lt, 499
Bearcat, Grumman, 223
Beaver, de Havilland, 482
Bee-line, Project, 303
Belknap, USS, 257
Bellinger, Lt (Jg) P. N. L., 96
Bennett, Floyd, 39
Bennington, USS, 347
Bettis, Lt Cyrus, 125
Bird Dog, Cessna, 419
Birmingham, USS, 1, 92
Blue Angels, 235, 239
Boeing
  Model C, 51
  314, 416
Boxer, USS, 218, 345
Bradley, Lt B. G., 278
Bronco, North American, 355
Brow, Lt H. J., 121, 125
Buccaneer, Brewster, 73
Buckeye, North American, 359
Buffalo, Brewster, 71

Bulldog, Bristol, 478
Bureau of Aeronautics, 2
Burgess
  Model I., 418
  -Dunne, AH, 417
Byrd, Cdr Richard E., 39

C

Cactus Kitten, Curtiss, 120, 425
Camel, Sopwith, 494
Caproni Ca 44, 479
Catalina, Consolidated, 80
Chambers, Capt W. I., 1
Cleveland, USS, 148
Colours of aircraft, 19
Constellation, Lockheed, 272
Constitution, Lockheed, 447
Coral Sea, USS, 265, 346
Coronado, Consolidated, 85
Corsair, Vought, 390, 393, 403
Corsair II, LTV, 292
Cougar, Grumman, 233
Crusader, Vought (LTV), 289
Cuddihy, Lt G. T., 122, 126
Curtiss, Glenn, 91
Curtiss Marine Trophy Race, 118
Curtiss
  18-T, 117
  A, 91, 92, 93
  AH, 94
  AB, 96
  AX, 93
  C, 96
  Dunkirk Fighter, 424
  F-Boat, 95
  F-5L, 114
  H-12, 106
  H-16, 106
  HA-1, 424
  Jennie, 100
  JN series, 100
  R-3, -6, -9, 103
  NC, 328
  Triad, 92
  Twin JN, 100
  TS, 321
Cutlass, Vought, 408

544